Innovative Processing and Manufacturing of Advanced Ceramics and Composites

Innovative Processing and Manufacturing of Advanced Ceramics and Composites

Ceramic Transactions, Volume 212

A Collection of Papers Presented at the 8th Pacific Rim Conference on Ceramic and Glass Technology
May 31–June 5, 2009
Vancouver, British Columbia

Edited by
Zuhair A. Munir
Tatsuki Ohji
Yuji Hotta

Volume Editor
Mrityunjay Singh

A John Wiley & Sons, Inc., Publication

Published by John Wiley & Sons, Inc., Hoboken, New Jersey.
Published simultaneously in Canada.

Library of Congress Cataloging-in-Publication Data is available.

ISBN 978-0-470-87646-6

Printed in the United States of America.

10 9 8 7 6 5 4 3 2 1

Contents

NOVEL, GREEN, AND STRATEGIC PROCESSING

ADVANCED POWDER PROCESSING

Preface

With continued discoveries and innovations, the field of materials synthesis and processing remains as it has been for many decades, a vibrant and fertile area for research and development. It comes, therefore, as no surprise that every Pac Rim conference has had considerable emphasis on this topic with many symposia devoted to various aspects of this field.

This Ceramic Transactions volume represents selected papers based on presentations in four symposia during the 8th Pacific Rim Conference on Ceramic and Glass Technology, held in Vancouver, British Columbia, May 31-June 5, 2009. The symposia and their organizers are:

Synthesis and Processing of Materials by the Spark Plasma Method, organized by Zuhair A. Munir, University of California-Davis, USA, Manshi Ohyanagi, Ryukoku University, Japan, Enrique J. Lavernia, University of California-Davis, USA, Masao Tokita, SPS SYNTEX INC, Japan, and Javier E. Garay, University of California-Riverside, USA.

Innovative Processing and Manufacturing, organized by Tatsuki Ohji, National Institute of Advanced Industrial Science and Technology (AIST), Japan, Juergen G. Heinrich, Clausthal University of Technology, Germany, Dongliang Jiang, Shanghai Institute of Ceramics, China, Takashi Goto, Tohoku University, Japan, Richard D. Sisson, Jr., Worcester Polytechnic Institute, MA, USA, and Junichi Tatami, Yokohama National University, Japan.

Advanced Powder Processing and Manufacturing Technologies, organized by Koji Watari, National Institute of Advanced Industrial Science and Technology (AIST), Japan, George V. Franks, University of Melbourne, Australia, Jianfeng Yang, Xi'an Jiaotong University, China. Guo-Jun Zhang, Shanghai Institute of Ceramics, China, Yoshio Sakka, National Institute for Materials Science, Japan, Junichi Tatami, Yokohama National University, Japan, Satoshi Tanaka, Nagaoka University of Technology, Japan, Hae Jin Hwang, Inha University, Korea, Lennart Bergstrom, Stockholm University, Sweden, Christopher B. DiAntonio, Sandia National Laboratories, USA, and Yuji Hotta, National Institute of Advanced Industrial Science and Technology (AIST), Japan.

We are grateful for the help of all of our co-organizers and for the support of the Pac Rim organizers and the American Ceramic Society. We want to especially acknowledge the help of Mr. Gregory Geiger of the Society. We also acknowledge the financial support from SPS SYNTEX, Inc. of Japan to the symposium on the Synthesis and Processing of Materials by the Spark Plasma Method.

ZUHAIR A. MUNIR, University of California-Davis, USA
TATSUKI OHJI, National Institute of Advanced Industrial Science and Technology (AIST), Japan
YUJI HOTTA, National Institute of Advanced Industrial Science and Technology (AIST), Japan

Introduction

The 8th Pacific Rim Conference on Ceramic and Glass Technology (PACRIM 8), was the eighth in a series of international conferences that provided a forum for presentations and information exchange on the latest emerging ceramic and glass technologies. The conference series began in 1993 and has been organized in USA, Korea, Japan, China, and Canada. PACRIM 8 was held in Vancouver, British Columbia, Canada, May 31–June 5, 2009 and was organized and sponsored by The American Ceramic Society. Over the years, PACRIM conferences have established a strong reputation for the state-of-the-art presentations and information exchange on the latest emerging ceramic and glass technologies. They have facilitated global dialogue and discussion with leading world experts.

The technical program of PACRIM 8 covered wide ranging topics and identified global challenges and opportunities for various ceramic technologies. The goal of the program was also to generate important discussion on where the particular field is heading on a global scale. It provided a forum for knowledge sharing and to make new contacts with peers from different continents.

The program also consisted of meetings of the International Commission on Glass (ICG), and the Glass and Optical Materials and Basic Science divisions of The American Ceramic Society. In addition, the International Fulrath Symposium on the role of new ceramic technologies for sustainable society was also held. The technical program consisted of more than 900 presentations from 41 different countries. A selected group of peer reviewed papers have been compiled into seven volumes of The American Ceramic Society's Ceramic Transactions series (Volumes 212-218) as outlined below:

- **Innovative Processing and Manufacturing of Advanced Ceramics and Composites, Ceramic Transactions, Vol. 212,** Zuhair Munir, Tatsuki Ohji, and Koji Watari, Editors; Mrityunjay Singh, Volume Editor
  Topics in this volume include Synthesis and Processing by the Spark Plasma

Method; Novel, Green, and Strategic Processing; and Advanced Powder Processing

- **Advances in Polymer Derived Ceramics and Composites, Ceramic Transactions, Vol. 213,** Paolo Colombo and Rishi Raj, Editors; Mrityunjay Singh, Volume Editor
  This volume includes papers on polymer derived fibers, composites, functionally graded materials, coatings, nanowires, porous components, membranes, and more.
- **Nanostructured Materials and Systems, Ceramic Transactions, Vol. 214,** Sanjay Mathur and Hao Shen, Editors; Mrityunjay Singh, Volume Editor
  Includes papers on the latest developments related to synthesis, processing and manufacturing technologies of nanoscale materials and systems including one-dimensional nanostructures, nanoparticle-based composites, electrospinning of nanofibers, functional thin films, ceramic membranes, bioactive materials and self-assembled functional nanostructures and nanodevices.
- **Design, Development, and Applications of Engineering Ceramics and Composite Systems, Ceramic Transactions, Vol. 215, Dileep Singh, Dongming Zhu, and Yanchum Zhou; Mrityunjay Singh, Volume Editor**
  Includes papers on design, processing and application of a wide variety of materials ranging from SiC SiAlON, ZrO_2, fiber reinforced composites; thermal/environmental barrier coatings; functionally gradient materials; and geopolymers.
- **Advances in Multifunctional Materials and Systems, Ceramic Transactions, Vol. 216,** Jun Akedo, Hitoshi Ohsato, and Takeshi Shimada, Editors; Mrityunjay Singh, Volume Editor
  Topics dealing with advanced electroceramics including multilayer capacitors; ferroelectric memory devices; ferrite circulators and isolators; varistors; piezoelectrics; and microwave dielectrics are included.
- **Ceramics for Environmental and Energy Systems, Ceramic Transactions, Vol. 217,** Aldo Boccaccini, James Marra, Fatih Dogan, and Hua-Tay Lin, Editors; Mrityunjay Singh, Volume Editor
  This volume includes selected papers from four symposia: Glasses and Ceramics for Nuclear and Hazardous Waste Treatment; Solid Oxide Fuel Cells and Hydrogen Technology; Ceramics for Electric Energy Generation, Storage, and Distribution; and Photocatalytic Materials.
- **Advances in Bioceramics and Biotechnologies, Ceramic Transactions, Vol. 218;** Roger Narayan and Joanna McKittrick, Editors; Mrityunjay Singh, Volume Editor
  Includes selected papers from two cutting edge symposia: Nano-Biotechnology and Ceramics in Biomedical Applications and Advances in Biomineralized Ceramics, Bioceramics, and Bioinspiried Designs.

I would like to express my sincere thanks to Greg Geiger, Technical Content Manager of The American Ceramic Society for his hard work and tireless efforts in the publication of this series. I would also like to thank all the contributors, editors, and reviewers for their efforts.

MRITYUNJAY SINGH
Volume Editor and Chairman, PACRIM-8
Ohio Aerospace Institute
Cleveland, OH (USA)

Synthesis and Processing by the Spark Plasma Method

SIMULATION OF CONTACT RESISTANCES INFLUENCE ON TEMPERATURE DISTRIBUTION DURING SPS EXPERIMENTS

A. Cincotti[1], A.M. Locci[1,*], R. Orrù[1], G. Cao[1,2,*]

[1]Dipartimento di Ingegneria Chimica e Materiali, Unità di Ricerca del Consorzio Interuniversitario Nazionale di Scienza e Tecnologia dei Materiali (INSTM), Unità di Ricerca del Consiglio Nazionale delle Ricerche (CNR), Università degli Studi di Cagliari, Piazza d'Armi, 09123 Cagliari, Italy
[2]CRS4 – Centro di Ricerca, Sviluppo e Studi Superiori in Sardegna, Parco Scientifico e Tecnologico, POLARIS, Edificio 1, 09010 PULA (CA), Italy

* Authors to whom correspondence should be addressed

ABSTRACT

The behavior of the Spark Plasma Sintering/Synthesis (SPS) apparatus, which represents an effective tool for sintering/synthesizing advanced materials, is simulated in this work. A step-by-step heuristic procedure is proposed since several, concomitant physico-chemical phenomena, for example heat transfer and generation, electric current transport, and stress-strain mechanics along with chemical transformation and sintering, take place during SPS processes. In this work we consider the SPS behavior of specific sample configurations characterized by the absence of powders. This approach permits to determine the electric and thermal resistances experimentally evidenced in the horizontal contacts between stainless steel electrodes and graphite spacers as functions of temperature and applied mechanical load. Horizontal contact resistances between graphite elements are experimentally found to be negligible and, accordingly, they are not modeled. Model reliability is tested by comparing numerical simulations with experimental data obtained at operating conditions far from those adopted during fitting procedure of unknown parameters. The proposed model can be successfully compared from a quantitative point of view to the measured temperature, voltage once rms current, geometry are taken into account.

INTRODUCTION

SPS is an effective process for the sintering/synthesis of advanced materials like ceramics, metals, polymers and semiconductors.[1] Basically, it consists in heating up the powder sample shaped into a die inserted between two water cooled electrodes (rams) by means of a pulsed electric DC forced to pass through, while uniaxially pressing the system in order to facilitate sintering processes and guarantee electric circuit closure.

In the technical literature, SPS is considered a thermo efficient sintering process since highly dense products in relatively short times are attainable.[1-8] A volumic heating rate due to joule effect, in contrast to the conductive heat transport applied in conventional sintering systems, permits a quick temperature rise able to enhance the mass transport mechanisms responsible for sintering phenomena, thus improving consolidation rate and minimizing grain coarsening. The latter aspect leads to improved mechanical, physical and optical properties of final sintered products.[2]

While an updated review of modeling approaches adopted to simulate the behavior of SPS apparatus is reported elsewhere[1], a reliable mathematical model of SPS can be obtained in our view by separately analyzing an increasing complex system behavior in the framework of a step by step procedure, where physico-chemical phenomena, previously excluded, are gradually introduced along with their unknown model parameters. This approach allows one to independently fit the complete set of unknown parameters of the comprehensive SPS model, thus avoiding the masking effect given by the various phenomena involved in the whole process. In this work, the first step of this ideal approach is carried out by taking into account heat transfer phenomena, and current distribution. In particular, the evaluation of the predominant electric and thermal contact resistances is carried out by comparing model results with experimental data obtained when appropriate samples characterized by the absence of powders are used. Specifically, explicit dependence of horizontal electric and thermal contact resistances on applied load and local temperature is obtained.

EXPERIMENTAL SECTION

A SPS apparatus 515S model (Sumitomo Heavy Industries Ltd., Japan) is used for the experimental runs. The power supply is a DC pulse generator which is reported to provide a maximum current and voltage equal to 1500 A and 10 V, respectively, while the mechanical load applied through an hydraulic system can be varied between 3 and 50 kN. Specifically, current pulses of 3.3 ms fixed duration are generated. Operator is free to select the pulse sequence, i.e. number of ON pulses (from 1 to 99) vs. number of OFF pulses (from 1 to 9) that represent periods of time with zero current. Typically, a 12/2 sequence is adopted (as prescribed by Sumitomo). This choice corresponds to the repetition of a sequence of 12 ON pulses followed by 2 OFF pulses for a total sequence period of 46.2 ms (i.e. 3.3x14 ms). It should be mentioned that no specifications are available regarding the measured current and voltage, i.e. average or rms values. Referring to Figure 1, the sample is inserted into a die placed between two plungers that are not in direct contact with the stainless steel rams, but spacers are typically inserted in between. From the electric point of view, the end parts of the rams are connected to an electric generator through copper bars and wires. Spacers, plungers and die are made of AT101-grade graphite (ATAL, Italy) which guarantees relatively high electric and thermal conductivities, i.e. lower power dissipation, higher heat transfer to powder specimen, and quicker cooling step. The use of graphite limits the attainable pressure level to a value less than 100 MPa, while the vacuum chamber permits to avoid chemical oxidation of graphitic elements. As it may be seen in Figure 1, the vacuum chamber is made of two coaxial cylinders both jacketed with cooling water circulation. A vacuum level down to 10 Pa is attainable with the SPS 515S model. Rams are made of stainless steel (AISI 304) and cooling water flows through them, as depicted in Figure 1, where the corresponding horizontal section a-a of the water circuit is also shown.

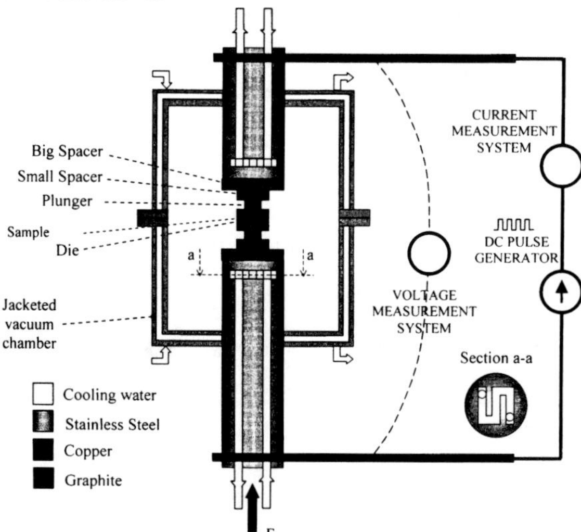

Figure 1: Schematic representation of SPS experimental set-up (not in scale).

A new data acquisition system has been designed and installed for independently measuring instantaneous (pulsed) values of electric current and voltage, from which calculating average and rms values. In particular, referring to Figure 1, an open loop Hall effect current transducer has been used (LEM HAX 2500-S, nominal primary current 2500 A rms, maximum primary current 5500 A, bandwidth 25 kHz, accuracy 1 % at nominal current) along with a voltage isolation amplifier (DATAFORTH DSCA41-09, Input range -40 to + 40 V, bandwidth 3 kHz, accuracy 0.03 % of full scale). The latter one is connected to the copper bars right close to the stainless steel electrodes. The output signals of these transducers are fed to a data acquisition board (200 kS/s, 12-Bit, 16 Analog Input Multifunction, National Instruments) connected to a PC, where a specifically designed Labview (National Instruments) virtual instrument is installed. This data acquisition system is able to collect instantaneously current and voltage measurements and calculate the corresponding average and rms values (sampling time $\tau = 0.5$ s), along with all the other variables typically measured in SPS processes (i.e. time, temperature, displacement, load, and gas pressure).

Specific sample configurations characterized by the absence of powders are considered in this work. In particular, we used the graphite cylindrical samples reported in Table I, along with the size of upper and lower stainless steel electrodes provided with SPS 515S model. Graphite samples have been inserted between rams during experimental runs. It should be noted that sample IV consists of two big spacers, two small spacers and one monolithic block in order to avoid vertical contact resistances, mimicking two plungers slid into a die.

MODELLING SECTION

Due to heat losses by radiation from lateral surfaces as well as heat removal by cooling water in axial direction, along with variations of cross sections, a 2D model for the energy balance of SPS technique is proposed, while radial symmetry is considered. Vertical symmetry cannot be assumed due to different heights of stainless steel electrodes and the corresponding cooling circuits. Although isotropic materials are considered, temperature variation in radial and axial directions induces spatial gradients of thermophysical properties like electric and thermal conductivities and coefficient of thermal expansion.

The energy balance in cylindrical coordinates (r,z) related to the stainless steel rams as well as the graphite samples depicted in Figure 1 and Table I is given by:

$$\rho_i\, C_{p,i}\frac{\partial T}{\partial t}=\frac{1}{r}\frac{\partial}{\partial r}\left(r\,k_i\frac{\partial T}{\partial r}\right)+\frac{\partial}{\partial z}\left(k_i\frac{\partial T}{\partial z}\right)+\frac{1}{\rho_{el,i}}\left[\left(\frac{\partial \varphi}{\partial r}\right)^2+\left(\frac{\partial \varphi}{\partial z}\right)^2\right];\ i=\begin{cases}\text{Stainless Steel}\\ \text{Graphite}\end{cases} \tag{1}$$

with the initial condition $T = T_0$ at $t=0$, while boundary conditions are reported in Figures 2 and 3. The meaning of the other symbols is reported in the Nomenclature Section. Only contact resistances at stainless steel-graphite interfaces are considered, with a local joule heat (q_e) due to electric contact resistance,[9] which has been equally split between the materials at the interface (i.e. $f = 0.5$ is considered when solving the model). The following steady-state conduction model:

$$\frac{1}{r}\frac{\partial}{\partial r}\left(\frac{1}{\rho_{el,i}}\,r\frac{\partial \varphi}{\partial r}\right)+\frac{\partial}{\partial z}\left(\frac{1}{\rho_{el,i}}\frac{\partial \varphi}{\partial z}\right)=0\quad ;\quad i=\begin{cases}\text{Stainless Steel}\\ \text{Graphite}\end{cases} \tag{2}$$

coupled with the boundary conditions reported in Figures 2 and 3, is adopted for describing the electrical behavior inside the SPS system. Only contact resistances at stainless steel-graphite interfaces are considered, while equipotential conditions for electrode surfaces in contact with copper bars (cf.

Figure 1) have been adopted. The resistive portion of the rms voltage (φ) is given by the equation $\varphi = R\ I_{RMS}$, where resistance is determined from the measured average voltage and electric current ($R = \bar{V}/\bar{I}$). The pseudo isostatic equilibrium model adopted to simulate the mechanical behavior of SPS systems is reported elsewhere[10] for sake of brevity.

Table I: Samples configurations and dimensions (not in scale) of graphite and stainless steel elements of the SPS system investigated.

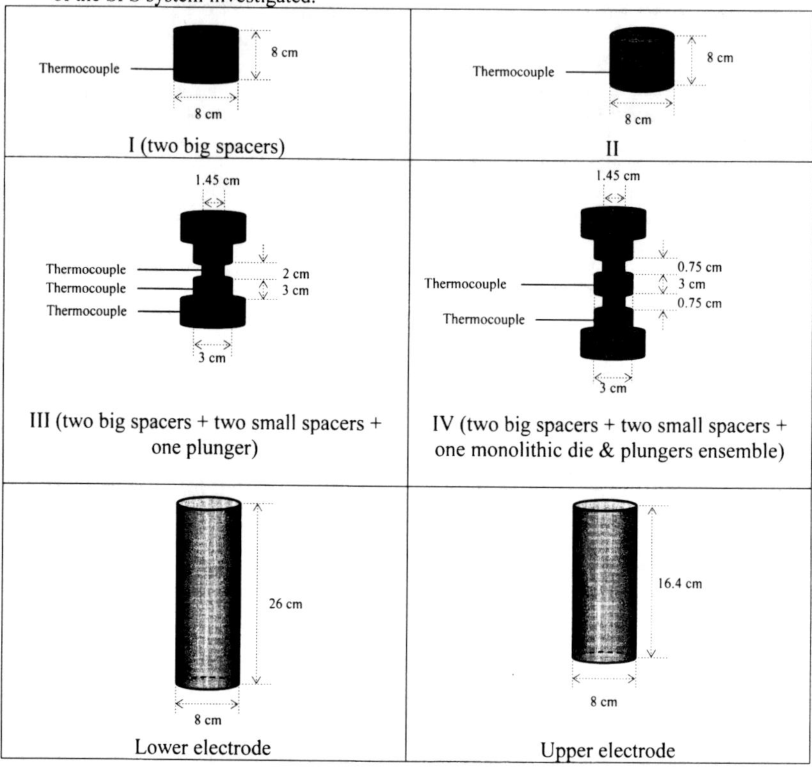

The resulting system of differential-algebraic-integral equations has been solved by FEM numerical technique. In particular, the commercial software COMSOL MULTYPHISYCS® 3.2 has been adopted (numbers of DOF and elements equal to about 25000 and 3000, respectively). Parameters used for computations are reported in Table II and Figure 4. In particular, heat transfer coefficient (h) is calculated as discussed elsewhere while heat graphite thermal conductivity determination deserves a comment, since different data are reported in the literature. In this work, $k_G(T)$ is evaluated by averaging the temperature dependences given by the two different references available in the literature consistent with the only value (100 W m^{-1} °C^{-1} at 25 °C) provided by ATAL vendor for the graphite AT101. The only unknown model parameters remain therefore thermal and electric conductances, C_T and C_E, at the horizontal contacts between stainless steel electrodes and graphite spacers (cf. see boundary conditions in Figures 2-3). The determination of the dependence of these two parameters on temperature and applied mechanical load is described in the following section.

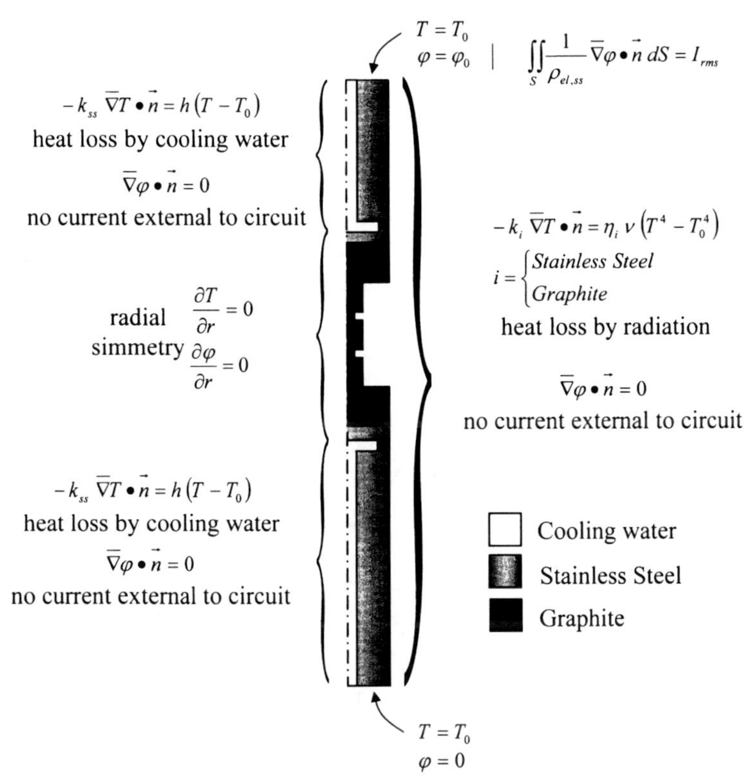

Figure 2: Initial and boundary conditions (except those involving contacts resistances) for the energy balance and steady-state electric conduction model equation (not in scale).

RESULTS AND DISCUSSION

In what follows, the results related to the comparison between experimental data and model results will be illustrated by considering both the fitting procedure adopted to evaluate the unknown parameters and the prediction capability of the proposed SPS apparatus model.

Figure 5 reports the direct comparison between temperature and voltage temporal profiles when sample I and sample II are used under the same operating conditions, i.e. a rectangular profile of rms current (amplitude 1200 A, 35 min duration) at 12/2 pulse sequence, and an initial mechanical load equal to 3 kN. It clearly follows that horizontal contact resistances between graphite elements can be neglected. Figure 6 reports the same comparison when a graphite foil (0.13 mm thick, Alfa Aesar) is inserted at the contacts of sample I with stainless steel electrodes.

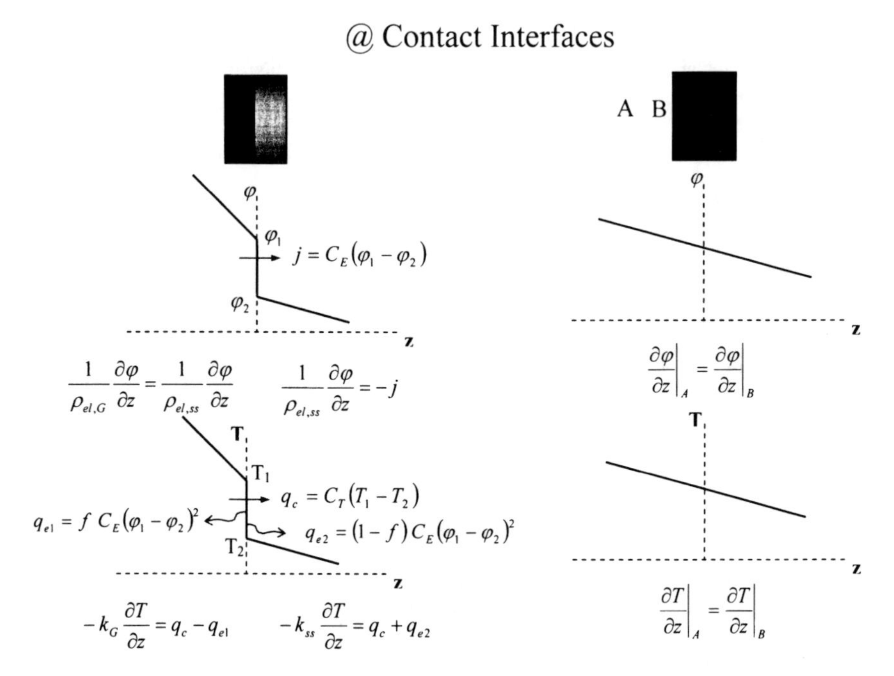

Figure 3: Boundary conditions involving contacts resistances for the energy balance and the steady-state electric conduction model (not in scale).

The experimental runs are repeated several times as reported in Figures 5 and 6 in order to appreciate the reproducibility level obtainable with SPS systems. Since significant differences are found in terms of both temperature and voltage temporal profiles, it is apparent that both thermal and electric contact resistances between graphite and stainless steel elements need to be taken into account. Indeed, according to Madhusadana,[21] conducting interstitial or filler material inserted in the gaps left by actual solid-solid point contact, increases the real surface of contact between interfacing elements, thus reducing constriction resistances. Therefore, in our case the presence of graphite foil reduces the relevant horizontal contact resistances, so that lower temperature and voltage temporal profiles are obtained. Presumably, machined graphite parts used in this work possess a lower surface roughness than that of stainless steel electrodes provided with the SPS 515S model. It is worth noting that, in order to experimentally highlight the presence of thermal and electric contact resistances in horizontal position, the lowest applicable mechanical load and the higher nominal contact surface among the available samples (cf. Table I) have been used.

According to Madhusadana[21] and Babu *et al.*,[22] thermal and electric constriction conductances are related to temperature and applied mechanical load as follows:

$$C_T(T,P) = \alpha_T\, k_{Harm}\, (P/H_{Harm})^{\beta_T} \tag{3}$$

$$C_E(T,P) = \alpha_E \, \sigma_{el,Harm} \, (P/H_{Harm})^{\beta_E} \tag{4}$$

where *P (=F/S)* represents the mechanical pressure uniformly applied at the contact surface area (*S*) between graphite and stainless steel.

Table II: Model parameters.

Parameter	Value	Reference
$C_{p,G}$	see Figure 4	11
$C_{p,ss}$	see Figure 4	12
E_G	1.1 10^4 [N mm^{-2}]	ATAL vendor
E_{ss}	19.3 10^4 [N mm^{-2}]	13
f	0.5	9
h	4725 [W m^{-2} K^{-1}]	This work
H_G	3.5 10^9 [Pa]	ATAL vendor
H_{ss}	1.92 10^9 [Pa]	14
k_G	see Figure 4	ATAL vendor; 12; 15
k_{ss}	see Figure 4	16
T_0	298.15 [K]	This work
α_G	see Figure 4	17
α_{ss}	see Figure 4	13
η_G	0.85	6; 8; 9; 18
η_{ss}	0.4	19
ν	5.67 10^{-8} [W m^{-2} K^{-4}]	20
ρ_G	1750 [kg m^{-3}]	ATAL vendor
ρ_{ss}	8000 [kg m^{-3}]	12
$\rho_{el,G}$	see Figure 4	ATAL vendor
$\rho_{el,ss}$	see Figure 4	12
υ_G	0.33	This work
υ_{ss}	0.29	13

The parameters k_{Harm} and $\sigma_{el,Harm}$ take into account the temperature dependence of contact conductances, and are expressed as follows using the harmonic mean of the individual thermal and electric conductivities of graphite and stainless steel:

$$k_{Harm}(T) = \frac{2\, k_G(T)\, k_{ss}(T)}{k_G(T) + k_{ss}(T)} \tag{5}$$

$$\sigma_{el,Harm}(T) = \frac{2\, \sigma_{el,G}(T)\, \sigma_{el,ss}(T)}{\sigma_{el,G}(T) + \sigma_{el,ss}(T)} \tag{6}$$

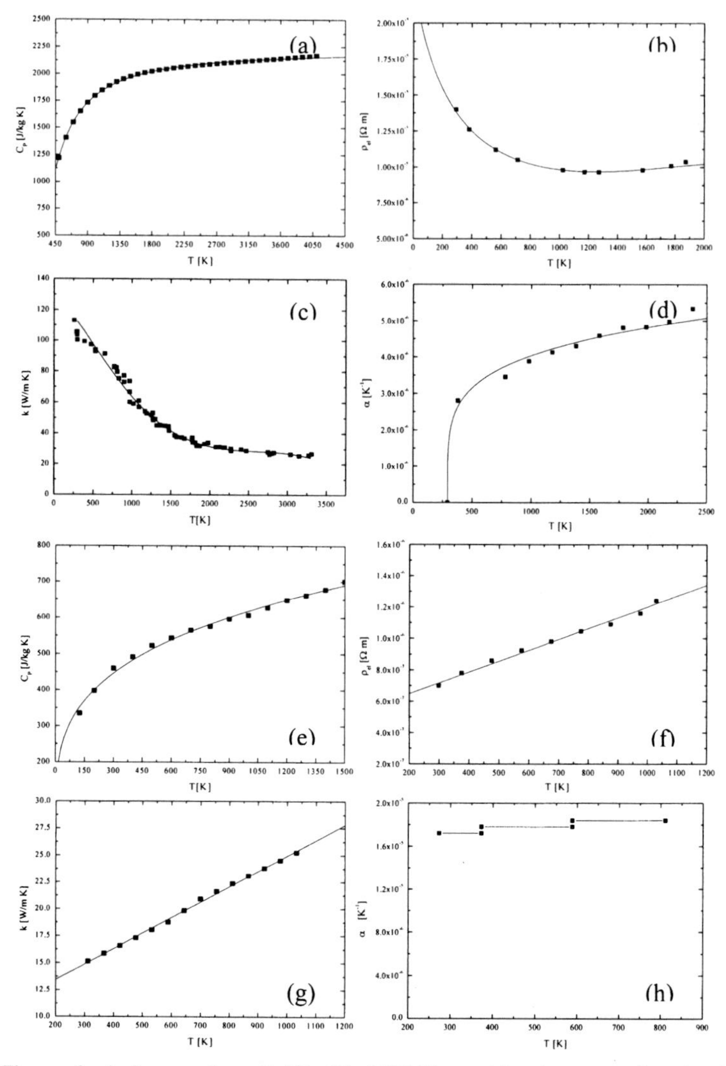

Figure 4: Thermophysical properties of AT 101 (ATAL) graphite: heat capacity a), electrical resistivity b), thermal conductivity c), and coefficient of thermal expansion d), and AISI 304 stainless steel: heat capacity e), electrical resistivity f), thermal conductivity g), and coefficient of thermal expansion h). Dots represent experimental data taken from the literature as specified in Table II, while lines are the corresponding fitting curves adopted for the numerical simulation.

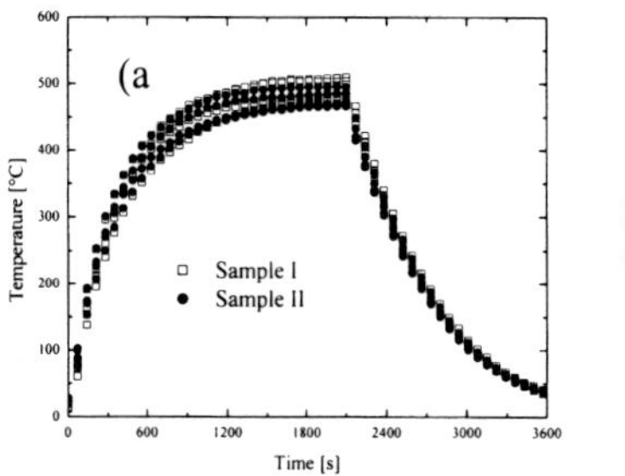

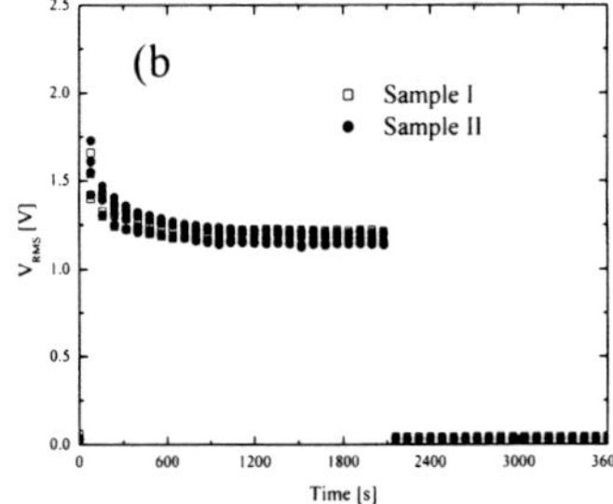

Figure 5: Comparison between sample I and sample II in terms of temporal profiles of measured temperature (r = 0, z = 28 cm) a) and rms voltage b), when a rms current step of 1200 A is applied for 35 min with a initial load of 3 kN.

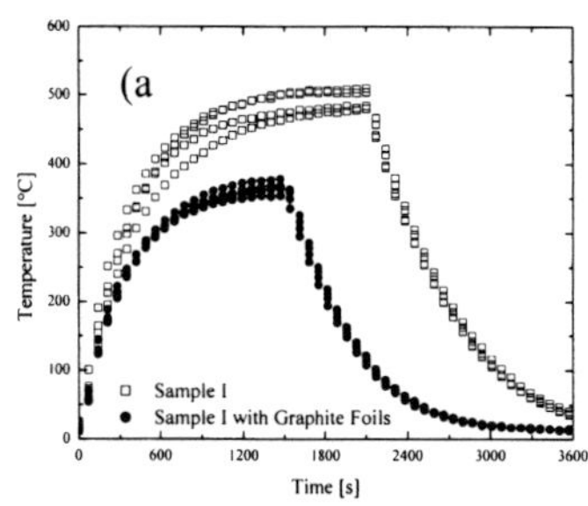

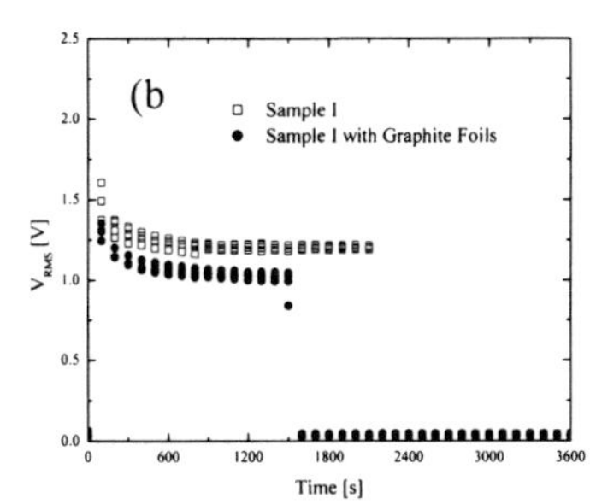

Figure 6: Comparison between sample I with and without graphite foils at stainless steel/graphite interfaces in terms of temporal profiles of measured temperature (r = 0, z = 28 cm) a) and rms voltage b), when a rms current step of 1200 A is applied for 35 min and 25 min, correspondingly, with a initial load of 3 kN.

Basically, the use of harmonic mean is a consequence of the series combination of the two interfacing materials, which characterize the real contact. Analogously, in this work the parameter H_{Harm} is the harmonic mean between graphite and stainless steel hardnesses:

$$H_{Harm} = \frac{2\, H_G\, H_{ss}}{H_G + H_{ss}} \tag{7}$$

The remaining parameters appearing in Eqs. 3-4 (i.e., α_T, β_T, α_E, and β_E,) are the only adjustable ones in our model. They depend on surface roughness and plastic behavior of contact between graphite and stainless steel interfaces. These four parameters have been fitted by direct comparison between modeling results and experimental data in terms of temporal profiles of resistive portion of rms voltage measured between electrodes' ends, φ_0, and temperature measured by a thermocouple placed in axial position inside sample I, as depicted in Table I. In particular, experimental runs conducted under applied initial mechanical loads in the range 3-50 kN, at a constant I_{RMS} of 1200 A at 12/2 pulse cycle, have been performed. Experimental data for the case of 3, 20 and 50 kN are compared with modeling results in Figures 7-9, respectively, while fitted parameters are reported in Table III along with the 95% confidence band and correlation coefficients.

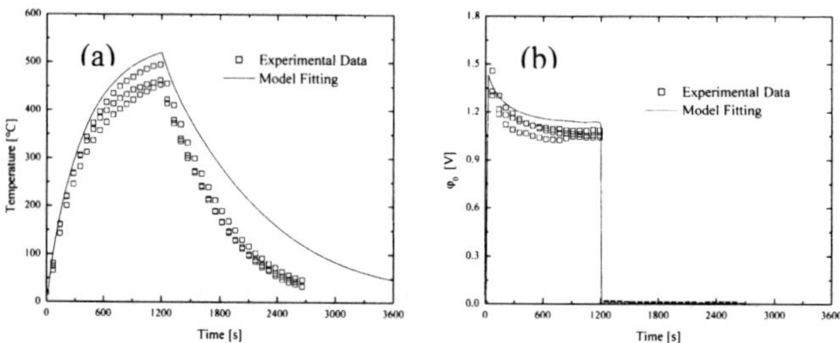

Figure 7: Comparison between experimental data and model results in terms of temporal profiles of temperature a) taken at the center of the lower big spacer (r = 0, z = 28 cm), resistive portion of rms voltage b), when sample I is submitted to a rms electric current step of 1200 A for 20 min, with a load of 3 kN.

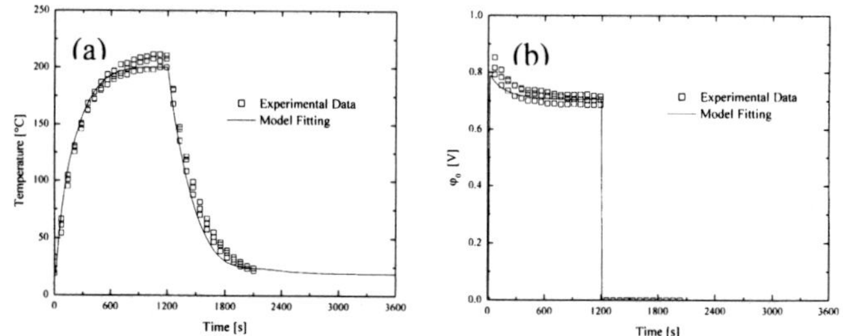

Figure 8: Comparison between experimental data and model results in terms of temporal profiles of temperature a) taken at the center of the lower big spacer (r = 0, z = 28 cm), resistive portion of rms voltage b), when sample I is submitted to a rms electric current step of 1200 A for 20 min, with a load of 30 kN.

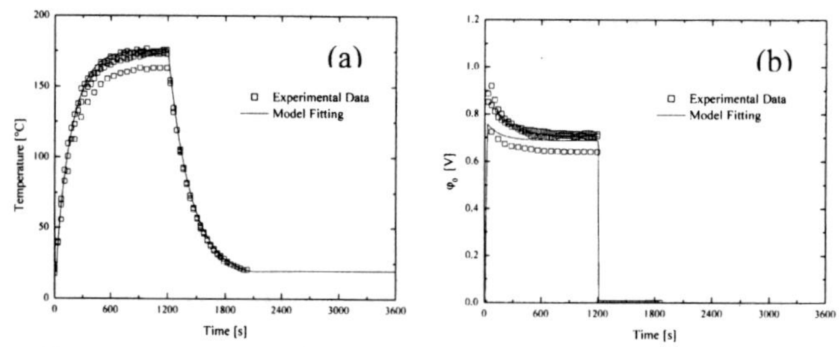

Figure 9: Comparison between experimental data and model results in terms of temporal profiles of temperature a) taken at the center of the lower big spacer (r = 0, z = 28 cm), resistive portion of rms voltage b), when sample I is submitted to a rms electric current step of 1200 A for 20 min, with a load of 50 kN.

Table III: Fitted parameters of contact conductances along with 95% confidential band, and correlation coefficient R.

Parameter	Value ± Error	R
α_T	22810 ± 4.6 [m^{-1}]	0.9368
β_T	1.08 ± 0.23	
α_E	64 ± 1.38 [m^{-1}]	0.9609
β_E	0.35 ± 0.05	

The corresponding values of electric and thermal conductances obtained in the range of temperature and mechanical load investigated in this work are 0.25÷1.7 10^6 Ω^{-1} m^{-2} and 0.007÷2 10^3 W m^{-2} K^{-1}, respectively. It is worth noting that constant values for electric and thermal contact conductances in horizontal position equal to 1.25 10^7 Ω^{-1} m^{-2} and 2.4 10^3 W m^{-2} K^{-1}, respectively, were obtained by Zavaliangos *et al.*[9]. These discrepancies may be ascribed to the fact that Zavaliangos *et al.*[9] considered the electrodes made of graphite (instead of stainless steel), whose thermophysical properties are rather different from those used in the present paper. A satisfactory simulation of SPS system behavior in terms of temperature, voltage is obtained, especially at higher applied mechanical loads, when temperature and voltage decrease due to lower contact resistances.

Model reliability is further tested by comparing model results with experimental data obtained using samples III and IV and adopting operating conditions significantly far from those used during fitting procedure. In particular, the I_{RMS} value is decreased from 1200 A to 670 A and 980 A. It is worth noting that also sample geometry is changed and, accordingly, a different joule heat generation is expected to produce different temperature distribution, since electric resistance varies as function of cross section surface area. For the same reason, voltage changes as well. The obtained results are shown in Figures 10-11.

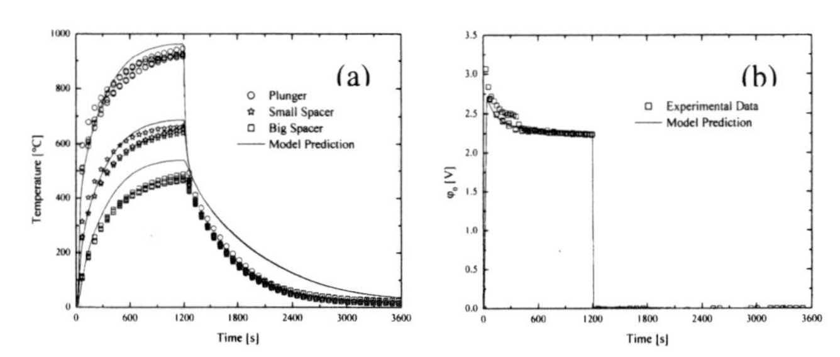

Figure 10: Comparison between experimental data and model predictions in terms of temporal profiles of temperature a), resistive portion of rms voltage b), when sample III is submitted to a rms electric current step of 670 A for 20 min, with a load of 3.5 kN. Temperature is taken at three different locations, i.e. center of lower big spacer (r = 0, z = 28 cm), center of lower small spacer (r = 0, z = 31.5 cm), and center of plunger (r = 0, z = 34 cm).

It is seen that the developed model is able to entirely predict SPS system behavior in terms of temperature, voltage, and displacement temporal profiles. In order to further highlight the importance of taking into account contact resistances, Figure 11 reports model predictions when the fitted contact resistances are neglected (i.e. model equations are solved by setting very high contact conductances).

In this latter case, temporal profiles of temperatures and voltage are underestimated. As expected, the larger discrepancy is related to temperature taken in axial position of lower graphite small spacer, i.e. the one closer to the relevant contact resistances. Moreover, cooling dynamics, occurring once the electric current is turned off, is overestimated so that displacement rises relatively too quickly to the initial value. It is apparent that sample IV has been chosen to make these considerations, since, among those investigated in this work, is the most similar in size to the actual configuration used when sintering/synthesis experiments are performed.

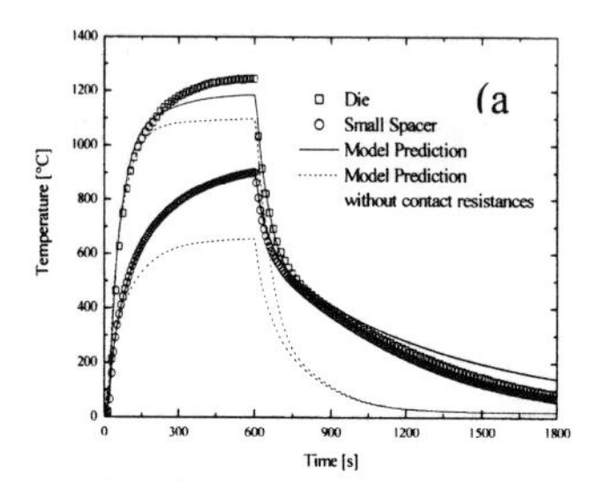

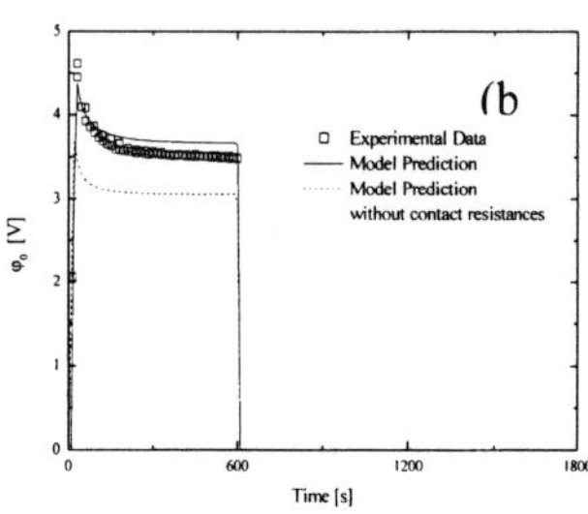

Figure 11: Comparison between experimental data and model predictions in terms of temporal profiles of temperature a), resistive portion of rms voltage b), and displacement c), when sample IV is submitted to a rms electric current step of 980 A for 10 min, with a load of 3.5 kN. Temperature is taken at two different locations, i.e. center of lower small spacer (r = 0, z = 31.5 cm) and inside the die (r = 1.1 cm, z = 32.25 cm).

CONCLUDING REMARKS

In this paper, the quantitative description of SPS system behavior in the case of the absence of powders is addressed in the framework of a novel heuristic approach. In particular, electric and thermal resistances in horizontal contacts as functions of temperature and applied mechanical load have been obtained by direct comparison between model results and experimental data. The electric behavior of the system is described in details by highlighting the importance of considering rms electric current and voltage (whatever pulse cycle is adopted) when joule effect needs to be quantitatively determined. Model reliability is for the first time tested by comparing model predictions with experimental data obtained at operating conditions significantly far from those ones considered during fitting procedure. The proposed model has been successfully compared from a quantitative point of view to the measured temperature, voltage and displacement when rms current, geometry and initial mechanical load are set. This work represents the first step towards the development of a complete model for the simulation of SPS system behavior where sintering/synthesis of powders sample takes place.

ACKNOWLEDGMENTS

IM (Innovative Materials) Srl, Italy is gratefully acknowledged for granting the use of SPS apparatus.

NOMENCLATURE

C_E	electric contact conductance, [Ω^{-1} m^{-2}]
$C_{p,G}$	graphite heat capacity, [J kg^{-1} K^{-1}]
$C_{p,ss}$	stainless steel heat capacity, [J kg^{-1} K^{-1}]
C_T	thermal contact conductance, [W m^{-2} K^{-1}]

E_G	graphite Young's modulus, [N mm^{-2}]
E_{ss}	stainless steel Young's modulus, [N mm^{-2}]
F	mechanical load, [N]
f	fraction of localized joule heat due to electric contact resistance
H_G	graphite hardness, [N m^{-2}]
H_{Harm}	harmonic mean between graphite and stainless steel hardness, [N m^{-2}]
H_{ss}	stainless steel hardness, [N m^{-2}]
h	heat transfer coefficient, [W m^{-2} K^{-1}]
I	instantaneous electric current, [A]
$\bar{I}$	temporal average electric current, [A]
I_{rms}	temporal root mean squared electric current, [A]
j	rms current density, [A m^{-2}]
k_G	graphite thermal conductivity, [W m^{-1} K^{-1}]
k_{Harm}	harmonic mean between graphite and stainless steel thermal conductivities, [W m^{-1} K^{-1}]
k_{ss}	stainless steel thermal conductivity, [W m^{-1} K^{-1}]
L	self-inductance, [H]
$\vec{n}$	surface area unit vector in outward direction
P	applied mechanical pressure, [N m^{-2}]
q_c	heat flux due to thermal contact resistance, [W m^{-2} K^{-1}]
q_e	localized joule heat flux due to electric contact resistance, [W m^{-2} K^{-1}]
R	electric resistance, [Ω]
T	temperature, [K]
T_0	initial ambient temperature, [K]
t	time, [s]
r	radial coordinate, [m]
S	surface area, [m^2]
V	instantaneous voltage, [V]
$\bar{V}$	temporal average voltage, [V]
V_L	instantaneous inductive voltage, [V]
V_R	instantaneous resistive voltage, [V]
V_{rms}	temporal root mean squared voltage, [V]
z	axial coordinate, [m]

Greek letters

α_E	adjustable parameter in Eq. 4, [m]
α_G	graphite coefficient of thermal expansion, [K^{-1}]
α_{ss}	stainless steel coefficient of thermal expansion, [K^{-1}]
α_T	adjustable parameter in Eq. 3, [m]
β_E	adjustable parameter in Eq. 4
β_T	adjustable parameter in Eq. 3
φ	resistive rms voltage, [V]

φ_0	resistive rms voltage between electrodes ends, [V]
η_G	graphite emissivity
η_{ss}	stainless steel emissivity
ν	Stefan-Boltzmann constant, [W m^{-2} K^{-4}]
ρ_G	graphite density, [kg m^{-3}]
ρ_{ss}	stainless steel density, [kg m^{-3}]
$\rho_{el,G}$	graphite electric resistivity, [Ω m]
ρ_{Harm}	harmonic mean of graphite and stainless steel electric resistivities, [Ω m]
$\rho_{el,ss}$	stainless steel electric resistivity, [Ω m]
$\sigma_{el,G}$	graphite electric conductivity $\left(=\dfrac{1}{\rho_{el,G}}\right)$, [Ω^{-1} m^{-1}]
$\sigma_{el,Harm}$	harmonic mean of graphite and stainless steel electric conductivities, [Ω^{-1} m^{-1}]
$\sigma_{el,ss}$	stainless steel electric conductivity $\left(=\dfrac{1}{\rho_{el,ss}}\right)$, [Ω^{-1} m^{-1}]

REFERENCES

[1]R. Orrù, R. Licheri, A.M. Locci, A. Cincotti, G. Cao, Consolidation/synthesis of materials by electric current activated/assisted sintering, *Mat. Sci. Eng. R: Reports*, **63**, 127-287 (2009).
[2]M. Omori, Sintering, consolidation, reaction and crystal growth by the spark plasma sintering system (SPS), *Mat. Sci. Eng A*, **287**, 183-188 (2000).
[3]J.R. Groza, A. Zavaliangos, Sintering activation by external electrical field, *Mat. Sci. Eng. A*, **287**, 171-177 (2000).
[4]W. Yucheng, F. Zhengyi, Study of temperature field in spark plasma sintering, *Mat. Sci. Eng. B*, **90**, 34-37 (2002).
[5]G. Xie, O. Ohashi, K. Chiba, N. Yamaguchi, M. Song, K. Furuya, T. Noda, Frequency effect on pulse electric current sintering process of pure aluminum powder, *Mat. Sci. Eng. A*, **359**, 384-390 (2003).
[6]R.S. Dobedoe, G.D. West, M.H. Lewis, Spark plasma sintering of ceramics: understanding temperature distribution enables more realistic comparison with conventional processing, *Adv. Appl. Ceram.*, **104**, 110-116 (2005).
[7]W. Chen, U. Anselmi-Tamburini, J.E. Garay, J.R. Groza, Z.A. Munir, Fundamental investigation on the spark plasma sintering/synthesis process. I. Effect of dc pulsing on reactivity, *Mat. Sci. Eng. A*, **394**, 132-138 (2005).
[8]K Vanmeensel, A. Laptev, J. Hennicke, J. Vleugels, O.Van der Biest, Modelling of the temperature distribution during field assisted sintering, *Acta Mater.*, **53**, 4379-4388 (2005).
[9]A. Zavaliangos, J. Zhang, M. Krammer, J.R. Groza, Temperature evolution field activated sintering. *Mat. Sci. Eng. A*, **379**, 218-228 (2004).
[10]A. Cincotti, A.M. Locci, R. Orrù, G. Cao, Modelling Spark Plasma Sintering/Synthesis: Horizontal Contact Resistances Determination, *AIChE Journal*, **53**, 703-719 (2007).
[11]I. Barin, *Thermochemical data of pure substances*. New York: VCH, 1993.
[12]Y.S. Touloukian, *Thermophysical properties of high temperature solid materials*. New York: MacMillan Co. 1967.
[13]ASM International. *High-temperature property data: ferrous alloys*. USA: ASM International, 1988.
[14]CRC, *Materials science and engineering handbook*, (3rd Edition). Boca Raton: CRC Press LLC, 2001.

[15]IIT Research Institute. *Handbook of thermophysical properties of solid materials.* New York: Macmillan, 1961.
[16]R.L. Peters, *Materials data nomographs.* New York: Reinhold Publishing Corporation, Chapman & Hall, Ltd., 1965.
[17]A. Goldsmith, T.E. Waterman, H.J. Hirschhorn. *Handbook of thermophysical properties of solid materials.* Oxford: Pergamon Press, 1961.
[18]U. Anselmi-Tamburini, S. Gennari, J.E. Garay, Z.A. Munir, Fundamental investigation on the spark plasma sintering/synthesis process. II. Modeling of current and temperature distributions. *Mat. Sci. Eng. A*, **394**, 139-148 (2005).
[19]F.P. Incropera, D.P, DeWitt, *Fundamentals of heat and mass transfer.* New York: Jon Wiley & Sons, 1981.
[20]CRC, *Handbook of chemistry and physics* (80th Edition). Boca Raton: CRC Press LLC, 2000.
[21]C.V. Madhusadana, *Thermal contact conductance.* New York: Springer-Verlag, 1996.
[22]S.S. Babu, M.L. Santella, Z. Feng, B.W. Riemer, J.W. Cohron, Empirical model of effects of pressure and temperature on electrical contact resistance of metals, *Sci. Technol. Weld. Joining,* **6**, 126-132 (2001).

SPARK PLASMA SINTERING (SPS) PROCESSSING OF HIGH STRENGTH TRANSPARENT $MgAl_2O_4$ SPINEL POLYCRYSTALS

Koji Morita[†], Byung Nam Kim, Hidehiro Yoshida and Keijiro Hiraga
National Institute for Materials Science, Nano Ceramics Center
1 2 1 Sengen, Tsukuba, Ibaraki 305 0047, Japan.
[†] Corresponding Author: MORITA.Koji@nims.go.jp

ABSTRACT

In order to fabricate high strength transparent $MgAl_2O_4$ spinel polycrystals by employing a low heating rate SPS technique, the effect of the sintering temperature was examined. For a constant heating rate of 10°C/min, fine grained and high density spinel can be attained at about 1300°C for only a 20 min soak. For the short sintering time, the sintering temperatures lower than 1300°C are not enough to achieve high density spinel. The temperatures higher than 1350°C, on the other hand, cause rapid grain growth, resulting in the coalescence into larger pores and the precipitation of second phases. The spinel SPSed at the optimum temperature of 1300°C exhibits a good combination of in line transmission (50 70%) and three point bending strength (>500 MPa), simultaneously. The good optical and mechanical properties can be ascribed to the superimposed effects of the fine grain size, fine pore size and low density of the defects (pore and second phase), which are related closely to the sintering temperature during the low heating rate SPS processing. This suggests that a sintering temperature of 1300°C is effective for fabricating high strength transparent spinel polycrystals.

INTRODUCTION

Spark Plasma Sintering (SPS) processing is one of the novel sintering techniques and is now widely used to fabricate several polycrystalline materials including metal, ceramic and composite. The noteworthy feature of the SPS technique is available high heating rates, and hence, the SPS processing can complete the powder densification in a short processing time. Since the short processing time has been regarded as one of the primary advantages over the well known HIP and HP techniques, the SPS processing has usually been employed at high heating rates of typically ≥100°C/min [1 4].

In contrast to the previous studies, Kim et al. [5 7] have recently demonstrated that for attaining dense and fine grained oxide ceramics, a low heating rate SPS processing is more effective than the widely used high heating rate processing. For a widely used high heating rate of α = 100°C/min, alumina (α Al_2O_3) is low density and almost opaque, whereas for a low heating rate of α = 8°C/min, it has high density enough to show a good in line transmission even for only a 20 min soak and at a low temperature of 1175°C. Likewise, by employing the low heating rate SPS processing, a transparent spinel ($MgAl_2O_4$) having a in line transmission of T_{in} = 47% at a wavelength of λ = 550 nm can successfully be fabricated for only a 20 min soak at 1300°C and at α = 10°C/min [8 10]. For the present spinel, although the transmission can be improved with decreasing the heating rate ranging from 10 to 100°C/min, it shows the maximum value of $T_{in} \approx 50\%$ at $\alpha = 10$°C/min and is unchanged at $\alpha < 10$°C/min [8,9]. This suggests that the low heating rate of $\alpha = 10$°C/min can be regarded as an effective condition for attaining a transparent spinel in short processing time.

However, the attained light transmission, particularly in visible light wavelength range, is not enough for industrial applications, and hence, further improvement of the light transmission is necessary. The present study was therefore performed to examine an optimum SPS temperature condition in order to obtain high strength transparent spinel polycrystals.

EXPERIMENTAL PROCEDURES

Spark Plasma Sintering

The raw material used was a high purity $MgAl_2O_4$ spinel powder (TSP 15, Taimei Chemical Co., Ltd., Japan) with a purity of >99.97% (30 SiO_2, 50 Fe_2O_3, 100 NaO, <10 K_2O, 70 CaO in wt ppm). The average particle size of the raw powder is about 360 nm.

The densification was performed with a SPS machine (SPS 1050, SPS Syntex Inc., Japan) as described elsewhere [8]. Briefly, the powder was placed in a graphite die with a 30 mm inner diameter. Under the conditions of vacuum and at a uniaxial pressure of about 80 MPa, the powder was heated to the desired temperatures of 1250 1500°C at heating rates ranging from 2 to 100°C/min and held at the respective soaking temperatures for 20min under loading. In order to relieve any residual strain, the applied load was released to less than 5 MPa and then the specimens were subsequently held at 1150°C for 10 min. During the SPS processing, the temperature was measured on the surface of the graphite die. By this procedure, we fabricated a disk with a 30 mm diameter and 3 mm thickness.

Optical Properties

For optical characterization, square plates with dimensions of 12×12×1.8 mm were machined from the center of the SPSed circular disks. Both surfaces of the plates were carefully mirror polished with diamond pastes. The in line transmission T_{in} measurement was conducted in the visible and near IR wavelength range (λ = 240 2500 nm) using a double beam spectrophotometer (SolidSpec 3700DUV, Shimazu Co. Ltd., Japan) equipped with an integrating sphere.

Microstructural Examination

The microstructures were examined at around the center of the circular disks using a scanning electron microscope (SEM) and a transmission electron microscope (TEM).

The porosity measurement was performed using a SEM. For the SEM observations, the surface of the specimen was mechanically removed and mirror polished with a colloidal silica suspension. For the pore observations, the polished specimen was lightly thermal etched at 50 100°C below the sintering temperature for 30 min in air. The pore diameter, $2r$, the pore density, N and the average area fraction of the pore A_p and of the second phase A_s were measured using image analysis software (Image Pro plus, Media Cybernetics, USA) in the SEM images. In this study, N was defined as the number of pores per unit area. After the pore observations, the specimen was additionally thermal etched at 50 100°C below the sintering temperature for 30 min in order to more clearly reveal the grain boundaries. Using the SEM micrographs, the grain size was determined by counting the number of grains. Assuming the grains to be spherical, the average grain size, $\bar{d}$, was determined to be 1.225 times the apparent grain size [11].

For the TEM observations, thin sheets with a thickness of about 500 μm were cut using a low speed diamond cutter, and mechanically polished to a thickness less than 100 μm in thickness, followed by further thinning with an Ar ion milling machine. The TEM observations were performed using a JEOL 2010F operated at 200kV.

Fracture Strength

The fracture strength σ_f at room temperature was measured by three point bending tests using an Instron type machine. For the bending test, the rectangular bars with a cross section of 2.5 mm width and 1.8 mm thickness were machined from the fabricated disks by simply grinding using a diamond wheel. The bending test was performed at a 16 mm outer span length and a crosshead displacement rate of 0.5 mm/min. The fracture strength was evaluated by the following equations:

$$\sigma_{3f} = \frac{3PL}{2wt^2} \quad (1)$$

where P is the peak load, L is the outer span length and w and t are the width and the thickness of the bending specimen, respectively. In this study, the average fracture strength at each temperature was measured at five bending specimens.

EXPERIMENTAL RESULTS

Sintered Microstructures

Figure 1 shows the sintering temperature dependent microstructures of spinel SPSed at a constant heating rate of 10°C/min [9]. It is apparent from the SEM images that the sintering temperature strongly affects the microstructure of spinel. As shown in **Fig. 2(a)**, the grain size $\bar{d}$ increases from 0.4 to 1.2 μm as the temperature increases. The rate of the grain growth becomes significant at the sintering temperature of ≥1350°C.

Figures 2(b) and **(c)** shows the residual porosity A_p and the pore size distribution, respectively [9,10]. Both A_p and the pore size distribution are not simple function of the sintering temperature. A_p monotonously decreases from 0.5 to less than 0.1% with temperature at 1275 1400°C and it takes almost a constant value of about 0.1% at >1400°C. The pore size distribution, on the other hand, tends to decrease with the temperature at 1275 1325°C. Above 1350°C, where rapid grain growth occurs, the density of the residual pores is much lower as compared to that of the lower temperatures of ≤1325°C,

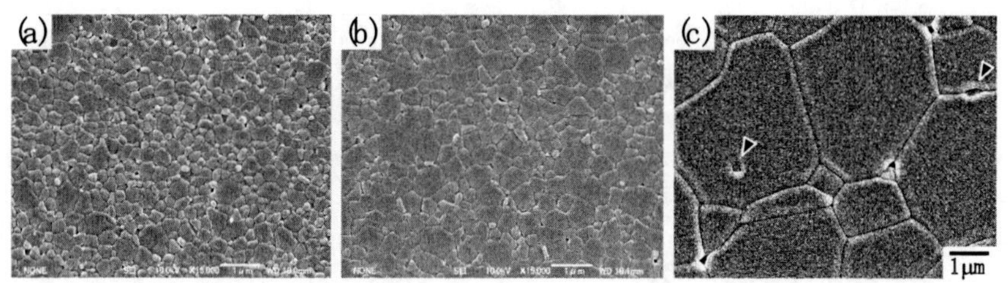

Figure 1. Typical SEM images of spinel SPSed for a 20min soak at (a) 1275°C, (b) 1325°C and (b) 1500°C [9].

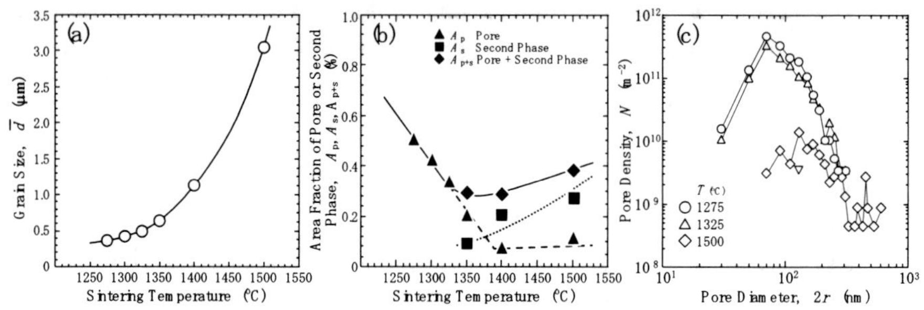

Figure 2. Sintering temperature dependent (a) grain size d, (b) average area fraction of pore A_p, second phase A_s and pore + second phase A_{s+p}, and (c) pore size distribution [9].

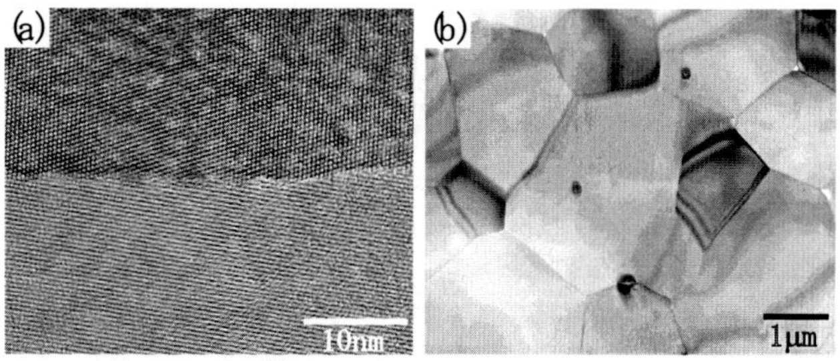

Figure 3. (a) High resolution TEM and (b) bright field TEM images of spinel SPSed at 1500°C for 20min [9].

but the number of pores larger than 300 nm increases with the temperature.

Another important microstructural feature is the precipitation of fine particles as indicated by the triangles in **Fig. 1(c)**. As shown in a high resolution TEM image of **Fig. 3(a)** [9], no amorphous phase is observed along the grain boundaries. At the limited grain boundaries and grain interiors (**Fig. 3(b)**), however, rounded fine particles were observed to precipitate at ≥1350°C. From TEM EDS analysis [9], it is found that most of the particles were composed of both Mg and Cl, but some particles are composed of either Mg or Cl; the reason why Cl exists in the present material is not clear at the present time. It is reasonable to conclude that the fine particles are the second phases precipitated due to the rapid grain growth for sintering at ≥1350°C.

The average area fraction of the second phase A_s tends to increase with the temperature as shown in **Fig. 2(b)**. If the area fractions of A_p and A_s are added together, the total area fraction of defects A_{p+s} ($= A_p + A_s$) shows a minimum at around 1300 1325°C. For the present study, therefore, the spinel with fine grain size and low defects can be attained at around 1300 1325°C for $\alpha = 10$°C/min.

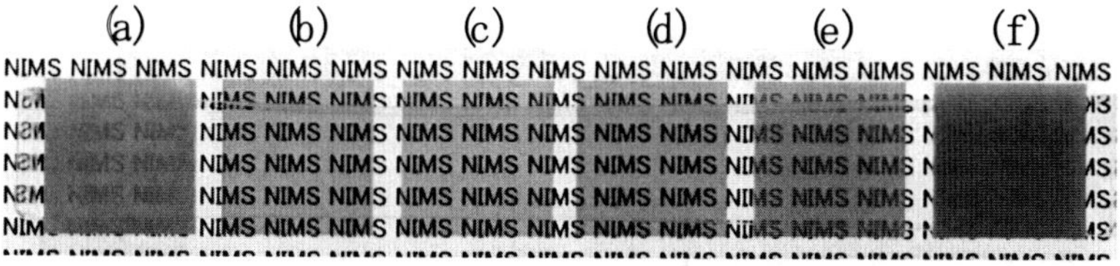

Figure 4. Photographs of the spinel plates SPSed for a 20min soak at (a) 1275°C, (b) 1300°C, (c) 1325°C, (d) 1350°C, (e) 1400°C and (f) 1500°C [9]. The 12×12×1.8 mm square samples are placed 10 mm above the text.

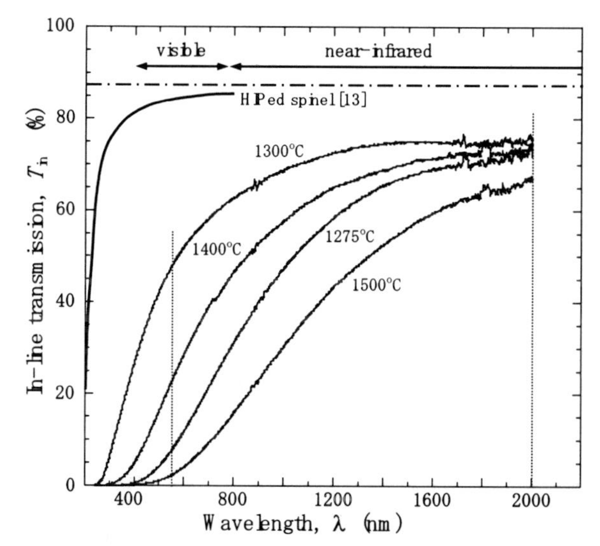

Figure 5. Temperature dependent in line transmission T_{in} of spinel plotted as a function of wavelength λ [9]. The theoretical transmittance, T_{th}, expected for a single crystalline $MgAl_2O_4$ spinel [12] and the highest T_{in} reported for a submicro grained spinel with $d \approx 0.4\mu m$ after HPing at 1360°C [13] are also shown for comparison. The data of the submicro grained spinel [13] with a thickness of w 5mm were normalized at the same thickness of w 1.8mm using the following equation of $T_{in,1}$ $(1 \; R)^2 \, [T_{in,2}/(1 \; R)^2]^{w_1/w_2}$, where R is the reflection loss (R 0.068 [12]) and $T_{in,1}$ and $T_{in,2}$ are the in line transmission for the specimen thickness of w_1 and w_2, respectively.

Optical Property

Figure 4 shows the temperature dependent optical properties of spinel SPSed at α = 10°C/min [9]. The transparency apparently depends on the sintering temperature. The spinel exhibits good transparency in the temperature range of 1300 1400°C, but is almost opaque at 1275 and 1500°C.

The in line transmission T_{in} is plotted as a function of wavelength λ in **Fig. 5**. For comparison, the theoretical transmission T_{th} [12] and the highest transmission reported by a fine grained spinel with $d \approx 0.4\mu m$ [13] are shown by dash dotted and broken lines, respectively. The sintering temperature strongly affects T_{in}, particularly in the visible light range. Although the attained transmission T_{in} approaches the theoretical value T_{th} in the IR wavelength range, T_{in} is much lower than the highest value reported in HIPed spinel [13] in visible light wavelength range. When the in line transmissions was compared at visible and IR wavelengths of λ = 550 and 2000nm ($T_{in,550}$ and $T_{in,2000}$), the influence becomes more clear. In the IR wavelength range, $T_{in,2000}$ is relatively insensitive to the temperature and exceeds 60%. As compared to the IR wavelength range, however, $T_{in,550}$ is highly sensitive to the temperature and has a peak of about 50% at around 1300°C [9]. This suggests that for a heating rate of α = 10°C/min, a sintering temperature of about 1300°C is an effective condition for attaining good transmission in spinel.

Fracture Strength

Figure 6 shows the temperature dependent three point bending strength σ_{3f}. For comparison, four point bending strength σ_{4f} [10] and the fracture strength of the spinel with several grain sizes ($\bar{d} \approx$ 0.4 100μm) [14 16] are also plotted.

The fracture strength σ_{3f} is relatively insensitive to the sintering temperature and exceeds

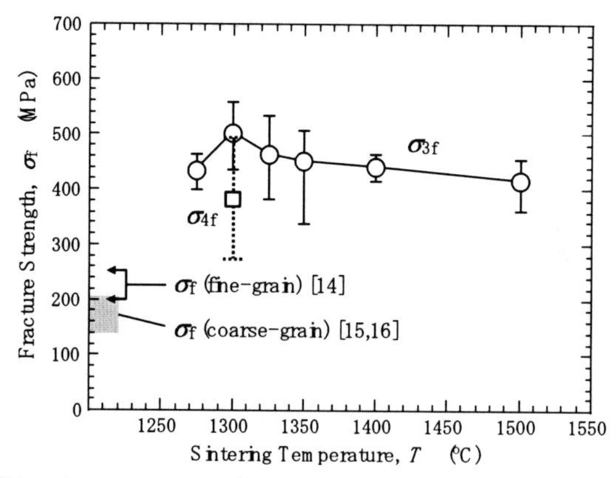

Figure 6. Sintering temperature dependent three and four point bending strengths, σ_{3f} and σ_{4f} [10]. The arrows and hatching show the earlier data for the fine and coarse grained transparent spinel ($d \approx 0.4\mu m$ and 100μm) [14 16], respectively.

400MPa though it exhibits a small peak of 500MPa at around 1300°C. Irrespective of the temperature, the σ_{3f} value is much greater than those (250MPa) of the conventional coarse and finr grained spinel [14 16]. It is well known, however, that σ_{3f} generally provides higher value than σ_{4f} because the measuring maximum load does not necessarily correspond to the load acting at the fracture point for the three point bending test. We have recently confirmed that for the spinel SPS processed at 1300°C, σ_{3f} exhibited about 20% higher values than σ_{4f} as shown in **Fig. 6** [10]. Even if the value of σ_{4f} is 20% lower than that of σ_{3f} at all the temperatures, the value of σ_{4f} can be expected to be larger as compared with the earlier data [14 16]. This suggests that the low heating rate SPS processing can improve the fracture strength of the spinel by a factor of 1.5 2.0, particularly for the sintering at 1300°C.

DISCUSSION

The present result suggests that a high sintering temperature is not necessarily effective for fabricating high strength transparent spinel polycrystals. For a low heating rate of 10°C/min, the sintering at 1300°C can be regarded as the optimum condition for attaining high transparency ($T_{in,550} \approx$ 47% and $T_{in,2000} >$ 70%) and high strength ($\sigma_{3f} \approx$ 500MPa) simultaneously, and this is quite consistent with the microstructural feature (**Fig. 2(b)**). Since both the optical and mechanical properties are known to be sensitive to microstructural factors, such as the porosity and grain size [11,17], the good properties can be ascribed to the microstructures, such as fine grain size and high density (**Figs. 1** and **2**) that is closely related to the sintering temperature.

For spinel with a symmetric cubic crystal structure, the absorption loss and the birefringent scattering loss at the grain boundaries can be neglected, and hence, the in line transmission T_{in} is not affected by grain size. Thus, for high purity spinel, it is reasonable to assume that T_{in} can be ascribed to the light scattering caused mainly by the residual pore using the following equation [11],

$$T_{in} = (1 \quad R)^2 \exp(\quad w\gamma) \qquad \text{and} \tag{2}$$

$$\gamma = \left(p \Big/ \frac{4}{3}\pi r^3 \right) C , \tag{3}$$

where R is the reflection loss at the sample surface, w is the sample thickness, γ is the total scattering coefficient, p is the total porosity, r is the radius of the pores, and C is the scattering cross section of one spherical pore.

Figure 7 shows the effects of pore size on in line transmission calculated by Eqs. (2) and (3) using the computer program given by Bohren and Huffman [18]. The calculation clearly shows that in line transmission depends on the pore diameter. The transmission significantly decreases with the increasing the pore diameter and show a minimum when the wavelength of the incident light becomes as large as the pore diameter [9,11]. Furthermore, the effect of the residual pore on the transmission becomes significant with porosity [9]. For the present material, most of the residual pores have the

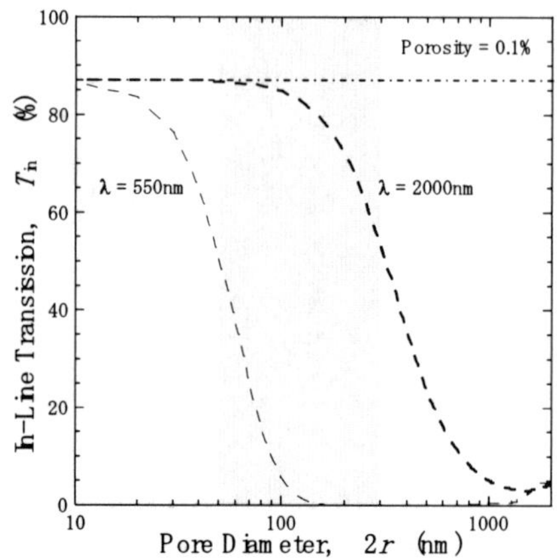

Figure 7. Pore size dependent in line transmission, $T_{in,550}$ and $T_{in,2000}$, calculated for wavelengths of λ = 550 and 2000nm and a porosity of p = 0.1% [9].

pore diameter $2r$ ranging from 50 to 300nm (hatching in **Fig. 7**), suggesting that the residual pores significantly affect the visible right transmission ($T_{in,550}$), but barely affect the IR transmission ($T_{in,2000}$). For the present spinel, the degradation of T_{in} is more remarkable in the visible wavelength range ($T_{in,550}$) than in the IR wavelength range ($T_{in,2000}$) as shown in **Fig. 5**. The trend of the present result is reasonably associated with the theoretical prediction.

For the present materials, however, the second phases also act as the light scattering source in addition to the residual pores, and hence, the quantitative comparison is not easy. Nevertheless, an approximate discussion of the scattering arising from the defects, such as the pores and second phases, may be possible. For the present material, the pore size and the density of defects (pore and second phase) depend on the sintering temperature (**Fig. 2**). With an increase in the temperature, the pore size and porosity decrease at ≤1325°C, but the pore size and the number of second phase increase at >1350°C. Thus, the spinel with fine pore size and low defects can be attained at around 1300°C. The transmission also shows a peak of about 50% at around 1300°C and the trend of the result is reasonably associated with the microstructural feature. This suggests that the pore and second phase act as a primary source of the light scattering for the spinel sintered at ≤1325°C and at >1350°C, respectively.

Fracture strength is also related to microstructural factors, such as the porosity and grain size [10,19]. According to an earlier study on polycrystalline Al_2O_3 [19], if large intrinsic flaws are eliminated during sintering, fracture is often initiated from one of the surface machining damages that are induced depending on the grain size. Since the surface damage tends to increase with grain size, fracture strength decreases with grain size [19]. In addition, we have recently indicated that for the higher density spinel, the residual pores also act as a possible factor for the machine induced flaw formation and are reflected in the fracture strength σ_f [10]. For the present spinel sintered at several temperatures, the grain size increases significantly at >1350°C as shown in **Fig. 1**. The porosity and pore size, on the other hand, decrease with temperature and shows the minimum at around 1300°C. Therefore, it is reasonable to conclude that for the present spinel, since both the grain size and porosity influence the surface damage formation, the high strength attained at 1300°C can be ascribed to the simultaneous reduction of the grain size and porosity.

Krell *et al.* [13,14] have recently attained a fine grain size of $\bar{d} \approx$ 0.4μm and a high transmittance of $T_{in} \approx$ 87% in a high purity spinel by HIPing at 1360°C for 2h. The fine grained spinel exhibited excellent T_{in} close to the theoretical value ($T_{th} \approx$ 87%) [13,14]. Although this suggests that the spinel attained fully density close to the theoretical value, the fracture strength (σ_{4f} = 200 250 MPa) is similar to that of the coarse grained spinel [15,16] and is much lower than σ_{4f} = 400MPa predicted in the present fine grained spinel ($\bar{d} \approx$ 0.4μm). Since the size of the present four point bending specimen is slightly smaller (outer span length is L = 30 mm and inner span length is l = 10 mm) than that of their study (L = 40 mm and l = 20 mm), the different sample size may affect the bending strength. At the present time, however, the origin of the lower strength is not clear in their high density sample.

The present study shows that for a constant heating rate of 10°C/min, fine grained and high density microstructure can be attained at about 1300°C for only a 20 min soak. The superimposed effects of the fine grain size, fine pore size and low density of defects (pore and second phase) can realize a good spinel polycrystal that simultaneously possesses high strength and good light transmission. Further improvement of light transmission, particularly in visible light wavelength range, is necessary in the present spinel. Nevertheless, the simple low heating rate and low temperature SPS processing, which can fabricate the high density and fine grained spinel by short time processing, has an advantage over the previous powder processing using the high temperature sintering for long times and/or the doping of sintering aids.

The enhanced sinterability during the low heating rate SPS processing can be explained as follows. First, the applied field during the SPS processing assists the enhanced sinterability by accelerating the mass transport. Conrad [20] indicated that an electric field can lower the high temperature flow stress of oxide ceramics and the lowered flow stress can be ascribed to enhanced diffusion. Thus, I think that the applied field also plays an important role for the enhanced sinterability of the spinel.

Second, the microstructural evolution during the heating process affects the sinterability. I confirmed that for the high heating rates, many large closed pores were observed to form at the grain boundary junctions, whereas for the low heating rates, sintering was almost completed during the heating process [9]. For the high heating rates, that is high densification rates, open pores rapidly changed into closed pores before shrinkage of the pore takes place by diffusional processes. Once the large closed pores are formed, they are difficult to remove because further sintering for long times and/or at high temperatures results in grain growth accompanied by the coalescence of the pores. For the high heating rate SPS processing, consequently, many large pores remain around the grain junctions. In contrast, for slow densification rates, the shrinkage of the pores takes place during the heating process due to the contribution of the diffusion related processes. Thus, for the low heating rates, the residual pores are limited and the sinterability of spinel becomes high.

A comprehensive understanding of the enhanced sinterability, of course, is difficult in the present study, and another examination is necessary. I think, however, that in the low heating rate SPS processing, the above two factors work simultaneously and enable the high sinterability of the spinel.

CONCLUSION

The effect of sintering temperature during the low heating rate spark plasma sintering (SPS) processing was examined to fabricate fine grained transparent $MgAl_2O_4$ spinel polycrystals. For a constant heating rate of 10°C/min, a sintering temperature of about 1300°C is an effective in attaining high strength transparent spinel polycrystals and the high sintering temperature is not necessarily effective owing to the rapid grain growth causes the coalescence into larger pores and the precipitation of second phases. For 1300°C, the in line transmissions evaluated at visible and IR wavelengths of λ = 550 and 2000nm ($T_{in,550}$ and $T_{in,2000}$) reaches 47% and 70%, respectively. The three point bending strength σ_{3f} exhibits the maximum value of 500MPa, which can be expected to be about two times larger than that reported in coarse grained spinel. The present study shows that the superimposed effects of the submicro grain size, fine pore size and low defect density can realize a good spinel polycrystal that possesses a high strength and good light transmission simultaneously.

ACKNOWLEDGEMENT

This study was financially supported by the Amada Foundation (AF 2008016) and the Grant in Aid for Young Scientists (B) (19760497) from the Ministry of Education, Culture, Sports, Science and Technology (MEXT), Japan and World Premier International Research Center Initiative (WPI Initiative), MEXT, Japan.

REFERENCES

[1] Z. A. Munir, U. Anselmi Tamburini, and M. Ohyanagi, The Effect of Electric Field and Pressure on the Synthesis and Consolidation of Materials: A Review of the Spark Plasma Sintering Method, *J. Mater. Sci.*, **41**, 763 77 (2006).

[2] R. Chaim, J. Z. Shen, and M. Nygren, Transparent Nanocrystalline MgO by Rapid and Low Temperature Spark Plasma Sintering, *J. Mter. Res.*, **19**, 2527 31 (2004).

[3] U. Anselmi Tamburini, J. N. Woolman, Z. Munir, Transparent Nanometric Cubic and Tetragonal Zirconia Obtained by High Pressure Pulsed Electric Current Sintering, *Adv. Funct. Mater.*, **17**, 3267 73 (2007).

[4] D. T. Jiang, D. M. Hulbert, U. Anselmi Tamburini, T. Ng, D. Land, and A. K. Mukherjee, Optically Transparent Polycrystalline Al_2O_3 Produced by Spark Plasma Sintering, *J. Am. Ceram. Soc.*, **91**, 151 54 (2008).

[5] B. N. Kim, K. Hiraga, K. Morita, and H. Yoshida, Spark Plasma Sintering of Transparent Alumina, *Scripta Mater.*, **57,** 607 10 (2007).

[6] B. N. Kim, K. Hiraga, K. Morita, and H. Yoshida, Effects of Heating Rate on Microstructure and Transparency of Spark Plasma Sintered Alumina, *J. Euro. Ceram. Soc.,* **29,** 323 27 (2009).

[7] B. N. Kim, K. Hiraga, K. Morita, H. Yoshida, T. Miyazaki, and Y. Kagawa, Microstructure and optical properties of transparent alumina, *Acta Mater.,* **57,** 1319 26 (2009).

[8] K. Morita, B. N. Kim, K. Hiraga and H. Yoshida, Fabrication of Transparent $MgAl_2O_4$ spinel polycrystal by Spark Plasma Sintering Processing, *Scripta Mater.,* **58,** 1114 17 (2008).

[9] K. Morita, B. N. Kim, K. Hiraga, and H. Yoshida, Spark plasma sintering (SPS) condition optimization for producing transparent $MgAl_2O_4$ spinel polycrystals, to be published to *J. Am. Ceram. Soc.,* **92,** 1208 16 (2009).

[10] K. Morita, B. N. Kim, K. Hiraga, and H. Yoshida, Fabrication of high strength transparent $MgAl_2O_4$ spinel polycrystals, to be published (2009).

[11] R. Apetz and M. P. B. van Bruggen, Transparent Alumina: A light Scattering Model, *J. Am. Ceram. Soc.*, **86,** 480 86 (2003).

[12] A. F. Dericioglu, and Y. Kagawa, Effect of Grain Boundary Microcracking on the Light Transmittance of Sintered Transparent $MgAl_2O_4$," *J. Euro. Ceram. Soc.*, **23** 951 59 (2003).

[13] A. Krell, J. Klimke, and T. Hutzler, Advanced spinel and sub μm Al_2O_3 for transparent armour applications. *J. Euro. Ceram. Soc.,* **29,** 275 81 (2009).

[14] A. Krell, J. Klimke, and T. Hutzler, Transparent Compact Ceramics: Inherent Physical Issues, *Optical Mater.*, **31**, 1144 50 (2009).

[15] D.W.Roy, J.L.Hastert, L.E.Coubrough, K.E.Green, and A.Trujillo, Method for producing transparent polycrystalline body with high ultraviolet transmittance, U.S. Patent **No.**5244849 (1993).

[16] P. J. Patel, G. A. Gilde, P. G. Dehmer, and J. W. McCauley, Transparent Armor, AMPTIAC Newsletter **4**, 1 6 (2000).

[17] A. Krell, and P. Blank, Grain Size Dependence of hardness in Dense Submicrometer Alumina, *J. Am. Ceram. Soc.,* **78,** 1118 20 (1995).

[18] C. F. Bohren, and D. R. Huffman, Absorption and Scattering of Light by Small Particles, Wiley VCH, New York, pp. 477 482 (1983).

[19] A. Krell, Fracture Origin and Strength in Advanced Pressureless Sintered Alumina, *J. Am. Ceram. Soc.,* **81,** 1900 1906 (1998).

[20] H. Conrad, Electroplasticity in metals and ceramics, *Mat. Sci. Eng. A*, **287**, 276 87 (2000).

CONSOLIDATION OF CARBON WITH AMORPHOUS GRAPHITE TRANSFORMATION BY SPS

Naoki Toyofuku[1], Megumi Nishimoto[1], Kazuki Arayama[1], Yasuhiro Kodera[1], Manshi Ohyanagi[1] and Zuhair Munir[2]

[1]Department of Materials Chemistry, Innovative Materials and Processing Research Center, Ryukoku University, Ohtsu, Japan

[2]Department of Materials Science and Chemical Engineering, College of Engineering, University of California, Davis, CA, USA

ABSTRACT

Amorphous carbon with 5 at% boron powders prepared by a mechanical grinding of graphite and boron under an argon atmosphere were consolidated by spark plasma sintering (SPS). When the powders were consolidated at 2200°C for 10 min under applied pressure of 70 MPa, the amorphous structure was transformed into graphite structure after the consolidation. The sintered compact showed a relative density of 96.2%. The c axis direction of the hexagonal lattice of the sintered compact, prepared from crystalline graphite powder, preferably aligns in parallel with the pressure direction. On the other hand, using amorphous powder led the alignment of the c axis toward perpendicular to the pressure direction. The electrical conductivity of the sintered compact was 8.48×10^4 S/m and much higher than that of commercial graphite.

INTRODUCTION

Graphitic carbon has many excellent properties such as high heat resistance, high corrosion resistance, and high electrical and high thermal conductivity.[1, 2] It has been used in a crucible and a bearing. The graphite, h BN and related materials consist of hexagonal crystal structure with basal plane stacking. This structure induces anisotropy in the morphology of grains; for example, graphite generally exhibits flake shape and the thickness direction of flakes corresponds to the direction of the c axis of crystal structure. The crystal orientation in a sintered compact have been observed by XRD analysis.[3, 4] Also, micro structural analysis, such as SEM and optical microscope, has been used to observe the crystal structural orientation. The investigation of the structural orientations is important due to the significant influence of the orientation on mechanical, thermal and electrical properties for bulk material.[5 8] In addition, the dense fiber reinforced graphitic carbon has been studied for aerospace materials.[9, 10] However, because carbon atoms have the covalent nature bonding and low self diffusion coefficient, graphitic carbon is not easily consolidated without high heat treatment for a long time.[11, 12]

High dense graphitic carbon has been sintered from coke powder, which is a disordered material, with boron or boron carbide (B_4C) by using a high temperature (2200°C) for a long time (2h).[13 15] Although graphite has a three dimensional structure,[16, 17] the degree of graphitization of the sample was evaluated by using only the d space of the c axis, such as (002) and (004). In addition, the XRD pattern of the sintered compact did not show a diffraction peak of (101) and (112), which was considered when it had a completely graphite structure. Therefore, it is not revealed whether a sintered compact can be completely graphitized when the cokes were used.

Recent investigations have shown that a structural ordering during sintering can aid the densification process, making it possible to sinter such materials as nanostructured SiC to relatively high densities (> 99%) without the use of additives, e.g. Al_2O_3 Y_2O_3 and B C, or exceptionally high pressures (> 1 GPa).[18] We also showed that turbostratic boron nitride (t BN), prepared by a mechanical grinding (MG) method, could be consolidated by spark plasma sintering (SPS).[19] The consolidation of BN powder to high densities is due to the densification enhanced by a disorder order transformation from t BN to hexagonal boron nitride (h BN) when the t BN powder contains oxygen. In addition, the c axis direction of the hexagonal lattice of the sintered compact, prepared from crystalline h BN powder by hot pressing, usually aligns parallel along with the pressure direction. However, it was found the c axis direction of the sintered compact exhibited unusual orientation, which was relatively perpendicular to the pressure direction, when the sample was prepared from turbostratic boron nitride powder.[20]

We attempted consolidating an amorphous carbon (a C) powder, which was prepared by MG method of graphite under an argon atmosphere, by using SPS. However, a C powder did not transform into graphite structure and did not achieve a full consolidation despite sintering a C powder at 2200°C. To transform a C into graphitic carbon, we mixed 5 at% boron powder, which is well known as graphitization accelerator,[13 15, 21] with graphite powder by the MG method. When the powders were consolidated at 2200°C, the amorphous structure was transformed into graphite structure after the consolidation. The sintered compact showed a relative density of 96.2%. The c axis of the hexagonal lattice of the sintered compact was preferably aligned with the facet parallel to the pressure direction, which stood in contrast to the grain orientation in the graphite orientated from crystalline graphite powder. The electrical conductivity of the sintered compact was much higher that of commercial graphite. In this paper we describe these details.

EXPERIMENTAL

The elemental powders of graphite (ca. 7 μm, > 99%, Tokai Carbon Co. Ltd., Japan) and boron (300 mesh, > 99%, Furuuchi Chemical Co., Japan) were used in this experiment. The powders were milled by planetary ball milling (model: P 6, Fritsch, Japan) that was performed using 10 mm diameter silicon nitride balls (Nikkato, Japan) and silicon nitride vial with an inside diameter of 75 mm and a height of 70 mm (approximately 300 ml, Fritsch, Japan). A ball to powder mass ratio (B/P) of 40/1 was used with 7.5 g of the powders. Total milling time was 5 h. The milling speed was 560 rpm and the rotation speed of the vial was 459 rpm. All transfers of powders to and from the vials were handled in a glove box filled with Ar (O_2 concentration < 5 ppm and H_2O concentration < 24 ppm). The vials in the glove box were also covered with stainless pots filled with Ar. Residual H_2O and O_2 were removed from the Ar atmosphere by a recycle purification system during the handling of powders (model: MF 70, UNICO, Japan).

Consolidation of the milled sample was carried out using an SPS apparatus (model: 1050, SPS SYNTEX INC., Japan). The SPS apparatus is a uniaxial 100 kN press combined with a 15 V, 5,000 A DC power supply to simultaneously provide pulsed current and pressure to the sample. Pulsed cycles of approximately 36 ms on and of approximately 6 ms off were used. The milled powders were placed into a graphite die with an outside diameter of 40 mm, an inside diameter of 20.4 mm, and a height of 40 mm. Then, 70 MPa pressure was applied to the milled powers through the punches having a diameter of 19.6 mm and a length of 20 mm. The three spacers between the punch and the electrode of SPS were used; each spacer's size is ϕ120mm x 20mm, ϕ80mm x 40mm, and ϕ40mm x 10mm. The samples were heated at 60°C/min and then held at the consolidating temperature ranging from 1,100 to 2,200°C for 10 min. The consolidated samples were cooled to room temperature by turning off the power. The temperatures were measured by an optical pyrometer focused on a hole drilled into the die surface to a depth of 5 mm. An emissivity of 0.9 was used based on calibration with a two color pyrometer.

Structural analysis was performed by X ray diffractometer (model: RINT 2500, CuKa radiation operated at 100 mA and 40 kV, RIGAKU, Tokyo Japan). To calculate the Index of Orientation Preference (IOP) of the sintered compact, the XRD measurement of two surfaces with different dimensions, parallel or perpendicular to the pressure direction, was carried out. The value of IOP corresponds to a degree of the orientation of hexagonal plane. The detail of the calculation of IOP has been described in letter section on this paper. The step size was 0.02° in 2θ. Transmission electron microscope (TEM) analysis was performed on the sample (model: JEM 4000EX, operated at 400 kV, JEOL, Japan). The fracture surface was observed by scanning electron microscope (model: JSM 5200, JEOL, Japan). Mechanical property was characterized by Shore hardness (model: Shore D type, SHIMADZU, Japan). Electric conductivity was measured using four probe methods (model: ZEM 1, ULVAC, Japan) at 60°C.

RESULTS AND DISCUSSION

The milled C powder formed during milling of graphite powder for 5 h in a silicon nitride vial using balls of the same materials under an Ar gas atmosphere was consolidated by the SPS method at 2,200°C for 10 min under an applied pressure of 70 MPa. Figure 1 shows the XRD patterns of graphite powder, the milled C powder and its sintering compact. The pattern of unmilled powders shows the presence of crystalline graphite peaks only, as shown in Fig. 1 (a). When the graphite powder was milled for 5h, the XRD pattern of the powder showed the broadening peak of lattice plane (002) and the diffusing peak of lattice plane of lattice planes (100), (101) and (004), as shown in Fig. 1 (b). It is suggested that the structural ordering and hexagonal plane formed by a carbon atom is broken by ball milling. The XRD pattern of amorphous C is characterized by the presence of only two peaks corresponding to the (002) and (100) planes, where it is described as a hexagonal structure. Therefore, it can be said that the short time milling process slightly disarranges the hexagonal lattice of C, which is transformed to an amorphous structure.[22] Examination of the pattern of powder milled for 5 h reveals that it has a totally amorphous carbon (a C) structure.

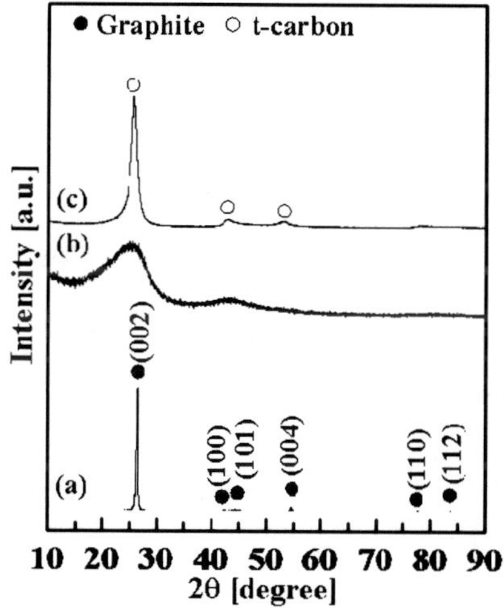

Figure 1. XRD patterns of (a) graphite powder and (b) milled carbon powder and (c) carbon compact consolidated by SPS under 70 MPa for 10 min hold at the holding temperature of 2200°C.

Fig 1 (c) is the XRD pattern of sintered compact of a C powder. There are only three peaks between 10° and 90°. The peak locations are nearly corresponding to ones of (002), (100), and (004) plains of the graphite structure. However, the XRD pattern of the sintered compact showed the defusing peaks of lattice planes (100) and (101) of graphite. Carbon often has a turbostratic structure without regularity of structure along the a and b axes.[22] Therefore, it can be said that the heat treatment at 2,200°C rearranges the hexagonal plane of C and leads to crystallization. However, the crystal stricture of the sample was the turbostratic structure with 2 dimensional ordering and didn't become a graphite structure. The sintered a C that was consolidated at 2,200°C had 1.68 g/cm^3 of density (relative density: 74.6%) and didn't achieve full consolidation. Generally, when the coke with B was consolidated above 2,000°C, the powder underwent ordering to graphite.[13 15, 21] Therefore, to transform into the graphite stricture, B powder was mixed in the a C powder by using the MG method.

a C with B powder was consolidated by the SPS method at a sintering temperature ranging from 1,100 to 2,200°C for 10 min under an applied pressure of 70 MPa. The XRD patterns of the samples are shown in Fig. 2. The XRD patterns show significant change as the temperature is increased from 1,100° to 2,200°C. The measurement of XRD patterns of the sintered compact was carried out on the surface, which is parallel to the pressure direction during SPS. The results showed a C forms at as low a temperature as 1,500°C. When the holding temperature was increased to 1,900°C, the XRD pattern mainly showed turbostratic structure. When the holding temperature was increased between 1,900° to 2,100°C, the XRD patterns still showed turbostratic structure.

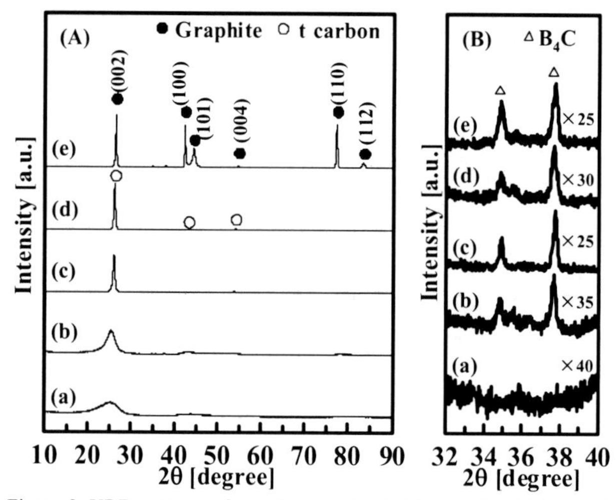

Figure 2. XRD patterns of sample consolidated from a C with B powder by SPS under 70 MPa for 10 min hold at the holding temperature: (a)1100°C, (b)1500°C, (c)1900°C, (d)2100°C, (e)2200°C. XRD patterns were magnified at 2θ value of (A) 10 90°, and (B) 32 40°.

However, the sample sintered at 2,200°C clearly shows the graphite structure, a higher level of order as seen by the increased symmetry of the main peak (at 2θ of about 26° in Fig. 2) and the existences of some peaks such as (101) and (112) peaks. The XRD pattern of sintered compact also considered several peaks appeared at the 2θ value of about 34° and 38° when the holding temperature was used at 1,200°C and higher (Fig.2 (B)). However, at 34° and 38°, there are small peaks which cannot be seen on the Figure 2 because of the high intensity at 26°. These peaks corresponded the (104) and (021) planes, where it is described as a B_4C, and they are the main peak of B_4C. It is reported that the maximum solubility of boron in graphite is approximately 2.35 at% at 2,350 °C.[21] Therefore, it is suggested that the part of boron in the a C with B powder becomes a B_4C because boron of 5 at% to graphite powder is mixed by using the MG method in this research. In the case of sintering cokes with B_4C, the boron atom in B_4C begins to enter the lattice of the hexagonal plane formed by the carbon atom when it is sintered above 2,000°C.[13] As a result, the value of the d space of (002) of the sample well rises to that of the graphite structure by influence of boron, which is a graphitization accelerator. Therefore, it is considered that the sample consolidated at 2,200°C becomes a graphite structure because boron is very well known material which takes the catalytic role for graphitization through the diffusion into carbon forming solid solution. Excess boron reacted with carbon to form B_4C. Those are the reason why we observed the graphitization and the formation of B_4C by increasing temperature. However, as shown in Fig. 2 (e), the intensity ratio between lattice planes (002) and (100) was different from those of the unmilled graphite powder (Fig.1 (a)) and the crystal database of graphite. The XRD pattern has been observed when the c axis of the hexagonal plane in sintered compact oriented perpendicular to the pressure direction. Generally, the c axis of the hexagonal plane in graphite material, which was prepared by uniaxial pressing during sintering, was oriented parallel to the pressure direction.[13] This difference suggested that the unique orientation of the hexagonal plane occurs during SPS with transformation from turbostratic carbon to graphite. These results were also observed on h BN, which is similar to graphite structure, prepared from turbostratic boron nitride. In Ref. [3, 23], IOP was used to quantify the orientation of the structure with hexagonal plane, such as h BN, in sintered compact. This method is also able to apply to the graphite having the crystal structure as similar as h BN. The IOP was estimated as

$$\mathrm{IOP} = (I_{100}/I_{002})_{\mathrm{par.}} \,/\, (I_{100}/I_{002})_{\mathrm{perp.}} \quad \cdots (1)$$

where $(I_{100}/I_{002})_{\mathrm{parallel}}$ and $(I_{100}/I_{002})_{\mathrm{perpendicular}}$ are the ratios between the intensities of the (100) and (002) diffraction peaks measured in direction parallel and perpendicular to the pressure direction, respectively. The value of IOP corresponds to a degree of the orientation of hexagonal plane. For example, the IOP shows 1, when the sample has an isotropic distribution of the c axis of the hexagonal plane, The IOP became over 1 and increase with the number of the c axis oriented toward perpendicular to the pressing direction. The IOP of the sample, which sintered at 2,200°C, was 128.4 by result of the calculation. Therefore, it is suggested that the c axis of the hexagonal lattice of the sintered compact was relatively perpendicular to the pressure direction.

Figure 3 shows the effect of the holding temperature on the d space and integral breadth of the sintered compact consolidated from a C with B powder by the SPS. The d space and integral breath were calculated from the peak at 2θ values of approximately 25° of the XRD results of the sample. The measurement of XRD patterns of the sintered compact was carried out on the surface, which is parallel to the pressure direction during SPS. The d space of the sintered compact changed by a small amount (from 3.51 Å to 3.49 Å) as the holding temperature increased from 1,100 to 1,500°C. However, the d space of the sample gradually degreased to 3.42 Å when sintering was carried out at 1,900°C. The d space of the sample almost didn't change when the holding temperature was increased to 2,100°C. At a holding temperature of 2,200°C, the d space of the sintered compact consolidated from a C with B abruptly decreased over again (3.37 Å). Its value was nearly consistent with that of JCPDS of graphite (3.376 Å) and natural graphite (3.354 Å). The reciprocal number of integral breadth of samples didn't change much when the holding temperature was increased from 1,100° to 1,500°C. However, when the holding temperature was increased to 1,900°C, the reverse of integral breadth of sintered compact gradually increased and flowed by flat value at 1,900° and 2,000°C. At the holding temperature above 2,100°C, the reciprocal number of integral breath of samples abruptly increased for the second time. The change of the number of the d space and the integral breadth corresponded well. In temperature ranges from 1,500° to 1,900°C and from 2,000° to 2,200°C, the change of the d space and integral breadth suggests that crystallization is drastically occurring.

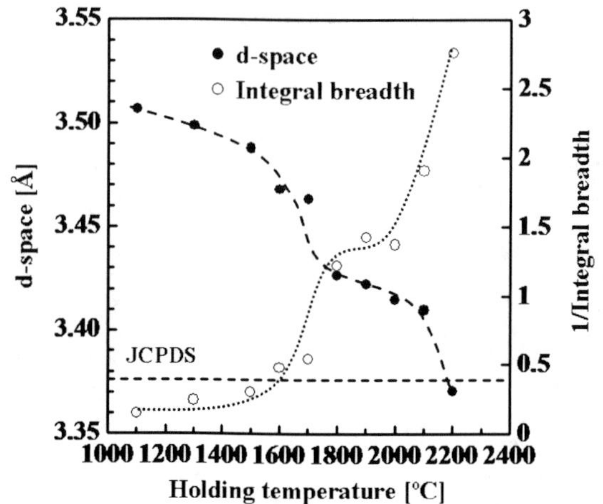

Figure 3. Effect of holding temperature on the d space and integral breadth of the sample consolidated from a C with B powder by SPS for 10 min hold under 70 MPa.

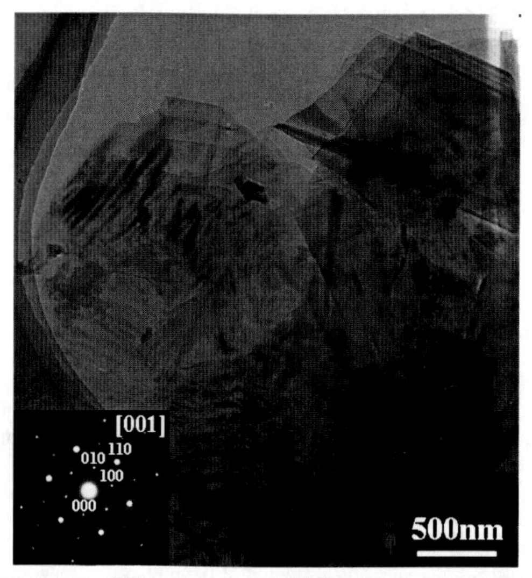

Figure 4. TEM image and SAD pattern of sample consolidated from a C with B powder by SPS under 70 Pa for 10 min at the holding temperature of 2200°C.

TEM analysis was used on the sample consolidated at 2,200°C in order to ascertain the grain size and the selected area diffraction (SAD) pattern, as shown in Figure 4. The inset shows the SAD pattern of its grain with [001] zone axes. Figure 4 shows the existence of relatively large graphite grains (about 1 μm in size). The SAD image is indexed based on the hexagonal unite cell for better understanding. The values of the d space calculated from diffraction spots of (100) and (110) were 2.14 Å and 1.26 Å, respectively. They nearly corresponded with those of d space of graphite of JCPDS (2.14 Å and 1.23 Å, respectively). It is reported that the value of d space of (110) increases when boron atoms enter the lattice of the hexagonal plane.[21] Therefore, it is considered that the value of d space of (110) of the sample increases because the boron atom substitutes for a part of the carbon atom in the hexagonal plane by heat treatment.

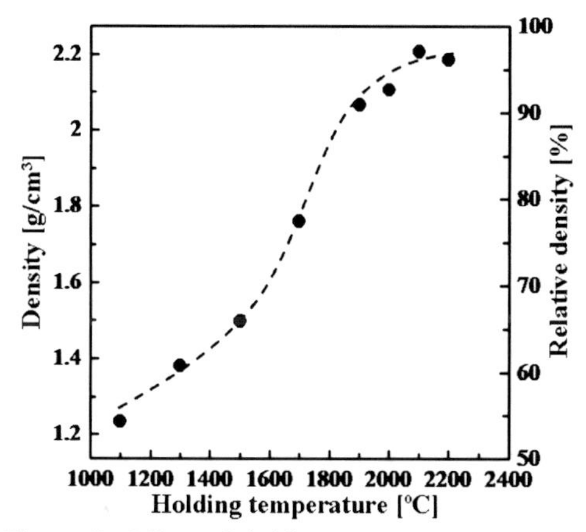

Figure 5. Effect of holding temperature on the density and relative density of carbon compact consolidated by SPS for 10 min hold under 70 MPa.

Figure 5 shows the effect of the holding temperature on the density and relative density of the sintered compact consolidated from a C with B powder by the SPS. At the holding temperature below 1,500°C, the samples were relatively porous with a 1.50 g/cm^3 or less density. At the holding temperature above 1,900°C, the density of the sample consolidated from a C with B was 2.06 g/cm^3 and higher. The relative density of sintered compacts sintered at 1,900°C and more was 91% and more. The change in the density between 1,500° and 1,900°C is relatively abrupt. The temperature changed gradually on the density of the sample corresponding with that of the drastic change of the number of the d space and the reciprocal number of integral breadth (Fig. 3). Moreover, the crystal structure of the sample transformed from a C to turbostratic carbon (t C) between these temperatures. Therefore, we have shown that the consolidation of a C with B powder to high densities is due to densification enhanced by disorder order transformation from a C structure to t C structure. In addition, it is suggested that the drastic consolidation doesn't occur from 2,000°C to 2,200°C because the densification of the sample consolidated from a C with B powder almost finishes at 1,900°C. In the case of a C without B powder, it is also considered that the sample didn't become high density in spite of using the high temperature (2,200°C) because the drastic change of structure didn't occur.

Figure 6 shows SEM photos of the sample consolidated from a C with B at (a) 1,900°C, (b) 2,100°C and (c) 2,200°C. The grains of the sintered compact with t C structure (Fig. 6 (a) and (b)) became large grains with an increase in the holding temperature and were much finer than that of the sintered compact with graphite structure (Fig. 6 (c)). In addition, the flake like particle of the sintered compact with graphite structure was arranged at random.

Figure 6. SEM photographs of sample consolidated from a C with B by SPS under 70MPa for 10 min at the holding temperature: (a) 1900°C, (b) 2100°C, (c) 2200°C.

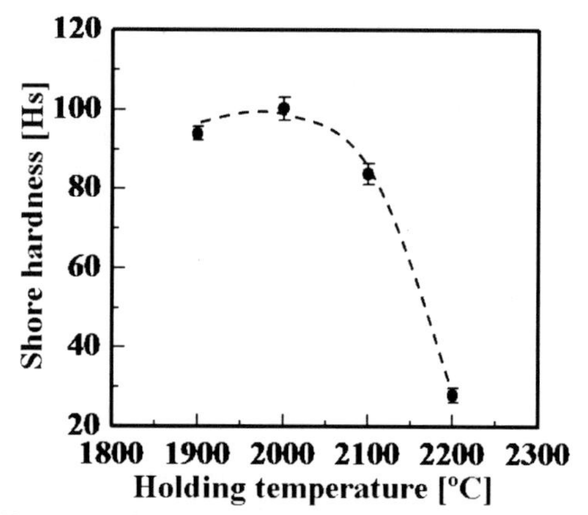

Figure 7. Effect of holding temperature on the shore hardness of sample consolidated from a C with B powder by SPS for 10 min hold under 70 MPa at the holding temperature of 1900° 2200°C.

Figure 7 shows the effect of holding temperature on the Shore hardness of the sample consolidated from a C with B by the SPS. The value of the Shore hardness slightly changed from 1,900° to 2,000 °C and, showed significant changed as the temperature was increased from 2,000° and 2,100°C. However, the Shore hardness of the sintered compact at 2,200°C abruptly decreased to 27.9 Hs in spite of the similar density (about 2.1 g/cm^3). It has been reported that t C and a C have sp^3 and sp^2 bonding.[24] In addition, it is well known that the carbon materials with sp^3 and sp^2 bonding, such as diamond like carbon (DLC), are very hard materials.[24] Moreover, from the SEM images of the sintered compact (Fig. 6), the grains of samples with t C structure were smaller than that of the sample with graphite structure. Therefore, it is suggested that the Shore hardness of the sintered compact with t C is much higher than that of samples with graphite structure because of having sp^2 and sp^3 bonding and small grains.

Finally, we measured the electrical conductivity of the sintered compact with graphite structure. It was 8.48×10^4 S/m, which is much higher than that of the commercial graphite material produced by STACKPOLE (grade: 2020) with 1.77 g/cm^3 of the density (4.93×10^2 S/m).

CONCLUSION

Amorphous carbon with 5 at% boron powders prepared by a mechanical grinding process from graphite and boron powder under an Ar atmosphere were consolidated by SPS. When the powders were consolidated at a holding temperature below 2,100°C for 10 min under an applied pressure of 70 MPa, the sample had a disordered carbon structure such as amorphous and turbostratic structure, which was characterized by XRD results. When the temperature was increased to 2,200°C, the turbostratic structure was transformed into graphite structure, which was characterized by XRD and TEM method. The sintered compact showed a relative density of 96.2%. The c axis direction of the hexagonal lattice of the sintered compact, prepared from crystalline graphite powder, preferably aligns in parallel with the pressure direction. On the other hand, using amorphous powder led the alignment of the c axis toward perpendicular to the pressure direction. The Shore hardness and electrical conductivity of the sintered compact with graphite structure were 27.9 ± 1.8 Hs and 8.48×10^4 S/m, respectively. The electrical conductivity of the sintered compact was much higher than that of commercial graphite.

References

[1]E. Fitzer and M. Heym, High temperature mechanical properties of carbon and graphite, *High Temp. High Press.*, **10**, 29 66 (1978).

[2]Y. Kuga, M. Shirahige, T. Fujimoto and A. Ueda, Production of natural graphite particles with high electrical conductivity by grinding in alcoholic vapor, *Carbon*, **42**, 293 300 (2004).

[3]M. Hubáček and M. Ueki, Effects of the Orientation of Boron Nitride Grain on the Physical Properties of Hot Pressed Ceramics, *J. Am. Ceram.*, **82**, 156 160 (1999).

[4]D. Gaies and K. T. Faber, Thermal properties of pitch derived graphite foam, *Carbon,* **40**, 1131 1150 (2002).

[5]M. Murakami, N. Nishiki, K. Nakamura, H. Okada, T. Kouzaki, K.Watanabe, T. Hoshi and S. Yoshimura, High quality and highly oriented graphite block from polycondensation polymer films, *Carbon,* **30**, 255 262 (1992).

[6]Y. Kaburagi, A Yoshida, Y. Hishiyama, Y. Nagata, M. Inagaki, Microtexture and crystallinity of highly crystallized graphite films prepared from aromatic polyimide films, *Tanso*, 166, 19 27 (1995).

[7]T. Oku, A. Kurumada, Y. Imamura, K. Kawamata and M. Shiraishi, Effects of prestresses on mechanical properties of isotropic graphite materials, *J. Nucl. Mat.,* **258-263**, 814 820 (1998).

[8]M. Inagaki, Y. Tamai and S. Naka, Texture and graphitizability of gilsonite coke, *Tanso,* 118 125 (1973).

[9]E. Bruneton, C. Tallaron, N. Gras Naulin and A. Cosculluela, Evolution of the structure and mechanical behavior of a carbon foam at very high temperatures, *Carbon,*. **40**, 1919 1927 (2002).

[10]M. Koyama, H. Hatta and H. Fukuda, Effect of temperature and layer thickness on these strengths of

carbon bonding for carbon/carbon composites, *Carbon,* **43**, 171 177 (2005).
[11]E. Fitzer and W. Hüttner, Structure and strength of carbon/carbon composite, *J. Phys. D: App. Phys.*, **14**, 347 371 (1981).
[12]E. I. Neroshin, T. A. Ostrovskaya, G. G. Sazonov, N. N. Samsonova, A. V. Kharitonov and N. N. Shipkov, Theory and technology of sintering, thermal, and chemicothermal treatment processes, *Sov. Powder Metall. Met. Ceram.*, **13,** 210 212 (1974).
[13]K. Miyazawa, T. Hagio, K. Kobayashi, T. Honda, Effect of boron carbide addition and Hot pressing Temperature on sintering and graphitization of coke powder, *Kyusyu kougyougijyutusikenjyo houkoku*, 1223 1231 (1978).
[14]T. Hagio, Ichitaro Ogawa and Kazuo Kobayashi, Effects of boron addition on some properties of hard type carbons, *J. Mater. Sci.*, **21,** 4147 4150 (1986).
[15]H. N. Murty, D. L. Biederman and E. A. Heintz, Apparent catalysis of graphitization. 3. Effect of boron, *Fuel,* **56,** 305 312 (1977).
[16]V. S. Babu and M. S. Seehara, Modeling of disorder and X ray diffraction in coal based graphitic carbons, *Carbon,* **34,** 1259 1265 (2005).
[17]M. Monthioux and J. G. Lavin, The graphitizabillity of fullerenes and related textures, *Carbon,* **32,** 335 343 (1994).
[18]T. Yamamoto, H. Kitaura, Y. Kodera, T. Ishii, M. Ohyanagi and Z. A. Munir, Consolidation of Nanostructured β SiC by spark plasma sintering, *J. Am. Ceram. Soc.*, **87**, 1436 1441 (2004).
[19]T. Yamamoto, N. Ishibashi, N. Toyofuku, Y. Kodera, M. Ohyanagi and Z. A. Munir, Consolidation of h BN with disorder order transformation, *Proceedings of Material Science and Technology 2006 on Innovative Processing and Synthesis of Ceramics, Glasses and Composites*, 531 538 (2006).
[20]K. M. Taylor, Hot pressed Boron Nitride, *Ind. Eng. Chem.*, **47,** 2506 2509 (1955).
[21]C. E. Lowell, Solid solution of boron in graphite, *J. Amer. Ceram. Soc.*, **50,** 142 144 (1967).
[22]S. Imada, H. Hasegawa, J. Ozaki and A. Oya, High temperature treatment and X ray diffraction analysis of electric plug resin residues for fire origin determination, *kanshiki kagaku,* **9,** 29 38 (2004)
[23]M. Hubáček, M. Ueki and T.Sato, Orientation and growth of grains in copper activated hot pressed hexagonal boron nitride, *J. Am. Ceram.*, **79**, 283 285 (1996).
[24]A. C. Ferrari and J. Robertson, Interpretation of Raman spectra of disordered and amorphous carbon, *Phys. Rev. B,* **61,** 14095 14106 (2000).

SPARK PLASMA SINTERING OF NANOSTRUCTURED CERAMIC MATERIALS WITH POTENTIAL MAGNETOELECTRICITY

C. Correas, R. Jiménez, T. Hungría, H. Amorín, J. Ricote, E. Vila, M. Algueró, A. Castro
Instituto de Ciencia de Materiales de Madrid, CSIC. Cantoblanco, 28049 Madrid. Spain

J. Galy
Centre d'Elaboration de Matériaux et d'Etudes Structurales, CNRS.
29 rue Jeanne Marvig, 31055 Toulouse. France

ABSTRACT

Multiferroic and magnetoelectric materials are of particular interest for technological exploitation, but few examples stable at atmospheric pressure are known. The $BiMnO_3$ $PbTiO_3$ and $BiFeO_3$ $PbTiO_3$ solid solutions are good candidates, because the $BiMnO_3$ and $BiFeO_3$ edge members are multiferroic, while $PbTiO_3$ helps the stabilization of the perovskite structure.

Based on the recent demands for high precision, high reliability and miniaturized ceramic electronic components, the grain size effect on the electrical properties of ferroelectric materials has attracted much attention. In this context, the technique of spark plasma sintering (SPS) enables a powder compact to be sintered at relative low temperature and short time, limiting grain growth, and allows thus nanostructured ceramics when nanoparticulate powders are used.

In this communication, results on the mechanosynthesis of the solid solutions x $BiBO_3$ (1 x) $PbTiO_3$ (B = Fe or Mn) are reported. Single perovskite phases have been synthesized at room temperature as nanoparticulate powders. The processing of ceramics of selected compositions by SPS has been investigated. The combination of mechanosynthesis and SPS permits the structure and particle size to be controlled.

INTRODUCTION

A multiferroic material is one that possesses at least two of the ferroic behaviours: ferroelectric, (anti)ferromagnetic and ferroelastic [1,2]. Nowadays, the interest in the multiferroic materials, and more exactly in magnetoelectrics, is increasing due to their potential applications in a wide range of novel technologies, which are based on the coupling between magnetic and electric order parameters [3-5].

Unfortunately, few multiferroic materials synthesized at atmospheric pressure are known. The bismuth perovskites with a magnetic active element in the B position are investigated nowadays. The Bi^{3+} cation contributes to the ferroelectric effect due to its typical stereochemistry, due to the lone pair of electrons, while a magnetic active cation, as Fe^{3+} or Mn^{3+}, generates the ferromagnetic effect [5,6]. For example, $BiMnO_3$ has been considered as a good candidate because it shows ferroelectric ordering, below the Curie temperature of Tc ~ 632 °C; however this compound cannot be prepared at room pressure and its conductivity is very high [6,7]. On the other hand, another good candidate is $BiFeO_3$ that has been studied in powder and bulk forms. This material exhibits a perovskite type structure at room temperature, with rhombohedral distortion and R3c space group, it is a ferroelectric compound, with T_C ~ 830 °C, and it is also an antiferromagnetic, with T_N ~ 370 °C [5,9-12].

The main problems for the application of these materials are: first, its high electric conductivity; and second the formation of secondary phases ($Bi_2Fe_4O_9$ and $Bi_{24}FeO_{39}$) during the synthesis [9,11-13]. One possible solution is the addition of more stable compounds and materials with less electric conductivity. The $PbTiO_3$ is a tetragonal perovskite with ferroelectric properties at room temperature and combined with the $BiMnO_3$ or $BiFeO_3$ might maintain the multiferroic properties [8,9,13]. For this reason, the systems x $BiBO_3$ (1 x) $PbTiO_3$ (B = Mn or Fe) are good candidates to exhibit good multiferroic properties.

Moreover, based on the recent demands for high precision, high reliability and miniaturized electronic components, the grain size effect on dielectric and magnetic properties of fine grained materials has attracted much attention. As consequence, different synthesis methods have been studied in order to obtain reactive precursors with smaller particle size. Mechanochemical routes have been very useful to prepare suitable precursors for electroceramics [14-23].

In this communication, results on the synthesis by solid state and mechanochemical activation routes of the x $BiMnO_3$ (1 x) $PbTiO_3$ and (1 x) $BiFeO_3$ x $PbTiO_3$ solid solutions are reported. The processing of selected compositions for these solid solutions by spark plasma sintering has also been investigated.

EXPERIMENTAL

Conventional ceramic route has been applied for the preparation of different oxides of the system (1 x) $BiFeO_3$ x $PbTiO_3$. About 2 g of analytical grade Bi_2O_3, Fe_2O_3, PbO and TiO_2 for x= 0, 0.1, 0.3, 0.4, 0.5, 0.7 and 1 compositions were homogenized by hand in an agate mortar and heated at increasing temperatures from 400°C up to the melting point, maintaining each temperature during 12 h. The heated powders were cooled by quenching, weighed, reground by hand in an agate mortar and analysed by X ray diffraction (XRD).

For the sake of comparison, members with x= 0, 0.3, 0.5, 0.7 and 1 were also prepared by a mechanochemical activation route. In this case, the stoichiometric mixture were mechanically activated in a planetary mill (Fritsch Pulverisette 6) using an 80 cm^3 stainless steel container, with five steel balls 2 cm in diameter and 35 g in weight. The grinding vessel was rotated at 300 rpm during different times, depending on the sample. The activations were carried out both under air atmosphere and oxygen atmosphere at 20 kPa. In order to investigate the influence of the milling medium on the mechanochemical process of the (1 x) $BiFeO_3$ x PbTiO system, compositions with x= 0 and 0.3 have also been activated using a tungsten carbide + cobalt vessel, with a volume of 250 cm^3 and seven tungsten carbide balls 2 cm in diameter and 63 g in weight. In this case, the vessel was rotated at 300 rpm during 24 hours.

In the case of the system $BiMnO_3$ $PbTiO_3$, stoichiometric mixtures of analytical grade Bi_2O_3, PbO, Mn_2O_3 and TiO_2 were mechanically activated in the same planetary mill using steel vessel and balls, in oxygen atmosphere, for times ranging between 35 and 140 h, depending on the Mn content. The members of the solid solution x $BiMnO_3$ (1 x) $PbTiO_3$ investigated were x = 0, 0.1, 0.2, 0.3, 0.4, 0.5, 0.6, 0.7 and 1.

The phase evolution during the mechanical activation and after thermal treatments of the different components has being investigated by XRD with a Bruker AXS D8 Advance diffractometer between 13° and 60° (2θ), with 2θ increments of 0.1° and counting time of 1.5 s per step. The Cu Kα doublet (λ = 0.15418 nm) was used in these experiments.

The thermal behaviour of the samples was investigated by differential thermal analysis (DTA) and thermogravimetry (TG). The measurements were made with a Seiko 320 system, with α Al_2O_3 as the inert reference material, between room temperature and 850°C, with heating cooling rates of 10°C min^{-1} and Ar atmosphere.

The morphology of the powder samples was examined by field emission scanning electron microscopy (FE SEM, FEI Nova NanoSEM 230 operating between 5 and 15 kV, equipped with an OXFORD INCA 250 electron dispersive X ray detector (EDX)) and transmission electron microscopy (TEM, CM20 FEG Philips working at 200 kV).

Ceramics sintering was accomplished in vacuum in an SPS 2080 Sumitomo apparatus. A cylindrical graphite die with an inner diameter of 15 mm was filled with 2.5 3 g of the powder. A pulsed direct current was then passed through the die to heat it up, while an increasing uniaxial pressure up to 75 MPa was applied. As a result, the sample was heated up to the final temperature (650 750 °C) at a rate of 100 °C min ^{1}. The soaking time at the final sintering conditions was 5 min for all the samples.

RESULTS AND DISCUSSION

STUDY OF THE (1 x) $BiFeO_3$ x $PbTiO_3$ SYSTEM:

Table 1 summarizes the identified phases after the thermal treatments corresponding to the conventional route, while Fig. 1 shows the X ray diffraction patterns of the final products resulted at the lowest synthesis temperatures.

Table 1. Phases identified after the thermal treatments carried out to obtain the different (1 x) $BiFeO_3$ x $PbTiO_3$ compositions and the lowest temperature of synthesis for each composition (rhom: rhombohedral perovskite, tet: tetragonal perovskite).

Composition (1 x) $BiFeO_3$ x $PbTiO_3$	T (°C) synthesis	Phases
x = 0	850	$BiFeO_3$ (rhom), $Bi_2Fe_4O_9$, $Bi_{25}FeO_{40}$
x = 0.1	900	Perovskite (rhom), $Bi_2Fe_4O_9$
x = 0.3	900	Perovskite (rhom), Perovskite (tet)
x = 0.4	950	Perovskite (tet), Perovskite (rhom)
x = 0.5	1050	Perovskite (tet), Perovskite (rhom)
x = 0.7	1000	Perovskite (tet)
x = 1	700	$PbTiO_3$ (tet)

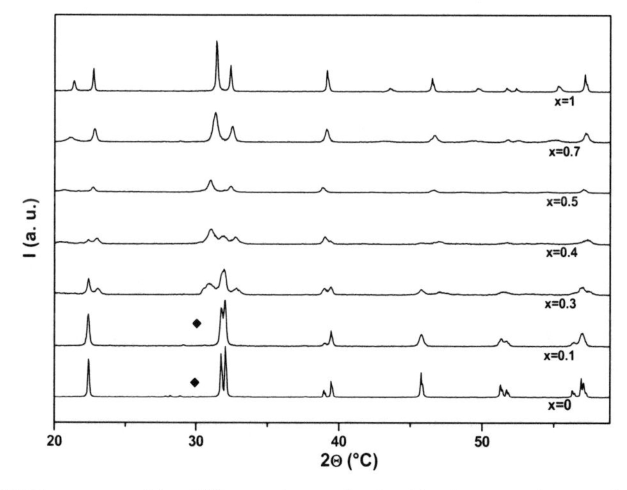

Figure 1. XRD patterns of the different phases obtained by the ceramic route for different (1 x) $BiFeO_3$ x $PbTiO_3$ compositions. (◆ : $Bi_2Fe_4O_9$ or $Bi_{25}FeO_{40}$ secondary phases)

For all the compositions the perovskites are formed. It can be found the perovskites together with secondary phases for the compositions with x = 0, 0.1; less crystalline perovskites in the range between 0.3 and 0.5 and single phases perovskites in the cases of the composites with x = 0.7 and 1. From XRD data, it can be concluded that oxides in the $BiFeO_3$ rich region present a rhombohedral perovskite structure, while in the $PbTiO_3$ rich region it is possible identify the tetragonal distortion. As it was previously reported, MPB region is located around x = 0.3 0.4 [5,9,10,13].

It is important to go into detail in the case of synthesis of the x = 0 and 0.1 compositions, due to the appearance of the impurities. Other authors have described that the preparation of

$BiFeO_3$ by the solid state reaction is quite difficult due to the formation of secondary phases as for example $Bi_2Fe_4O_9$ and $Bi_{24}Fe_2O_{39}$ [9,11-13]. For this reason, it is important to optimize the synthesis protocol. For example, Fig. 2 shows the difference between the products obtained after cumulative thermal treatment from 400 °C to 850 °C (a) and a direct triple treatment at 800 °C (b). The quantity of secondary phases is minimized by the direct heating at high temperatures.

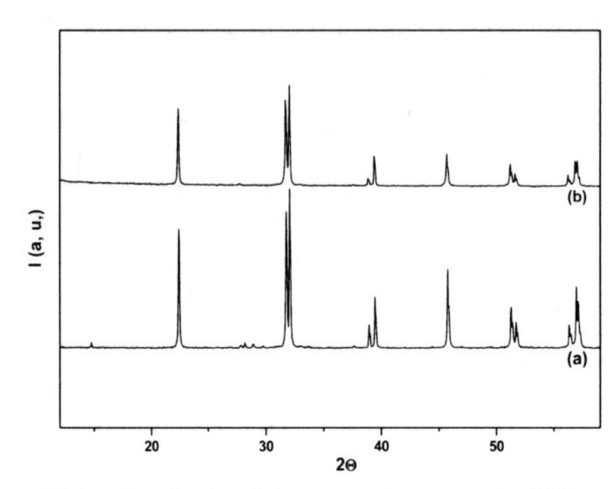

Figure 2. XRD patterns of $BiFeO_3$ obtained by ceramic route, in different forms. (a) cumulative thermal treatments from 400 to 850°C; (b) heated three times at 800°C.

The synthesis was also carried out by a mechanical activation route. Table 2 summarizes the conditions of the different mechanical activation processes. First, the mixtures with x = 0, 0.3, 0.5, 0.7 and 1 were milled in a stainless steel vessel in air and the XRD patterns are shown in the Fig. 3.

Table 2. Milling conditions of the mechanical activation processes for the different (1 x) $BiFeO_3$ x $PbTiO_3$ compositions in a stainless steel vessel.

Composition (1 x) $BiFeO_3$ x $PbTiO_3$	Milling time (h)	
	Air atmosphere	O_2 atmosphere
x = 0	140	70 and 140
x = 0.3	70	140
x = 0.5	140	70
x = 0.7	140	70
x = 1	35	

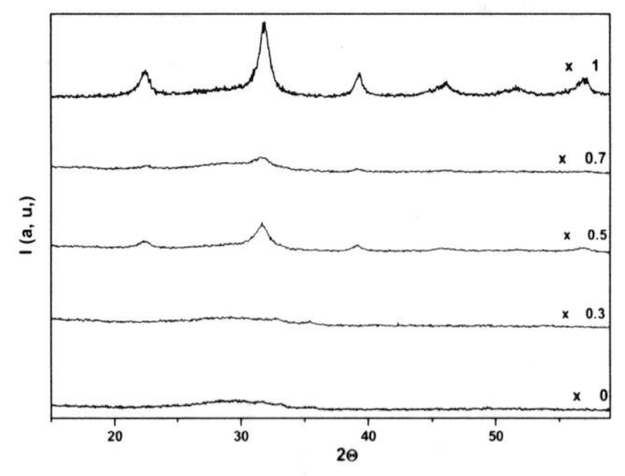

Figure 3. XRD patterns of the products obtained after mechanochemical activation in air for the different (1 x) $BiFeO_3$ x $PbTiO_3$ compositions.

The mechanosynthesis of the perovskite at room temperature can be observed in the case of the compositions close to the $PbTiO_3$. In the case of the $BiFeO_3$ rich compositions only the amorphization of the sample was detected. In some cases, as for example x = 0.7 (Fig. 4) it is possible to observe a XRD peak close to 2θ = 28° at 140h of milling time, together with the appearance of black powder in the mixture, corresponding to the Bi metal. For this reason, in order to avoid the reduction of the Bi oxides, mechanochemical activation was also carried out in O_2 atmosphere.

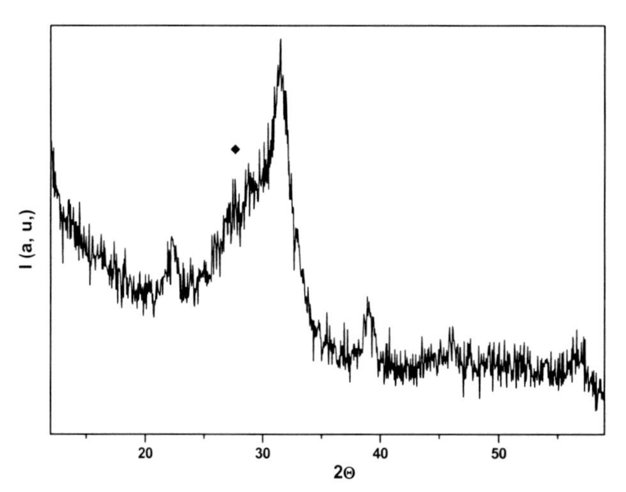

Figure 4. XRD pattern of the mixture corresponding to 0.3 $BiFeO_3$ 0.7 $PbTiO_3$ composition after 140h of mechanical treatment in air. (• : Bi metal)

The evolution of the mechanochemical activation of the mixtures in O_2 atmosphere was very similar to the mechanoactivation in air. However, the main difference was observed after the subsequent thermal treatments as can be clearly observed in Fig. 5 for x = 0.3. Unlike the sample milled in O_2 atmosphere, perovskite could not be isolated from the mixture activated in air, even after the heating at 790 °C. Instead of the perovskite, a fluorite phase was obtained, standing out the influence of the milling atmosphere on the final products of the mechanochemical activation route.

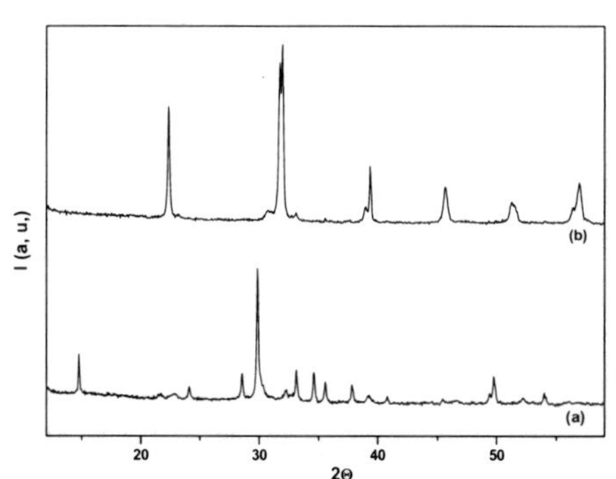

Figure 5. XRD patterns of the mixture corresponding to 0.7 $BiFeO_3$ 0.3 $PbTiO_3$ mechanochemically activated using different atmospheres (a) air and (b) O_2 and heated subsequently at 790 °C.

Mixtures corresponding to $BiFeO_3$ and 0.7 $BiFeO_3$ 0.3 $PbTiO_3$ compositions were also mechanically treated using a tungsten carbide vessel and balls. The results of the grinding during 24 hours are shown in Fig. 6. It is worth nothing that in this case it was possible to obtain $BiFeO_3$ rich perovskites by mechanosynthesis at room temperature as single phases, which was not possible in the case of the activation in steel medium. This fact could be due to the higher energy supplied during the impacts of the tungsten balls to the sample that is caused by the higher density and hardness of the tungsten carbide compared to the steel.

Secondary phases were not observed even after the thermal treatment of the mechanosynthesized perovskite at quite low temperatures (Fig. 6b and 6d)

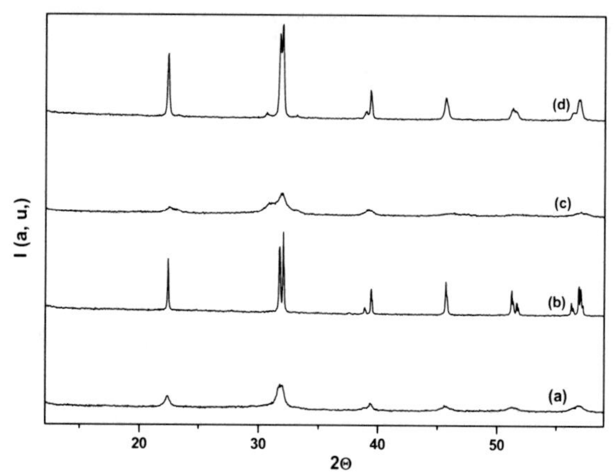

Figure 6. XRD patterns of the mechanosynthesized perovskites in a tungsten carbide vessel: $BiFeO_3$ (a) as milled and (b) treated at 700 °C after the mechanosynthesis and 0.7 $BiFeO_3$ 0.3 $PbTiO_3$ (c) as milled and (d) treated at 790 °C after the mechanosynthesis.

Fig. 7 shows the DTA results of the $BiFeO_3$ obtained by different methods: (a) conventional route and (b) mechanosynthesis in tungsten carbide medium. In both cases it is possible to observe two reversible transitions in the ranges of temperatures 364 366 and 795 820 °C, witch correspond at Néel temperature (T_N) and Curie Temperature (T_C) of the perovskite, respectively. In the DTA of the $BiFeO_3$ sample synthesized by ceramic route a different peak around 805 820 °C appears, this signal is attributed to the melting of the present $Bi_2Fe_4O_9$, $Bi_{25}FeO_{40}$ secondary phases. These results show that with mechanosynthesis it is possible to obtain a pure and stable perovskite phase.

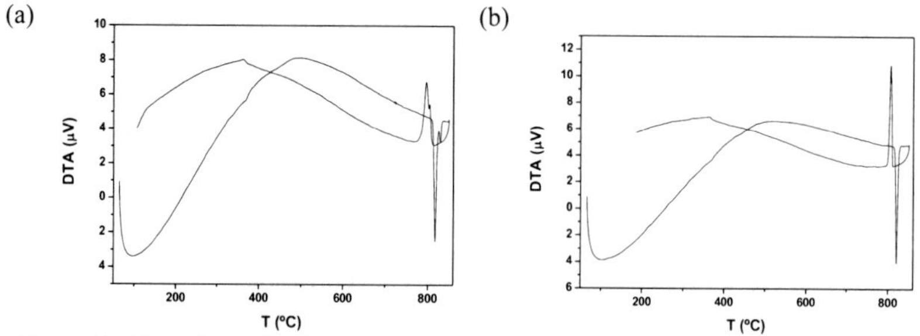

Figure 7. DTA of (a) $BiFeO_3$ synthesized by ceramic method and (b) mechanosynthesized $BiFeO_3$.

The morphological characteristics of $BiFeO_3$ and 0.7 $BiFeO_3$ 0.3 $PbTiO_3$ samples prepared by different synthesis methods were investigated by SEM. Fig. 8 depicts micrographs of different $BiFeO_3$ samples. The Fig. 8a shows the micrograph of the $BiFeO_3$ synthesized by ceramic method (at 800°C), while the micrographs in the Fig. 8b and 8c belong to $BiFeO_3$ mechanosynthesized in the tungsten carbide vessel, as milled and after annealing, respectively. In both cases, the particles of the oxides obtained by mechanical activation route were smaller that the particles of the $BiFeO_3$ synthesized by ceramic route. As can be observed in these SEM micrographs, all the samples are constituted by aggregates with different sizes, being smaller aggregates in the case of the mechanosynthesized powders. The EDX reveals that all the samples are homogeneous with a composition close to the stoichiometric one and there is neither tungsten nor cobalt contamination in the mechanosynthesized samples. TEM micrographs of mechanosynthesized $BiFeO_3$ (Fig. 9) show that the agglomerates observed by SEM (Fig. 8b) are constituted by nanocrystalline particles between 10 and 25 nm in size.

Figure 8. Scanning electron micrographs of (a) $BiFeO_3$ synthesized by ceramic method (b) $BiFeO_3$ mechanosynthesized and (c) $BiFeO_3$ mechanosynthesized and annealed at 700°C.

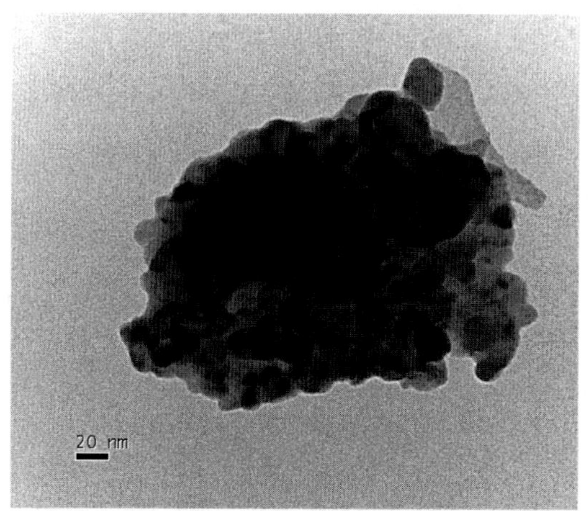

Figure 9. TEM micrographs of mechanosynthesized $BiFeO_3$ sample.

$BiFeO_3$ ceramics were processed by spark plasma sintering (SPS) from precursors obtained by different methods: solid state reaction, mechonoactivated in a steel vessel in O_2, and mechanosynthesized in tungsten carbide vessel. Table 3 summarizes the SPS conditions for the different precursors (being the final pressure in all the cases 75 MPa and the soaking time 5 minutes) and the composition of the ceramics. In the first case ceramics are constituted by perovskite and an important quantity of Bi secondary phases. This is not the case of ceramics processed from the mechanoactivated and mechanosynthesized oxide (Fig. 10), where a small quantity of reactives is detected. In the last case, ceramics are constituted mainly for perovskite and the density is close to the theoretical one. For this reason, 0.7 $BiFeO_3$ 0.3 $PbTiO_3$ ceramics were only processed from mechanosynthesized precursors.

Table 3. Spark Plasma Sintering (SPS) temperature conditions of the processed ceramics (1: Synthesized by ceramic route; 2: Mechanoactivated in steel vessel; 3: Mechanosynthesized in a tungsten carbide vessel; tr: traces).

Mixture (1 x) $BiFeO_3$ x $PbTiO_3$	Powder type	Sintering temperature (°C)	Composition	% theoretical density
x = 0	1	750	Perovskite + Bi secondary phases	97.77
x = 0	2	750	Perovskite + Reactives	89.78
x = 0	3	650	Perovskite + (tr) Reactives	98.89
x = 0.3	3	650	Perovskite + (tr) Reactives	98.23

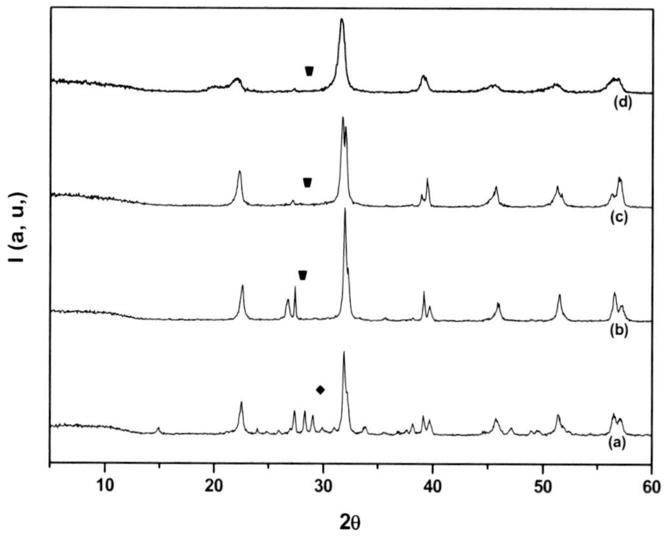

Figure 10. XRD patterns of different ceramics processed by SPS from: (a) $BiFeO_3$ obtained by solid state reaction; (b) mechanoactivated $BiFeO_3$; (c) mechanosynthesized $BiFeO_3$ and (d) mechanosynthesized 0.7 $BiFeO_3$ 0.3 $PbTiO_3$. (◆ : Bi secondary phases, ▼ : Reactives)

In the in situ shrinkage curves that are shown in Fig. 11, it can be observed two steps of compactation in the case of the powders of the $BiFeO_3$ synthesized by solid state reaction, while the shrinkage rate is almost constant in the case of the mechanosynthesized perovskite. The two steps, observed for the $BiFeO_3$ synthesized by ceramic route are attributed to the first compression of the powders due to the application of pressure and the subsequent sintering process. In the case of the mechanosynthesized powder, its higher reactivity due to the lower particle size and crystallinity, as well as the high quantity of defects, make possible the ceramic to be sintered at lower temperature, resulting in a higher density.

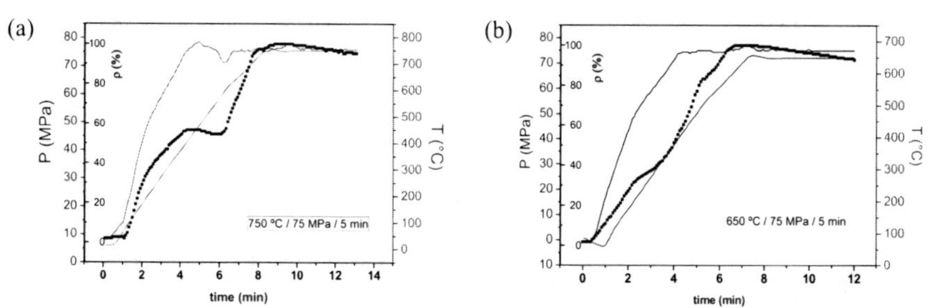

Figure 11. Shrinkage curves recorded during the SPS processing of the $BiFeO_3$ synthesized by (a) solid state reaction and (b) mechanosynthesis.

STUDY OF THE x $BiMnO_3$ (1 x) $PbTiO_3$ SYSTEM:

The study of the synthesis of the system with Mn by mechanical activation reveals that, as can be observed in Fig. 12, a perovskite type oxide was obtained by mechanosynthesis at room temperature for compositions in the range x $BiMnO_3$ (1 x) $PbTiO_3$ with $0 \leq x \leq 0.6$. It is worth noting that all mechanosynthesized phases show a pseudocubic symmetry, independently on the Mn/Ti ratio.

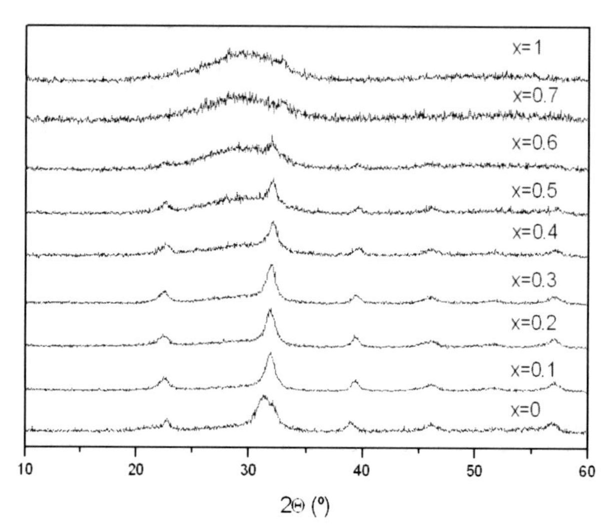

Figure 12. XRD patterns of the different Bi_2O_3, PbO, Mn_2O_3 and TiO_2 stoichiometric mixtures after mechanical activation to obtain the x $BiMnO_3$ (1 x) $PbTiO_3$ perovskites.

Similarly to that observed in the case of the $BiFeO_3$ $PbTiO_3$, the mechanosynthesized powders consist of nanosized particles as shown by the TEM image for 0.3 $BiMnO_3$ 0.7 $PbTiO_3$ (Fig. 13). Selected area electron diffraction (SAED, upper inset in the Fig. 13) indicates that these particles are tight aggregates of perovskite nanocrystals. SAED shows well defined rings with associated planar distances corresponding to those of the perovskite, in good agreement with XRD. Particle size distribution was obtained from the Feret's diameter measured for an ensemble of more than 100 particles, and an average size of 15 nm resulted (inner inset in Fig. 13).

Spark plasma sintering has been used to process ceramics for the composition 0.3 $BiMnO_3$ 0.7 $PbTiO_3$ using the same conditions than in the case of the mechanosynthesized $BiFeO_3$ $PbTiO_3$ oxides that are 650 °C/ 75 MPa. A transmission electron microscopy image of a ceramic and the corresponding grain size distribution are shown in Fig. 14. The combination of the highly reactive nanocrystalline powder and SPS, allow nanostructured ceramic with an average grain size of 30 nm and a high degree of densification (95 %) to be processed.

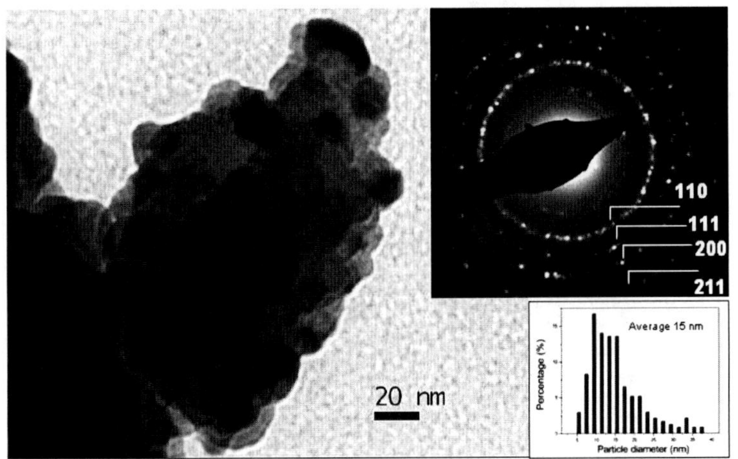

Figure 13. Transmission electron microscopy image of the 0.3 $BiMnO_3$ 0.7 $PbTiO_3$ oxide obtained by mechanosynthesis. (Upper inset: SAED of the sample, inner inset: particle size distribution of the sample).

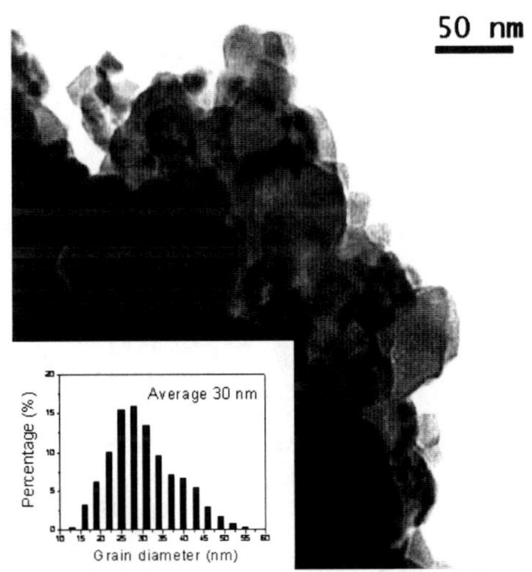

Figure 14. TEM Image and grain size distribution (inset) of a 0.3 $BiMnO_3$ 0.7 $PbTiO_3$ ceramic obtained by SPS of nanocrystalline powder.

Fig. 15 shows the shrinkage curve for the 0.3 $BiMnO_3$ 0.7 $PbTiO_3$ composition synthesized by mechanosynthesis. In this case it can be observed that saturation of compactation is obtained after 7 minutes at 650 °C, which corresponds with 95 % of densification, as in the case of the mechanosynthesized $BiFeO_3$.

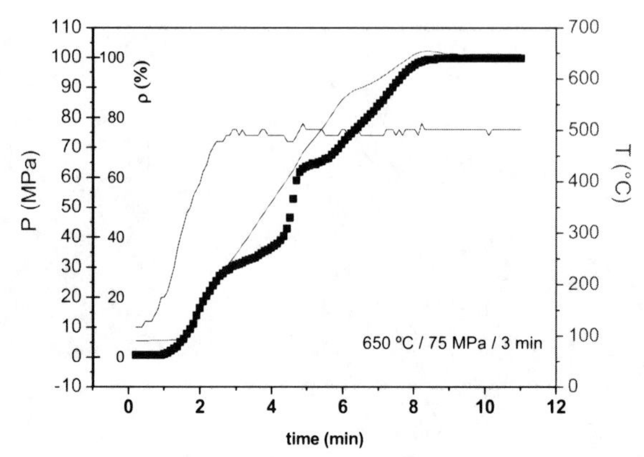

Figure 15. Compression curve recorded during the SPS processing of the 0.3 $BiMnO_3$ 0.7 $PbTiO_3$ synthesized by mechanochemical activation.

CONCLUSION

In summary, x $BiMnO_3$ (1 x) $PbTiO_3$ perovskite type oxides with a nanometer crystal size have been obtained by mechanosynthesis at room temperature, without need of high pressures. On the other hand, (1 x) $BiFeO_3$ x $PbTiO_3$ compounds, with high homogeneity, purity, stability and nanometer crystal size, can be synthesized by the same technique. In both cases, spark plasma sintering appears as a good method to process ceramics with a controlled grain size and high density at low temperature and short sintering time.

ACKNOWLEDGMENTS:
This work has been founded by the MICINN (Spain) through the MAT2007 61884 project. C.C., T.H. and H.A. thank the financial support by the Spanish MICINN by the BES 2008 005409, CSIC JAEDoc082 and R&C programs, respectively. Technical support by Ms. I. Martínez (ICMM) is also acknowledged.

REFERENCES

[1]W. Eerenstein, N. D. Mathur, and J. F. Scott, Multiferroic and magnetoelectric materials, *Nature*, **442**, 759 765 (2006).
[2]S. W. Cheong, and M. Mostovoy, Multiferroics: a magnetic twist for ferroelectricity, *Nature Mater.*, **6**, 13 20 (2007).
[3]O. E. González Vázquez, and J. Iñiguez, Pressure induced structural, electronic, and magnetic effects in $BiFeO_3$, *Phys. Rev. B*, **79**, 064102 (2009).
[4]D. Lebeugle, D. Colson, A. Forget, M. Viret, A. M. Bataille, and A. Gukasov, Electric field induced spin flop in $BiFeO_3$ single crystals at room temperature, *Phys. Rev. Lett.*, **100**, 227602 (2008).
[5]S. Bhattacharjee, V. Pandey, R. K. Kotnala, and D. Pandey, Unambiguous evidence for magnetoelectric coupling of multiferroic origin in $0.73BiFeO_3$ $0.27PbTiO_3$, *Appl. Phys. Lett.*, **94**, 012906 (2009).
[6]R. Seshadri, and N. A. Hill, Visualizing the role of Bi 6s "lone pair" in the off center distorsion in ferromagnetic $BiFeO_3$, *Chem. Mater.*, **13**, 2892 2899 (2001).
[7]D. I. Woodward, and I. M. Reany, A structural study of ceramics in the $(BiMnO_3)_x$ $(PbTiO_3)_{1-x}$ solid solution series, *J. Phys. Condes. Matter*, **16**, 8823 8834 (2004).

[8]J. Chen, R. Xing, G. R. Liu, J. H. Li, and T. Liu, Structure and negative thermal expansion in the $PbTiO_3$ $BiFeO_3$ system, *Appl. Phys. Lett.*, **89**, 101914 (2006).
[9]W. M. Zhu, H. Y. Guo, and Z. G. Ye, Structural and magnetic characterization of multiferroic $(BiFeO_3)_{1-x}(PbTiO_3)_x$ solid solutions, *Phys. Rev. B*, **78**, 014401 (2008).
[10]T. P. Comyn, T. Stevenson, M. Al Jawad, S. L. Tunner, R. I. Smith, W. G. Marshall, A. J. Bell, and R. Cywinski, Phase specific magnetic ordering in $BiFeO_3$ $PbTiO_3$, *Appl. Phys. Lett.*, **93**, 232901 (2008).
[11]R. Mazumder, D. Chakravarty, D. Bhattacharya, and A. Sen, Spark plasma sintering of $BiFeO_3$, *Mater. Res. Bull.*, **44**, 555 559 (2009).
[12]J. Chen, X. Xing, A. Watson, W. Wang, R. Yu, J. Deng, L. Yan, C. Sun, and X Chen, Rapid synthesis of multiferroic $BiFeO_3$ single crystalline nanostructures, *Chem. Mater.*, **19**, 3598 3600 (2009).
[13]T. P. Comyn, S. P. Bride, and A. J. Bell, Processing and electrical properties of $BiFeO_3$ $PbTiO_3$ ceramics, *Mater. Lett.*, **58**, 3844 3846 (2004).
[14] A. Castro, P. Millán, L. Pardo and B. Jiménez, Synthesis and sintering improvement of Aurivillius type structure ferroelectric ceramics by mechanochemical activation, *J. Mater. Chem.*, **9**, 1313 1317 (1999).
[15] J.G. Lisoni, P. Millán, E. Vila, J.L. Martín de Vidales, Th. Hoffmann and A. Castro, Synthesis of ferroelectric $Bi_4Ti_3O_{12}$ by alternative routes: wet no coprecipitation chemistry and mechanochemical activation, *Chem. Mater.*, **13**, 2084 2091 (2001).
[16] A. Castro and D. Palem, Study of fluorite phases in the system Bi_2O_3 Nb_2O_5 Ta_2O_5. Synthesis by mechanochemical activation assisted methods, *J. Mater. Chem.*, **12**, 2774 2780 (2002).
[17] A. Castro, P. Bégué, B. Jiménez, J. Ricote, R. Jiménez and J. Galy, New $Bi_2Mo_{1-X}W_xO_6$ solid solution: mechanosynthesis, structural study, and ferroelectric properties of the x=0.75 member, *Chem. Mater.*, **15**, 3395 3401 (2003).
[18]M. Alguero, J. Ricote, and A. Castro, Mechanosynthesis and thermal stability of piezoelectric perovskite $0.92Pb(Zn_{1/3}Nb_{2/3})O_3$ $0.08PbTiO_3$ powders, *J. Am. Ceram. Soc.*, **87**, 772 778 (2004).
[19]T. Hungria, M. Alguero, A. B. Hungria, and A. Castro, Dense, fine grained $xBaTiO_3$ (1 x)$SrTiO_3$ ceramics prepared by the combination of mechanosynthesized nanopowders and spark plasma sintering, *Chem. Mater.*, **17**, 6205 6212 (2005).
[20]M. Alguero, J. Ricote, T. Hungría, and A. Castro, High sensitivity piezoelectric, low tolerance factor perovskites by mechanosynthesis, *Chem. Mater.*, **19**, 4982 4990 (2007).
[21]T. Hungria, M. Alguero, and A. Castro, Grain growth control in $NaNbO_3$ $SrTiO_3$ ceramics by mechanosynthesis and spark plasma sintering, *J. Am. Ceram. Soc.*, **90**, 2122 2127 (2007).
[22]L.B. Kong, T.S. Zhang, J. Ma and F. Boey, Progress in synthesis of ferroelectric ceramic materials via high energy mechanochemical technique, *Prog. Mat. Sci.*, **53**, 207 322 (2008).
[23]M. Alguero, H. Amorin, T. Hungria, J. Galy, and A. Castro, Macroscopic ferroelectricity and piezoelectricity in nanostructured BiScO3 PbTiO3 ceramics, *Appl. Phys. Lett.*, **94**, 012902 (2009).

SINTERING AND PROPERTIES OF NANOMETRIC FUNCTIONAL OXIDES

Dat V. Quach[1], Sangtae Kim[1], Manfred Martin[2], and Zuhair A. Munir[1]
[1]Department of Chemical Engineering and Materials Science
University of California, Davis, CA 95616, USA.
[2]Institute of Physical Chemistry
RWTH Aachen University, Aachen, Germany.

ABSTRACT

The sintering of nanometric powders of functional oxides such as yttria stabilized zirconia (YSZ) and samaria doped ceria (SDC) was investigated by the pulsed electric current sintering (PECS) method. The effect of the applied uniaxial pressure and sintering time on the density and grain size of the consolidated oxides was studied. The pressure had a significant effect on the relative density with no apparent effect on grain size. Dense, bulk functional oxides with densities > 95% and a grain size of about 15 nm could be prepared by this method. The electrical conductivity of the nanometric oxides was found to be highly dependent on grain size. At low temperatures (<150 °C) the oxides are protonic conductors. The protonic conductivity is, however, associated with microstructure; samples with large grain size (~ 1 μm) did not exhibit protonic conduction at the corresponding temperatures.

INTRODUCTION

Fluorite structured oxides such as cubic yttria stabilized zirconia (YSZ) and samaria or gadolinia doped ceria are important functional ceramics which have great potential and existing applications as solid electrolytes in solid oxide fuel cells (SOFCs) due to their excellent electrical conductivity based on oxygen ion transport at high temperatures.[1] Recent observations brought forth unprecedented results showing protonic conductivity in these oxides when they are in the low nanometric range (<20 nm).[2,3] The discovery was made possible by the heretofore unachievable goal of sintering nanopowders of cubic YSZ to high densities and maintaining the grain size in low nanometric range. Samples with micrometric grain size did not exhibit protonic conduction.

Obtaining bulk, nanometric fluorite structured oxides from powders, especially cubic YSZ, is challenging due to significant grain growth during sintering. Nanometric cubic YSZ has been sintered by different methods including pressureless sintering,[4] hot pressing,[5] microwave sintering[6] and pulsed electric current sintering (PECS).[7] In most cases the sintering temperature is usually above 1300°C and excessive grain growth occurred due to a relatively low activation energy (288 $kJ.mol^{-1}$) of the cubic phase (compared to a higher activation energy for grain growth, 439 $kJ.mol^{-1}$, in the tetragonal modification of YSZ).[6] Even when nanometric powder is used, the final product often contains grains in the micrometric range, as has been reported by Dahl et al[5] where nanometric cubic YSZ powder (~ 10 nm) was sintered by hot pressing and by PECS. In both cases the average grain size was in submicrometer range after full densification. (0.21 and 0.37 μm for PECS and hot pressing, respectively.)

The recent success in consolidating nanometric YSZ was accomplished by a modified PECS method in which high pressure (up to 1 GPa) can be applied.[8] The application of the pressure has been shown to significantly lower the sintering temperature and thus reduce grain growth. In this paper we report on the sintering of cubic YSZ and SDC via this modified method and the effect of applied pressure and sintering time on densification and grain growth of these materials. In addition, the unique electrical properties of the bulk samples prepared by this method are also addressed.

EXPERIMENTAL

Yttria stabilized zirconia (8YSZ) and 20 mol% samaria doped ceria (20SDC) powders were synthesized using wet chemistry route. Yttrium nitrate hexahydrate ($Y(NO_3)_3.6H_2O$, 99.9%, Aldrich) and zirconyl nitrate hexahydrate ($ZrO(NO_3)_2.6H_2O$, 99%, Aldrich) were used for the synthesis of 8 mol% YSZ while samarium nitrate hexahydrate ($Sm(NO_3)_3.6H_2O$, 99.9%, Aldrich) and cerium nitrate hexahydrate ($Ce(NO_3)_3.6H_2O$, 99.9%, Aldrich) were used to obtain 20 mol% SDC. Precipitates formed as an aqueous solution of ammonia was added drop wise to an aqueous solution containing the proper amounts of the above salts. The precipitates were collected from centrifugation and washed with water, then a 50 50 vol% ethanol water solution, and finally with pure ethanol. The precipitates were dried at 120°C for 12 h, ground and then calcined in air at 450°C for 2 h. The phase composition of the resulting powders was determined by XRD, and for both 8YSZ and 20SDC the crystallite size was less than 10 nm, as determined by the Scherrer formula.

Sintering experiments were conducted in a SPS apparatus (Model 2050, Sumitomo Coal Mining Co., Japan) which is similar to a regular hot press but the graphite die and the powder compact were directly heated up by a pulsed DC electrical current. In a conventional SPS apparatus, the applied uniaxial pressure is usually constrained to about 100 MPa due to the limitation of the mechanical properties of graphite. A recently modified SPS assembly with the use of a double acting graphite die and SiC plungers allows for higher pressure application, up to 1 GPa. Dense samples with a diameter of 5 mm can be obtained from PECS at higher pressure and lower temperature. A more detailed description of this method can be found elsewhere.[8] Temperature in most experiments was measured by a shielded K type thermocouple inserted to the lateral wall of the external graphite die, and the actual temperature was determined from a calibration curve based on temperature reading from another thermocouple inserted into the internal graphite die, 2 mm away from the powder. The direct heating in the PECS method allows the use of a relatively high heating rate which was about 190 $^{o}C.min^{-1}$ in all experiments conducted in this study.

One set of samples was obtained from the sintering of 8YSZ powder at 980°C for 5 min under different applied pressure conditions in PECS, and another set was from the same temperature condition and a fixed applied pressure of 700 MPa but with different holding times. In a similar sintering experiment, 20SDC powder was consolidated at 820°C for 5 min under a pressure of 530 MPa. Density was determined by the Archimedes method and from geometric and gravimetric measurements. High resolution scanning electron microscopy (HRSEM) was utilized using a Philips FEI XL30 to observe fracture surfaces of these various samples with no conductive coating. The grain size was measured using the AnalySIS software (Soft Imaging System Corp., Lakewood, CO), and at least 100 grains were used to determine the average grain size of each sample. Samples were also ground for XRD analysis using a Scintag EDS 2000 diffractometer with Cu Kα radiation (λ = 1.544 Å) at a scan rate of 0.8 $degree.min^{-1}$. To compare with the grain size determined from SEM, the Williamson Hall method[9] was also applied to obtain the crystallite size based on peak broadening in the XRD pattern.

Bulk, dense 8YSZ and 20SDC samples sintered by the method described above having an average grain size of about 15 nm were used as an electrolyte in a custom designed water cell whose detailed description can be found elsewhere.[3] To study the dependence of protonic conductivity on grain size, the 8YSZ powder was sintered using PECS with sintering temperatures from 950 1150°C and applied pressure from 500 700 MPa. Secondary ion mass spectrometry (SIMS, ToF SIMS IV, ION ToF, Muenster, Germany) was performed on YSZ samples exposed to D_2O saturated air at 70 °C for 17 h using a beam of 25 keV Ga^+. The technique was sensitive enough to unambiguously distinguish D from H_2 .

RESULTS AND DISCUSSION

In conventional pressureless sintering or hot pressing, the temperature required to obtain high densification of nanometric zirconia powders is typically in the range from 1300 1500 °C. Anselmi Tamburini et al[10] used PECS at 1200°C with 141 MPa to consolidate 8YSZ powder with an initial particle size of about 20 nm and obtained dense ceramic specimens. In this study nearly full densification is achieved at temperature as low as 980 °C under an applied pressure of 700 MPa. The increase in pressure markedly reduces the sintering temperature at which dense samples (with density > 95%) are obtained. As shown in Figure 1 the pressure application has a great influence on the relative density with no apparent effect on grain size. At 980°C no significant densification occurs at 150 MPa, but samples with relative density ~ 97% can be obtained at 700 MPa. Although high pressure application facilitates fast densification, pressure appears to have no influence on grain size. Regardless of the applied pressure, samples sintered at 980°C in PECS have grain size of about 15 nm as determined from the fracture surfaces in SEM. This trend is confirmed by the Williamson Hall analysis based on XRD peak broadening of these samples, as can also be seen in Figure 1.

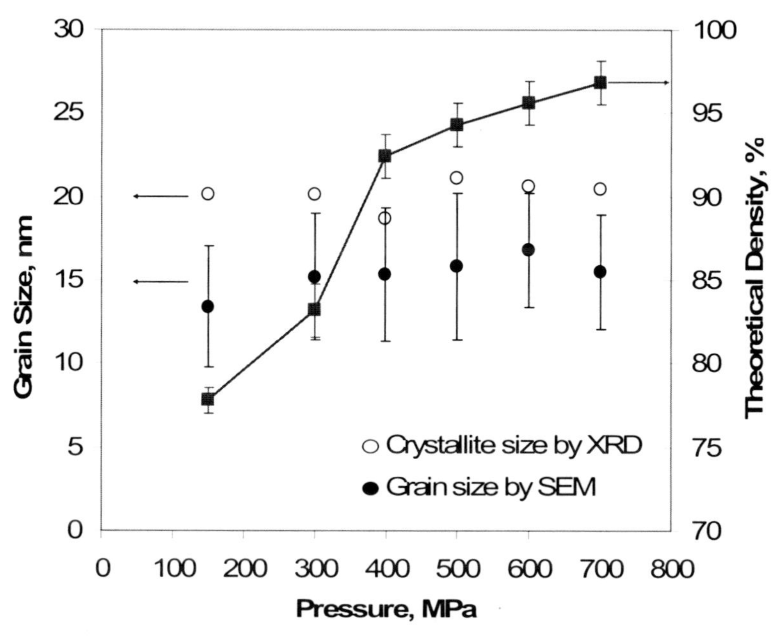

Figure 1. Effect of the applied uniaxial pressure on density and grain size of 8YSZ sintered in PECS at 980°C for 5 min.

The role of external pressure on sintering can be both extrinsic and intrinsic. The pressure helps break up agglomerates which often exist in nanometric powders and whose presence prevents complete densification. Pressure also can provide an additional driving to sintering aside from the sintering stress driven by surface curvature. There is a critical pressure above which the applied pressure can enhance densification. Skandan et al[11] studied the sintering of nanometric zirconia and found that for powders with 6 nm in size the critical pressure was 35 MPa. As the particle size was doubled, the critical pressure level was reduced to 10 MPa. The enhancement of sintering in PECS has been a topic of many discussions and investigations and has been attributed to various possible contributions, including electromigration, intensified diffusion under field application, cleansing effect on particle surfaces due to plasma, and high heating rates.[12 14] Recently Olevsky and Froyen[15] proposed thermal diffusion as another mechanism that can enhance densification during early stages of sintering in PECS. This mechanism can be applicable not only to electrically conductive materials but also to insulators such as zirconia. Although the mechanism for enhanced sintering in PECS is still not well understood, the use of PECS with high applied pressure (700 MPa) allows full densification in a shorter time at temperature that is at least 300°C lower than what is usually used in traditional pressureless sintering or hot pressing. By reducing sintering temperature and shortening sintering time, grain growth can be suppressed. Dense samples (> 95%) of 8YSZ with grain size of about 15 nm can be obtained.

The effect on density and grain size of holding time at 980°C under a pressure of 700 MPa for YSZ is shown in Figure 2. The errors in density measurements are relatively large due to the small sample size. Density is not significantly improved (96.8 to 98.8%) as the sintering time is increased from 5 to 60 min. The average grain size slightly increases with holding time, from 15 to about 20 nm, and fracture surfaces of several samples sintered under this condition are shown in Figure 3. Figures 1 and 2 indicate that at 980°C the external pressure is more effective than sintering time in obtaining high density. Pressure can dramatically enhance the densification of 8YSZ at this temperature with no influence on average grain size. Sintering time which is directly related to diffusion and grain growth, however, appears to have a very small effect on relative density and grain size under these conditions. Its effect at this low temperature may be insignificant because it takes an increase of more than one order of magnitude in holding time in order to notice a slight increase in both density and grain size that are still within the error limit of the measurements.

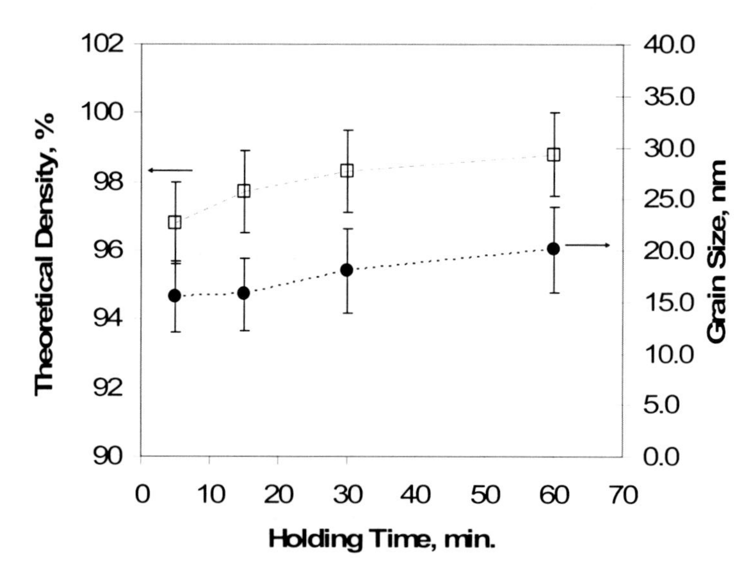

Figure 2. Density and average grain size of YSZ samples sintered at 980 °C for different holding time in PECS (applied pressure = 700 MPa).

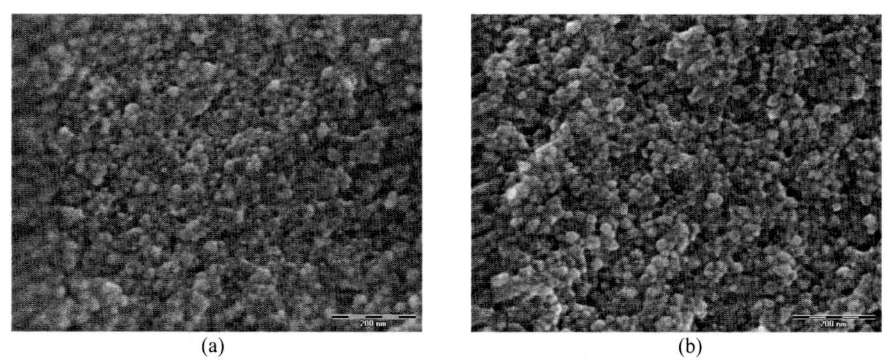

(a) (b)

Figure 3. High resolution SEM images of 8YSZ samples sintered at 980 °C under 700 MPa for (a) 5 min. and (b) 30 min. (Micron bar = 200 nm.)

Similarly, 20SDC powders were sintered by the same technique at 820°C for 5 min under a pressure of 530 MPa. The sample is fully dense with an average grain size of about 15 nm. Due to a significantly lower sintering temperature and shorter sintering time, bulk dense ceramic oxides with grain size as small as 15 nm can be obtained in PECS with high pressure application. The impact is enormous since it opens the door to the production of new nanometric materials, with the potential of providing totally new observations demonstrating the size effect on materials properties. Results from the water cell experiments with nanometric 8YSZ and 20SDC show a dependence of emf values on water partial pressure. Secondary Ion Mass Spectrometry (SIMS) analysis of the nanometric 8YSZ sample exposed to D_2O saturated air at 70°C for 17 h further proves the existence of deuterium in this nanometric zirconia, indicating protonic conduction within the sample. For the first time it was demonstrated that electrical power can be generated at room temperature from the use of bulk, dense nanometric 8YSZ and 20SDC prepared by PECS with an average grain size of about 15 nm. The power is low, however. A maximum power of about 6 and 42 $nW.cm^{-2}$ is obtained as the above 8YSZ and 20SDC, respectively, are used as the electrolyte in the cell. For comparison dense samples of 8YSZ and 20SDC with an average grain size in micrometric range (about 10 and 1 μm for 8YSZ and 20SDC, respectively) were prepared by conventional pressureless sintering. These micrometric samples, however, did not generate electrical power when they were used in the same water cell setup.

Bulk micrometric fluorite structured oxides such as YSZ and SDC have negligible protonic conductivity even at relatively high temperatures. Our results, however, show a different behavior when the oxides have a grain size in the nanometric range. They become protonic conductors at low temperatures (25°C). The grain size dependence of protonic conductivity in 8YSZ is shown in Figure 4. Above 200°C, nanometric 8YSZ behaves as an oxygen ion conductor with the conductivity decreasing sharply as the temperature is lowered. The slope of the data is 1.1 eV above 200°C, confirming the oxygen ion conduction mechanism.[16] However, the conductivity in wet air begins to increase below 150°C. As shown in Figure 4, although these 8YSZ samples have similar total conductivity above 150°C, their protonic conductivity values below 150°C are very different and grain size dependent. The smaller the grain size, the higher is the protonic conductivity. As the average grain size is increased from 13 to 100 nm, the conductivity drops by almost three orders of magnitude. Although grain boundaries appear to have a great influence on the proton conduction in these nanometric fluorite structured oxides,[17] further studies are needed before the conduction mechanism and the selective role of grain boundaries can be clearly understood.

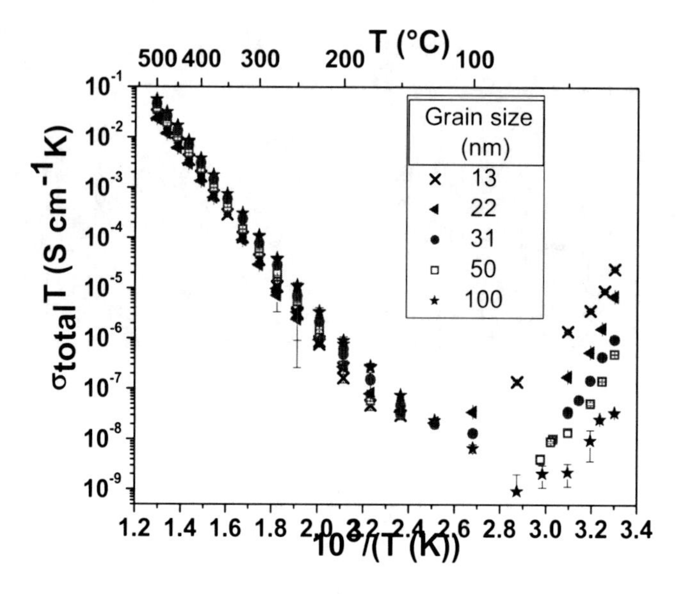

Figure 4. Total conductivity of 8YSZ samples with different average grain size in wet air, P_{H2O} = 23000 ppm.

CONCLUSION

The effect of sintering parameters such as applied pressure and sintering time on the relative density and average grain size of several nanometric fluorite structured oxides was investigated. The applied pressure had a marked influence on final density but virtually no effect on grain size. Sintering time at temperatures as low as 980°C for 8YSZ, however, had an insignificant impact on both density and average grain size of the sintered product. The use of high pressure (~ 700 MPa) helped lower sintering temperature, allowing full densification with very little grain growth. Bulk, dense 8YSZ and 20SDC samples with average grain size of about 15 nm could be produced by this technique. The electrical conductivity of these materials was shown to be microstructure specific at low temperatures: nanometric samples conduct protonically while their micrometric counterparts do not. Results clearly demonstrated that the protonic conductivity in 8YSZ was grain size dependent and that the smaller the grain size, the better the conductivity.

REFERENCES

[1]J. B. Goodenough, "Oxide ion electrolytes," *Annu. Rev. Mater. Res.*, 33, 91 128 (2003).

[2]U. Anselmi-Tamburini, F. Maglia, G. Chiodelli, P. Riello, S. Bucella, and Z. A. Munir, Enhanced low-temperature protonic conductivity in fully dense nanometric cubic zirconia, *Appl. Phys. Lett.*, 89, 163116/1-163116/3 (2006).

[3]S. Kim, U. Anselmi Tamburini, H. J. Park, M. Martin, and Z. A. Munir, Unprecedented room temperature electrical power generation using nanoscale fluorite structured oxide electrolytes, *Adv. Mater.*, 20, 556 559 (2008).
[4]M. C. Martin and M. L. Mecartney, Grain boundary ionic conductivity of yttrium stabilized zirconia as a function of silica content and grain size, *Solid State Ionics*, 161, 67-79 (2003).
[5]P. Dahl, I. Kaus, Z. Zhao, M. Johnsson, M. Nygren, K. Wiik, T. Grande, and M.-A. Einarsrud, Densification and properties of zirconia prepared by three different sintering techniques, *Ceram. Int.*, 33, 1603-1610 (2007).
[6]D. D. Upadhyaya, A. Ghosh, K. R. Gurumurthy, and R. Prasad, Microwave sintering of cubic zirconia, *Ceram. Int.*, 27, 415-418 (2001).
[7]X. J. Chen, K. A. Khor, S. H. Chan, and L.G. Yu, Preparation yttria-stabilized zirconia electrolyte by spark plasma sintering, *Mater. Sci. Eng.* A341, 43-48 (2003).
[8]U. Anselmi-Tamburinin, J. E. Garay, and Z. A. Munir, Fast low-temperature consolidation of bulk nanometric ceramic materials, *Scripta Mater.*, 54, 823-828 (2006).
[9]G. K. Williamson and W. H. Hall, X ray line broadening from filed aluminum and wolfram, *Acta Metall.*, 1, 22 31 (1953).
[10]U. Anselmi-Tamburini, J. E. Garay, Z. A. Munir, A. Tacca, F. Maglia, and G. Spinolo, Spark plasma sintering and characterization of bulk nanostructured fully stabilized zirconia: Part I. Densification studies, *J. Mater. Res.*, 19, 3255-3262 (2004).
[11]G. Skandan, H. Hahn, B. H. Kear, M. Roddy, and W. R. Cannon, The effect of applied stress on densification of nanostructured zirconia during sinter forging, *Mater. Lett.*, 20, 305 309 (1994).
[12]Z. A. Munir, U. Anselmi Tamburini, and M. Ohyanagi, The effect of electric field and pressure on the synthesis and consolidation of materials: A review of the spark plasma sintering method, *J. Mater. Sci.*, 41, 763 777 (2006).
[13]J. R. Groza, M. Garcia, and J. A. Schneider, Surface effects in field assisted sintering, *J. Mater. Res.*, 16, 286 292 (2001).
[14]E. A. Olevsky, S. Kandukuri, and L. Froyen, Consolidation enhancement in spark plasma sintering: Impact of high heating rates, *J. Appl. Phys.*, 102, 114913/1 114913/12 (2007).
[15]E. A. Olevsky and L. Froyen, Impact of thermal diffusion on densification during SPS, *J. Am. Ceram. Soc.*, 92, S122 S132 (2009).
[16]P. S. Manning, J. D. Sirman, and J. A. Kilner, Oxygen self-diffusion and surface exchange studies of oxide electrolytes having the fluorite structure, *Solid State Ionics*, 93, 125-132 (1997).
[17]S. Kim, H. J. Avila Paredes, S. Wang, C T. Chen, R. A. De Souza, M. Martin, and Z. A. Munir, On the conduction pathway for protons in nanocrystalline yttria stabilized zirconia, *Phys. Chem. Chem. Phys.*, 11, 3035 3038 (2009).

SPARK PLASMA SINTERING MECHANISMS IN Si_3N_4 BASED MATERIALS

M. Belmonte, J. Gonzalez Julian, P. Miranzo, M.I. Osendi
Institute of Ceramics and Glass (CSIC). Kelsen 5
Madrid, Spain

ABSTRACT

The spark plasma sintering (SPS) of Si_3N_4 powders plus Al_2O_3 and Y_2O_3 sintering additives, added in variable amounts from 2.5 to 7 wt%, is reported for the range of temperatures 1500 1675 °C. Same compositions hot pressed (HP) at higher temperatures (1600 1750 °C) only give dense materials for the composite with 7 wt% of additives. For lower additive contents, full densification is only possible by SPS. The comparison between the shrinkage rate curves of SPS specimens and published data for conventionally sintered Si_3N_4 of similar characteristics points out that an enhancement of the particle rearrangement stage by SPS occurs. A lack of homogeneity in the distribution of glassy phase is observed in the HP material with the lowest additive content, whereas the equivalent SPS material shows a completely wetting grain boundary phase. Based on these results, the role played by electrowetting in the first sintering stage of liquid phase sintering is signalled as responsible of the enhanced sintering rate.

INTRODUCTION

Although research on silicon nitride (Si_3N_4) ceramics started fifty years ago, the interest in these materials has progressively increased as well as their use in various technological applications,[1] such as engine components, ball bearings or metal cutting and shaping tools, due to their good thermomechanical and tribological properties. The densification of Si_3N_4 materials is quite complex and commonly entails the use of rare earth oxides additives and temperatures as high as 1700 1800 °C for promoting the liquid phase sintering process,[2] which includes particle rearrangement, solution precipitation and Ostwald ripening grain growth stages. That complexity in the sintering process can also be turn into advantage for tailoring microstructures and properties.[3]

The employ of conventional sintering techniques, such as hot pressing (HP) or hot isostatic pressing (HIP), simultaneously promotes $\alpha \rightarrow \beta$ phase transformation and grain growth during Si_3N_4 densification. However, the lately developed spark plasma sintering (SPS) technique,[4,5] based on a pressure assisted pulsed direct current sintering process, produces very fast densification of Si_3N_4 powders, which can even take place without any phase transformation or grain growth.[6 8]

The phenomena responsible of the enhanced sintering in SPS systems are still under debate.[5,9 11] Particularly controversial is the plasma generation during SPS process that some authors claim, which would cause the particle surface cleaning and the enhancement of the mass transport.[12] Nevertheless, the lack of experimental evidence of plasma formation has been pointed out.[11] Among the numerous SPS mechanisms proposed, there is a general consensus on the important role of both the Joule rapid heating[13] and the intrinsic field effects.[5,9] However, for non conductive powders, such as Si_3N_4, most of them are not entirely operative. For this particular ceramics, Shen et al.[14] proposed a dynamic ripening mechanism based on the enhanced motion of charged species by the electric field and rapid heating that promotes diffusion and homogenization of the formed liquid phase.

In the present work, we study the SPS process in Si_3N_4 materials varying the amount of sintering additives. The possible effect of the pulsed electric field on the wetting behaviour of the liquid phase during the densification is examined, analyzing the shrinkage rates, the microstructures and the morphological characteristics of the SPS materials. Furthermore, the results are compared with those got for similar hot pressed specimens and published data.

EXPERIMENTAL PROCEDURE

Three different batches of powders mixtures (Table I) containing α Si_3N_4 (SN E10 grade, UBE Industries) and distinct amounts of Al_2O_3 (SM8, Baikowski Chimie) and Y_2O_3 (Grade C, H.

C. Starck GmbH & Co.), as sintering additives, were attrition milled in ethanol for 2 h using Si_3N_4 grinding media. Afterwards, slurries were dried in a rotary evaporator and sieved through a 63 μm mesh. These powders mixtures were labelled as SN2A5Y, SN1A3Y and SN0.5A2Y (Table I).

Table I. Specimen label, weight composition, sintering additives ratio and total volume content of additives.

Composition	Si_3N_4 (wt%)	Al_2O_3 (wt%)	Y_2O_3 (wt%)	Y_2O_3/Al_2O_3	Additives (vol%)
SN2A5Y	93.0	2.0	5.0	2.5	4.85
SN1A3Y	96.0	1.0	3.0	3.0	2.74
SN0.5A2Y	97.5	0.5	2.0	4.0	1.70

Disc shaped samples of 3 mm thickness and 20 mm diameter were spark plasma sintered (SPS 510CE, SPS Syntex Inc.) at temperatures in the range 1500 1675 °C, using a heating rate of 133 °C·min^{1}, an uniaxial pressure of 50 MPa and holding time of 5 min, under 4 Pa of vacuum atmosphere. For comparative purposes, sintering runs were done in a hot press machine (HP W150/200 2200 100, FCT Systeme GmbH) at temperatures of 1600 1750 °C, 10 °C·min^{1} of heating rate, an uniaxial pressure of 50 MPa, holding times of 5 and 90 min, in 0.1 MPa of N_2 . The sintering parameters for the complete set of SPS and HP experiments are summarized in Table II. Temperature was controlled with pyrometers, and, in the case of SPS experiments, it was focused in a hole drilled in the middle of the external die surface through half the thickness of the die wall.

Table II. Sintering parameters, densities and α phase content for the SPS and HP specimens. T is maximum temperature; t is holding time; d and d_{rel} are the apparent and the relative densities, respectively; the later referred to the maximum density achieved.

Composition	Sintering	Sample ID	T (°C)	t (min)	d (g·cm^{3})	d_{rel} (%)	α phase (%)
SN2A5Y	SPS	SPS2A5Y 1500	1500	0	3.08	95.4	76
		SPS2A5Y 1600	1600	5	3.23	100	41
		SPS2A5Y 1650	1650	5	3.23	100	6
	HP	HP2A5Y 1600	1600	5	2.44	75.5	85
		HP2A5Y 1750	1750	90	3.23	100	0
SN1A3Y	SPS	SPS1A3Y 1600	1600	5	3.21	100	43
	HP	HP1A3Y 1750	1750	90	3.18	99.0	6
SN0.5A2Y	SPS	SPS0.5A2Y 1600	1600	5	2.96	92.8	39
		SPS0.5A2Y 1675	1675	5	3.19	100	15
	HP	HP0.5A2Y 1750	1750	90	3.12	97.8	13

The SPS parameters, including voltage, current, temperature, vacuum, axial pressure and total displacement (dz), were computer recorded during the whole sintering process.

Apparent density of the specimens was determined by the water immersion method. Crystalline phases and α/β transformation degree[15] were determined by X ray diffraction (XRD, Bruker D5000, Siemens) procedures. Microstructures of the specimens were observed using a field emission scanning electron microscope (FEM, Hitachi S 4700) on both fractured and polished samples. For the later, surfaces were previously etched in a CF_4/5 vol% O_2 plasma at 100 W of power for 40 s.

RESULTS AND DISCUSSION

Density and α phase content of each specimen are summarized in Table II. As theoretical density is difficult to set for these materials, the relative density (d_{rel}) given in the table is referred to the maximum density achieved for each composition. For the SN2A5Y composition, SPS technique gives materials with a high density (d_{rel} > 95 %) and reduced phase transformation (76 % α phase) at temperatures as low as 1500 °C, which become fully dense and yet with high α phase contents (41%) at 1600 °C. Conversely, at this temperature, highly porous specimens are obtained by HP, being necessary temperatures of 1750 °C and 90 min of holding time to get dense samples, then, leading to a complete α → β phase transformation. When the amount of additives is reduced to 2.74 vol% (SN1A3Y), similar results in terms of densification and α phase content are found for the SPS material, whereas the HP one begins to show small porosity (1 %) and residual amount of α phase (6 %). Finally, the composition with the lowest additive content, SN0.5A2Y (1.7 vol%), shows poor sinterability, requiring an increase in the SPS temperature. The same HP composition reaches a d_{rel} of ~ 98 % at 1750 °C, therefore, higher temperatures would be required for complete densification, which will prompt Si_3N_4 decomposition if nitrogen overpressure were not applied.

The present additive contents plus the SiO_2 proportion covering the Si_3N_4 particles, calculated from the O_2 amount in the powders provided by the supplier, were used to estimate the average composition of the grain boundary phase for each material. Those compositions were tentatively situated in the corresponding Al_2O_3 Y_2O_3 SiO_2 phase equilibrium diagram (Fig. 1).[16]

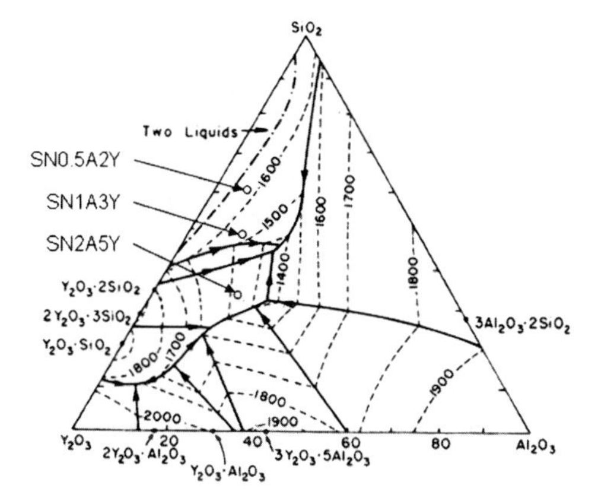

Figure 1. Al_2O_3 Y_2O_3 SiO_2 phase diagram[16] (in wt%) showing the location of the grain boundary composition for SN2A5Y, SN1A3Y and SN0.5A2Y.

Despite the fact that the grain boundary composition of SN2A5Y specimen is located in a different compatibility triangle from that of SN1A3Y and SN0.5A2Y, all three have the same invariant point and, therefore, the onset of liquid phase formation temperature would occur at 1345 °C.[2]

Fig. 2 shows examples of the displacement (dz, Fig. 2a) and the displacement rate (d(dz)/dt, Fig. 2b) curves as a function of temperature for the three different SPS compositions. The displacement curves evidence a different sintering behaviour between the SN2A5Y and the other two compositions (Fig. 2a), starting the densification at a lower temperature for the first one. The displacement rate curves for all the compositions exhibit the two typical peaks associated to the particle rearrangement (first maximum) and the solution precipitation mechanism (second maximum).[3] The first maximum in d(dz)/dt plots (Fig. 2b) indicates the onset of liquid formation and its shift from 1280 °C (SN2A5Y) to higher values for SN1A3Y and SN0.5A2Y compositions (1325 °C and 1342 °C, respectively) can be explained because these grain boundary compositions would need higher temperature for a complete melting than SN2A5Y, according to the phase equilibrium diagram (Fig. 1). One example of how the glassy phase surrounds Si_3N_4 grains can be clearly observed in the fracture surface of SPS2A5Y 1500 specimen (Fig. 3).

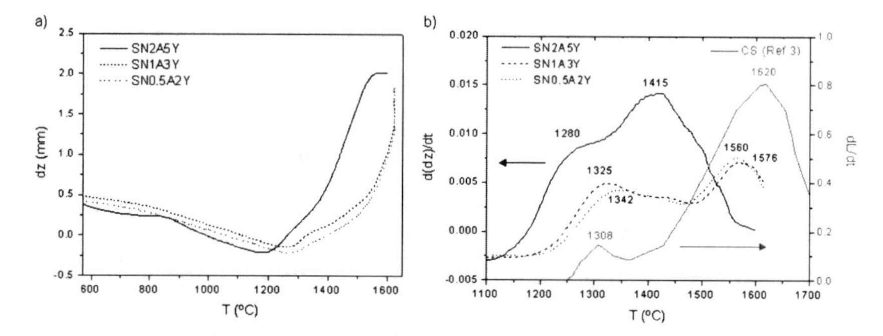

Figure 2. a) SPS displacement (dz) and b) SPS displacement rate (d(dz)/dt) curves as a function of the temperature for the different additives compositions. Petzow's data re plotted from ref. 3 for conventionally sintered Si_3N_4 (CS curve) with 6.0 vol% of additives in a 2:1 weight ratio (Y_2O_3:Al_2O_3) are included in Fig. 2b.

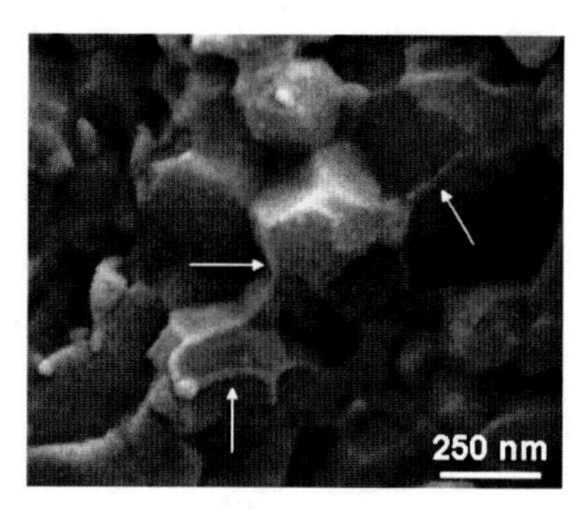

Figure 3. FEM micrograph of the fracture surface of SPS2A5Y 1500 specimen showing the glassy phase (pointed out by arrows) surrounding the Si_3N_4 grains.

Petzow and Herrmann[3] studied the sintering behaviour of Si_3N_4 ceramics containing mixtures of Al_2O_3 plus Y_2O_3 ranging from 3.5 to 8.5 vol%, using conventional sintering techniques. They observed peaks of maximum shrinkage at ~ 1320 °C and ~ 1600 °C, being the second peak, associated to the solution precipitation stage, about 4 times more intense than the first peak related to the rearrangement stage. For faster heating rates, typical of SPS runs, the same mechanisms should take place but shifted at higher temperatures and in a narrower temperature span.[17] If we plot in Fig. 2b Petzow's data[3] for similar Y_2O_3/Al_2O_3 additive ratio to that of the present SPS specimens (2.0 versus 2.5 for SN2A5Y specimen), we actually observe the opposite trend. That is, SPS process leads to slightly lower temperatures for the rearrangement stage and to a huge decrease (~ 200 °C) in the temperature associated to the solution precipitation stage, in spite of the smaller volume of additives in present materials (4.9 vol% versus 6.0 vol% for reference 3). Even more, the peak corresponding to the rearrangement process is strengthen in the SPS experiments, as the intensity ratio between the second and first peak is close to 2 instead of 4 observed for conventional sintering (Fig. 2b).

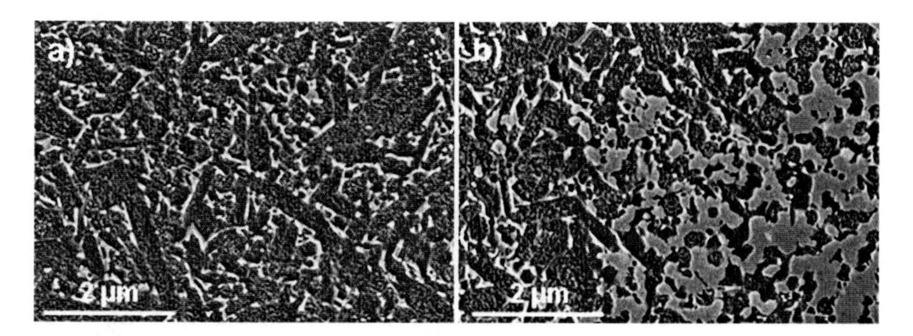

Figure 4. FEM micrographs of the polished and etched surfaces corresponding to: a) SPS0.5A2Y 1675 and b) HP0.5A2Y 1750 specimens, evidencing the differences in glassy phase distribution.

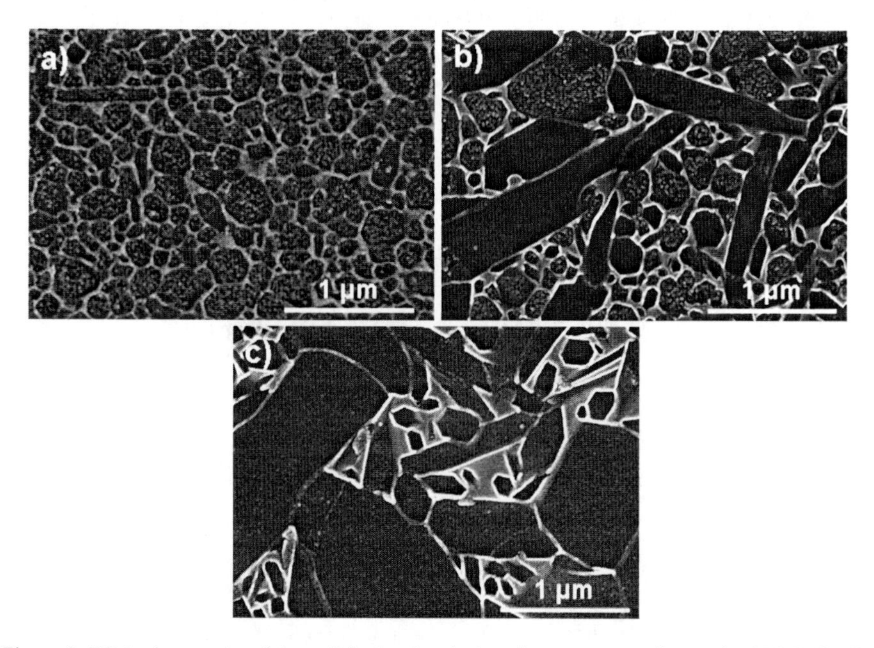

Figure 5. FEM micrographs of the polished and etched surfaces corresponding to the SPS SN2A5Y materials as a function of the SPS temperature: a) SPS2A5Y 1500, b) SPS2A5Y 1600 and c) SPS2A5Y 1650.

These phenomena can only be explained considering the effect of the applied electric field during SPS experiments on the sintering mechanisms, suggesting also that the rearrangement stage is enhanced under the presence of the electric field and that can only be due to an increased wetting by the liquid phase.[18] In this way, the so called electrowetting mechanism, i.e. a voltage induced wetting mechanism produced by an electric double layer building up at solid/liquid interfaces, is proposed as the responsible for that enhanced sintering rate during the first sintering stage. Notwithstanding the low voltage applied in SPS tests (~ 4 5 V), high values could be reached at the nano scale contacts between particles. Therefore, when voltage is applied, the solid liquid interfacial tension is reduced, lowering contact angle and increasing capillary forces at contacts between particles.[19,20] Electrocapillarity and electrowetting are well documented phenomena at room temperature for salt aqueous solutions[20] and they do not seem to be much dependent on the liquid nature. A similar effect could be envisaged for a high temperature dielectric fluid (sialon liquid) under the electric field (SPS voltage) applied on the graphite mold, where the Si_3N_4 particles would act as a dielectric layer.[20] This kind of mechanism would perfectly explain the enhancement in the particle rearrangement by the improved liquid wetting observed in SPS of Si_3N_4 materials as compared to conventionally sintered specimens.

As an example of above arguments, SPS SN0.5A2Y composition (Table II), having just 1.70 vol% of additives and relatively poor wettability due to its high Y_2O_3/Al_2O_3 ratio,[21] shows a well distributed grain boundary glassy phase as seen in Fig. 4a; while by contrast, the equivalent HP composition exhibits large pockets of glassy phase (Fig. 4b) confirming a poor wettability of the liquid when electric field is absent. This is an experimental evidence of the role of the electrowetting in the SPS process.

In the second sintering stage, the electric field can enhance the motion of charged species, favouring diffusion precipitation and grain growth processes through a dynamic ripening mechanism, as proposed by Shen et al.[14] This fact is observed in Fig. 5 for SN2A5Y specimens, where sub micron equiaxed microstructures with negligible grain growth are clearly visible at 1500 °C (Fig. 5a), and bimodal coarser microstructures are observed at 1600 °C and 1650 °C, respectively (Fig. 5b and c), in correspondence with a decrease in the α content phase 76% at 1500 °C and 6% at 1650 °C (Table II).

CONCLUSIONS

The application of an electric field, as occurs in the sintering of ceramics by the SPS technique, extraordinarily increases the sintering kinetics of Si_3N_4 materials. Displacement rate curves of SPS specimens showed in all cases two distinct peaks that corresponded to the common stages of the liquid phase sintering process, particle rearrangement and solution reprecipitation, but with a noticeable enhancement of the rearrangement step. The improved wetting of the liquid phase by the presence of an electric field is proposed, in similarity to the established electrowetting mechanism, to explain the enhanced sintering in Si_3N_4 ceramics.

ACKNOWLEDGEMENTS

Funding for this work was provided by Spanish Ministry of Science and Innovation (MICINN) and Spanish National Research Council (CSIC) under projects MAT2006 7118 and HA2007 0083. J. Gonzalez Julian acknowledges the financial support of the JAE (CSIC) fellowship Program.

REFERENCES

[1]F.L. Riley, Silicon nitride and related materials, *J. Am. Ceram. Soc.*, **83**, 245 265 (2000).

[2]D. Suttor and G. S. Fischman, Densification and sintering kinetics in sintered silicon nitride, *J. Am. Ceram. Soc.*, **75**, 1063 1067 (1992).

[3]G. Petzow and M. Herrmann, Silicon nitride ceramics, *Struct. Bond.*, **102**, 47 167 (2002).

[4]M. Tokita, Mechanism of spark plasma sintering and its application to ceramics, *Nyu Seramikkusu*, **10**, 43 53 (1997).

[5]Z. A. Munir, U. Anselmi Tamburini and M. Ohyanagi, The effect of electric field and pressure on the synthesis and consolidation of materials: a review of the spark plasma sintering method, *J. Mater. Sci.*, **41**, 763 777 (2006).

[6]T. Nishimura, M. Mitomo, H. Hirotsuru and M. Kawahara, Fabrication of silicon nitride nano ceramics by spark plasma sintering, *J. Mater. Sci. Lett.*, **14**, 1046 1047 (1995).

[7]M. Suganuma, Y. Kitagawa, S. Wada and N. Murayama, Pulsed electric current sintering of silicon nitride. *J. Am. Ceram. Soc.*, **86**, 387 394 (2003).

[8]Z. Shen, H. Peng, J. Liu and M. Nygren, Conversion from nano to micron sized structures: experimental observations, *J. Eur. Ceram. Soc.*, **24**, 3447 3452 (2004).
[9]E. A. Olevsky, S. Kandukuri and L. Froyen, Consolidation enhancement in spark plasma sintering: impact of high heating rates, *J. Appl. Phys.*, **102**, 114913 (2007).
[10]R. Chaim, Densification mechanisms in spark plasma sintering of nanocrystalline ceramics, *Mat. Sci. Eng. A*, **443**, 25 32 (2007).
[11]D. M. Hulbert et al., The absence of plasma in "spark plasma sintering", *J. Appl. Phys.*, **104**, 033305 (2008).
[12]J. R. Groza, M. Garcia and J. A. Schneider, Surface effects in field assisted sintering, *J. Mater. Res.*, **16**, 286 292 (2001).
[13]U. Anselmi Tamburini, S. Gennari, J. E. Garay, Z. A. Munir, Fundamental investigations on the spark plasma sintering/synthesis process: II. Modeling of current and temperature distributions, *Mat. Sci. Eng. A*, **394**, 139 148 (2005).
[14]Z. J. Shen, Z. Zhao, H. Peng and M. Nygren, Formation of tough interlocking microstructures in silicon nitride ceramics by dynamic ripening, *Nature*, **417**, 266 269 (2002).
[15]C. P. Gazzara and D. R. Messier, Determination of phase content of Si_3N_4 by X ray diffraction analysis, *Am. Ceram. Soc. Bull.*, **56**, 777 780 (1977).
[16]I. A. Bondar and F. Ya. Galakhov, *Izv. Akad. Nauk SSSR, Ser. Khim.*, **7**, 1325 (1963).
[17]O. Abe, Sintering process of Y_2O_3 and Al_2O_3 doped Si_3N_4, *J. Mater. Sci.*, **25**, 4018 4026 (1990).
[18]Fundamentals of ceramics, M.W. Barsoum, Publisher Taylor & Francis Group, NY pp. 337 341 (2003).
[19]M. W. J. Prins, W. J. J. Welters and J. W. Weekamp, Fluid control in multichannel structures by electrocapillary pressure, *Science*, **291**, 277 280 (2001).
[20]F. Mugele and J. C. Baret, Electrowetting: from basics to applications, *J. Phys. Condes. Matter.*, **17**, R705 R774 (2005).
[21]H. Lemercier, T. Rouxel, D. Fargeot, J. L. Besson and B. Piriou, Yttrium SiAlON glasses: structure and mechanical properties elasticity and viscosity, *J. Non Cryst. Solids*, **201**, 128 145 (1996).

CONSOLIDATION OF SIC WITH BN THROUGH MA SPS METHOD

Yasuhiro Kodera[1], Naoki Toyofuku[1], Ryousuke Shirai[1], Manshi Ohyanagi[1] and Zuhair Munir[2]

[1]Department of Materials Chemistry, Innovative Materials and Processing Research Center, Ryukoku University, Ohtsu, Japan

[2]Department of Materials Science and Chemical Engineering, College of Engineering, University of California, Davis, CA, USA

ABSTRACT

Two types of SiC/BN powder were prepared in Air and N_2 (O_2 concentration < 5 ppm and H_2O < 24 ppm) atmosphere. Both X ray diffraction and IR spectra analysis indicated that the mixture powder consists of the coexistence of stacking disordered SiC and turbostratic (t) BN. The successful consolidation of SiC/BN composite was achieved without sintering additives. The remarkable change of the crystal structure was obtained between the milled and sintered samples. The powder preparation atmosphere indicated negligible influence for the SiC phase, which exhibited the structural ordering from the stacking disorder to β SiC. In contrast, the atmosphere clearly affected BN phase. When the Air was used, sintered SiC/BN consisted of h BN as the result of the structural ordering from t BN. In N_2 atmosphere, sintered SiC/BN remained t BN due to the lack of H_2O and O_2 contamination. Homogeneous and small particles with the diameter of less than several hundred nanometers exhibited in SiC/BN composite. Selecting N_2 atmosphere led the SiC/BN to have not only t BN structure, but a also relatively small particle size. Without sintering aid, MA SPS method was able to prepare the composite with a higher Vickers hardness compared with the sample prepared by conventional process. The bending strength of approximately 388 and 137 MPa was obtained in the sample with SiC/BN=50/50 prepared in Air and N_2, respectively. These values were relatively high compared with the sample prepared by conventional process. These data indicates that MA SPS method is promised method for the preparation of SiC/BN composite.

NTRODUCTION

The addition of BN into SiC through making those composite extends the application field for SiC. SiC has been researched for present and potential technological applications due to its attractive properties, such as low density, chemical stability, low nuclear activation, and high strength at high temperatures.[1 4] However, SiC is a brittle material, which is a major impediment for engineering applications, including the difficulty of machining and cost. Also, SiC's low thermal shock resistance leads to a deterioration of its potential. Therefore, making a composite of SiC with other ceramics has been studied in order to improve its disadvantageous properties. h BN is top candidate material as a second phase owing to the lowest elastic modulus and excellent thermal shock resistance.[5,6] However, preparing SiC/BN composites is difficult due to the low sinterability of both SiC and h BN. [6,7]

The effectiveness of Mechanical Alloying Spark Plasma Sintering (MA SPS) technique has been reported for the consolidation of the material with low sinterability (e.g. SiC and BN).[7 11] With this technique, stacking disordered SiC was prepared from the elements by high energy ball milling and then consolidated by SPS. Without any sintering aids, highly dense β SiC (up to 99% relative density) was successfully obtained. In BN system, h BN was consolidated to have over 90% relative density without sintering aids through simultaneous structural ordering from t BN; it was prepared by mechanical grinding of h BN. MA SPS method was also extended to prepare bulk SiC/BN composite, and the method successfully led to effective consolidation without sintering additives.[12] In the previous paper, all powders of Si, C, and h BN were milled at once by high energy ball milling (as we called single step method) and then consolidated by SPS. During this method, the phase formation of β SiC and h BN from the amorphous like structure with Si C B N chemical bonding was suggested.

In this paper, we report the consolidation of SiC/BN composite by the MA SPS method, which is different than previous report.[12] As a powder preparation process, we developed double step method, in which only Si and C were milled to synthesize SiC and then milled with h BN to make SiC/BN composite. The double step method is able to have simple mechanism for making composite avoiding the phase formation of β SiC and h BN from the amorphous like structure with Si C B N chemical bonding. Also, the different atmosphere of Air and N_2 was selected during powder preparation. The influence of the atmosphere on the character of SiC/BN was discussed.

EXPERIMENTAL

The starting materials of Si (ca. 1.0 μm, 99.999% pure, Kojundo Chemical Co. Ltd., Japan), C (TGP 7, ca. 7.0 μm, >99.9% pure, Tokai Carbon Co. Ltd., Japan), and hexagonal BN (ca. 10 μm, >99% pure, Kojundo Chemical Co. Ltd., Japan) powders were selected. Si and C powder were blended to have a 50/50 mol ratio and then mechanically alloyed using a planetary ball mill (Pulverisette P5/2 Fritsch). Silicon nitride balls (10 mm diameter) and a vial (an inside diameter of 75 mm and a height of 70 mm) were used. The ball to powder mass ratio (B/P) was fixed to be 40/1 with 7.5 g of the mixed reactants. The revolution speed of the vial was 300 rpm, and a rotation/revolution ratio was 1.18/1. The Si and C were milled for 24 h to have mechanically alloyed (MA) SiC. The MA SiC and h BN were blended to have a SiC/BN = 100/0 to 0/100 ratio. This blended powder was mechanically grinded (MG) with same conditions for MA SiC preparation, but the different milling time, 12 h, was selected. Two atmospheric conditions of Air and N_2 were used during powder preparation. In the case of the former condition, all transfers of powders to and from the vials were handled in a glove box filled with N_2 (O_2 concentration < 5 ppm and H_2O concentration < 24 ppm). The vials in the glove box were also covered with stainless pots filled with N_2. Residual H_2O and O_2 were removed from the Ar atmosphere by a recycle purification system during the handling of powders (model: MF 70, UNICO, Japan). On the other hand, all powder preparation was done in Air for the latter condition.

Spark plasma sintering (SPS) apparatus (Model 1050, Sumitomo Coal and Mining Co. Japan) was utilized to consolidate samples. The milled powders were wrapped in a 0.2 mm thick graphite foil and placed in a cylindrical graphite die. Uniaxial pressure of 70 MPa was applied through graphite plungers. After the system with the sample was evacuated to a pressure of about 10 2 Torr, a DC pulsed current was supplied. The pulse cycle of the DC current was 12:2 i.e., 12 pulses of 3.6 ms on and 2 pulses of 3.6 ms off. The sample was heated to selected temperatures at rates of about 100°C/min and held for 10 min.

X ray diffraction (XRD) analyses were carried out using a RIGAKU RINT2500 (Rigaku Co., Ltd., Japan) diffractometer. To obtain structural information, an infrared absorption spectroscope (FTIR 660 Plus, JASCO, Japan) was used. Specimens were machined to 22 x 3 x 4 mm and polished for a 3 point bending strength measurement (PL 300, Marubishikagaku, Japan). The span length and crossed speed were 18 mm and 0.2 mm/min, respectively. The strength data were calculated based on the average of three measurements. The hardness was measured by Vickers method (HMV 2000, Shimazu, Japan) with a load of 98 N and a holding time of 15 s. The results of ten measurements were averaged and used for the hardness data. The fracture surfaces of the sintered specimens were observed using a Scanning Electron Microscope (SEM: JSM 5400, JEOL, Japan).

RESULTS AND DISCUSSION

In this research, two types of SiC/BN powder were prepared in Air and N_2 atmosphere with various SiC/BN ratios ranging from 100/0 to 0/100. There were the sample with SiC/BN ratio of 100/0 (mechanically alloyed SiC: MA SiC) and 0/100 (mechanically grinded BN: MG BN). Figure 1 and 2 shows the X ray diffraction patterns of MA SiC, SiC/BN, and MG BN powders prepared in N_2 and Air. MA SiC powders exhibited only the three broad peaks at 2θ values of about 36°, 60°, and 72°,

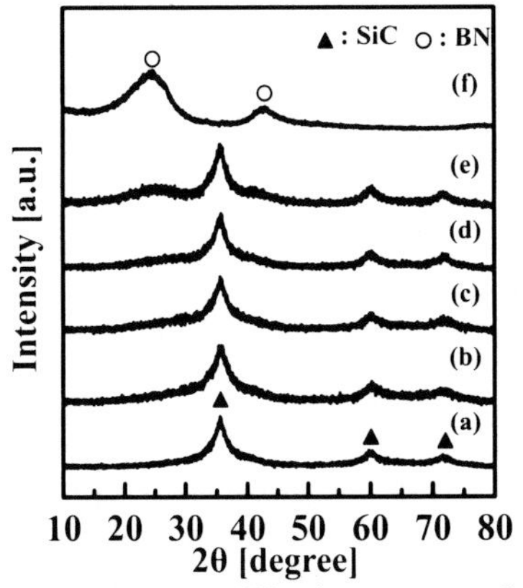

Figure 1. X ray diffraction patterns of sample powder prepared by MA process in Air atmosphere. SiC/BN vol% ratio is (a) 100/0:MA SiC, (b) 90/10, (c) 80/20, (d) 70/30, (e) 50/50, (f) 0/100: MG BN.

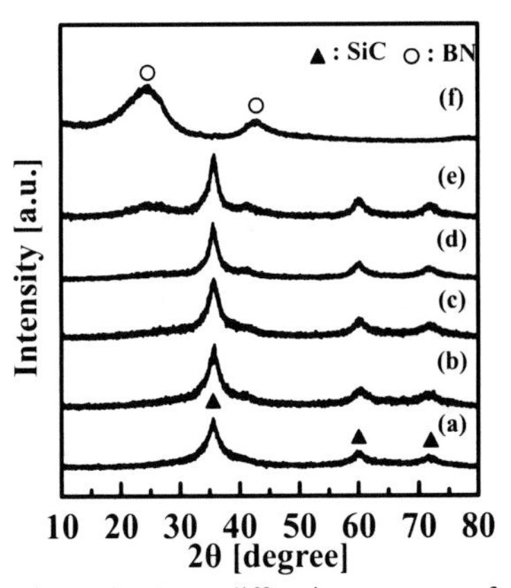

Figure 2. X ray diffraction patterns of sample powder prepared by MA process in N_2 atmosphere. SiC/BN vol% ratio is (a) 100/0:MA SiC, (b) 90/10, (c) 80/20, (d) 70/30, (e) 50/50, (f) 0/100: MG BN.

respectively. This peak pattern of MA SiC is totally different form that of α SiC (e.g. hexagonal, rhombohedra). Also, this pattern disagree with that of β SiC (cubic) due to the presence of considerably broad peaks and the absence of (200) peak. Silicon carbide exhibits considerable polytypism characterized by a one dimensional (stacking) disorder.[7,8,14 16] The X ray pattern of one dimensionally disordered SiC by the stacking sequence is characterized by the presence of only the three peaks of (111), (220), and (311) planes. [14] This pattern is consisted with that of MA SiC powders. In contrast, MG BN showed two broad diffractions from the (002) and (10l) planes of h BN peaks at 2θ values of about 24° and 42°, respectively. MG BN corresponded to turbostratic BN (t BN),[11,13] which has 2 dimensional order based on a B N six membered ring and its random stacking toward the c axis instead of the 3 dimensional order for h BN. During the milling process, this structural disordering was carried out with a sliding of the ring plane as a cleavage fracture. When the sample has highest BN ratio, the both peaks corresponding to MG BN and MA SiC coexisted. Additional peaks (e.g. Si_3N_4)

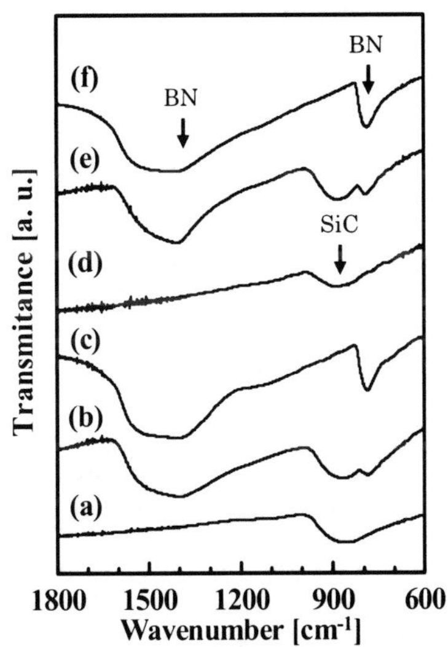

Figure 3. The IR spectra of the sample powder. (a) and (d) : MA SiC. (b) and (e) : SiC/BN, (c) and (d) : MG BN. (a)(b)(c) and (d)(e)(f) were prepared in Air and N_2, respectively.

were not detected. The influence of milling atmosphere was negligible in Fig. 1 and 2.

Figure 3 shows the IR spectra of the MA SiC, SiC/BN, and MG BN powder prepared in Air and N_2. In the range from 600 to 1800 cm ^{1}, SiC has a single peak at 800 cm ^{1} reflecting the stretching mode vibration of Si C bond position of Si C.[17] MA SiC showed a broad peak at approximately 820 cm ^{1} with FWHM over 150 cm ^{1}. The peak position agreed with other ball milled Si C result.[18,19] The FWHM of MA SiC was much larger than that of crystalline SiC with approximately 80 cm ^{1}. It is well known that the SiC with low crystalline order, which indicates the presence of amorphous structure and stacking defect, causes peak breading of Si C.[20] XRD peaks proved that MA SiC is not amorphous. Therefore, we considered that peak broadening of Si C was caused by the presence of the disordered structure.[17]

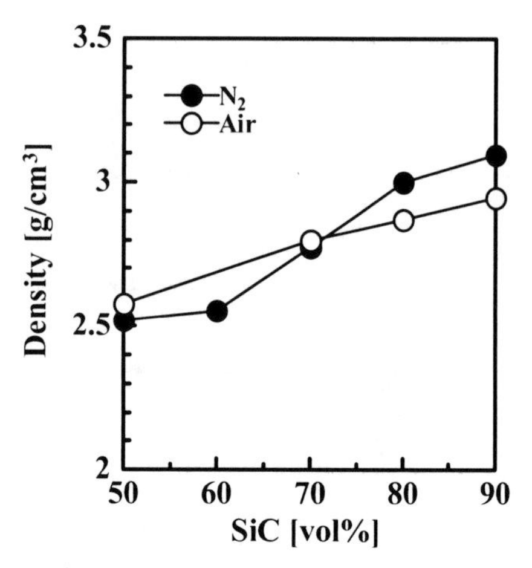

Figure 4. The density of the SiC/BN composite prepared in Air and N_2.

Generally, h BN showed two peaks at 817 and 1370 cm ^{1}, corresponded to out of plane and in plane vibrations, respectively. [21,22] The significant character of MG BN is exhibiting a broad peak at 1370 cm ^{1}, which agree with the reported data of t BN.[22] Thus IR data supported the XRD result suggesting that MG BN was t BN. In SiC/BN composite, the result of IR spectra indicates the characteristic peaks of MA SiC and MG BN. The mixture powder consists of the coexistence of stacking disordered SiC and t BN. The influence of milling atmosphere was also negligible in Fig. 3.

In the previous single step methods of powder preparation for SiC/BN composite, the X ray diffraction and IR spectra analysis suggested that the amorphous like structure and Si C B N chemical bonding existed as the result of planetary ball milling for 24 h with Si, C, and h BN.[12] However, during double step method, the influence of addition of h BN and milling for 12 h on the X ray diffraction and IR spectra analysis was not recognizable. Before h BN addition, the formation of strong Si C chemical bond of MA SiC was the key to have the coexistence of stacking disordered SiC and t BN in SiC/BN compost powder.

All powders were consolidated by SPS at 1900◦C for 10 min with the uniaxial pressure of 70 MPa. During SPS process, there was not clear effect of Air and N_2 atmosphere on the sintering behavior of SiC/BN composite. Figure 4 shows the density of the SiC/BN composite. Well known theoretical density of β SiC and h BN is 3.25 and 2.27 g/m^3, respectively. Traditionally, using sintering additives was unavoidable to achieve full consolidation for SiC/BN composite as well as SiC.[23 27] In this research, successful consolidation of SiC/BN composite was achieved without sintering additives, which is agreed with previous paper.[12] This consolidation was based on the mass transfer and the rearrangement of nano size grain, which was probably caused by the accelerated diffusion of defects and atoms in stacking disordered SiC and t BN. The bulk density decreased with the increasing of BN concentration due to the relative low density of BN. The influence of atmosphere during powder preparation on density was not clear in Fig. 4.

Figure 5 and 6 shows the X ray diffraction patterns of the consolidated specimens of MA SiC, SiC/BN, and MG BN prepared in Air and N_2. Remarkable peak change was obtained between the

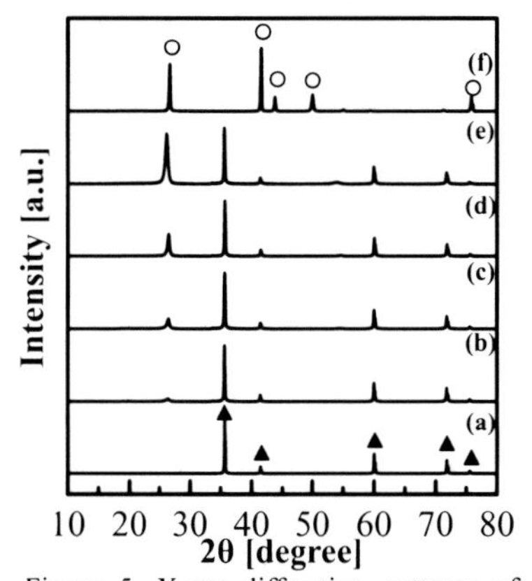

Figure 5. X ray diffraction patterns of sample consolidated by MA SPS process in Air atmosphere. SiC/BN vol% ratio is (a) 100/0:MA SiC, (b) 90/10, (c) 80/20, (d) 70/30, (e) 50/50, (f) 0/100: MG BN.

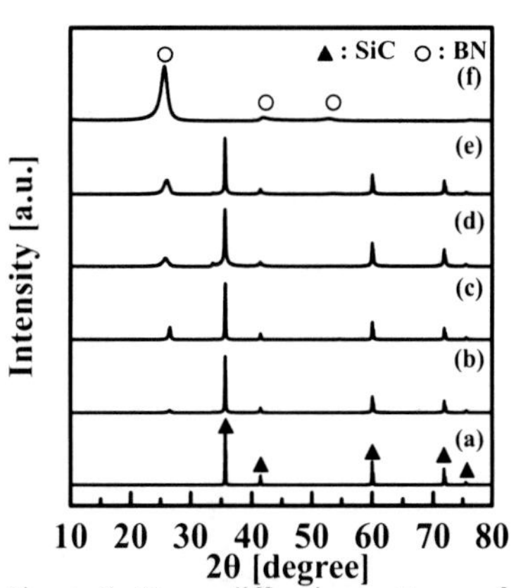

Figure 6. X ray diffraction patterns of sample consolidated by MA SPS process in N_2 atmosphere. SiC/BN vol% ratio is (a) 100/0:MA SiC, (b) 90/10, (c) 80/20, (d) 70/30, (e) 50/50, (f) 0/100: MG BN.

milled and consolidated samples. After the consolidation process, the five narrow peaks corresponding to β SiC at 2θ values of about 36°, 41°, 60°, 72°, and 76°, were observed instead of the three broad peaks of MA SiC powder (see Fig.1 and 2). This peak change was the result of structural ordering from the stacking disorder, and appeared in all sample, which containing SiC. All the data of X ray diffraction and IR spectra (shown in Fig. 1, 2, 3, 5, and 6) suggested that the powder preparation atmosphere indicated negligible influence for the SiC phase of the final product and the structural ordering from the stacking disorder to β SiC.

In contrast, the powder preparation atmosphere clearly affected BN peaks. The peak intensity of BN increased with increasing of BN concentration independent of the atmosphere. However, the peak intensity ratio of SiC/BN was different between the samples prepared in Air and N_2. The N_2 condition decreased the peak intensity of BN. During the SPS process, the decomposition of BN might be observed through the vaporization of the phase including B, N, and O. For example, the melting temperature of B O is lower than that of BN. Therefore, oxygen contamination in BN takes on an important role. However, the sample prepared in Air exhibited lower peak intensity ratio of SiC/BN, although the O_2 gas concentration in N_2 atmosphere was less than 5 ppm.

The h BN powder was milled and transferred to be t BN in the both of Air and N_2. In the case of Air atmosphere, sintered MG BN was identified as h BN structure by the peaks at 2θ values of about 26°, 41°, 44°, 50°, and 76°. The sintering process led the MG BN to be h BN through the structural ordering from t BN (see the both of MG BN samples in Fig. 1 and 5). However, when the MG BN was prepared in N_2 atmosphere, sintered MG BN exhibited two broad peaks at 2θ values of about 25° and 42°. Those peaks indicated that the sintered MG BN was still t BN: the structural ordering was not enough to form h BN. H_2O and O_2 contamination for BN accelerate BN atomic diffusion and phase transformation. [13,28] Therefore, the sintered MG BN prepared in N_2 atmosphere kept t BN structure by the lack of H_2O and O_2 contamination to be h BN. Consequently, N_2 atmosphere decreased peak

Figure 7. The SEM images of the fracture surface of the sintered sample. MG-BN prepared in Air, (b) MG-BN in N_2, (c)MA-SiC in Air, (d)SiC/BN = 90/10 in Air, (e) SiC/BN = 70/30 in Air, (f) SiC/BN = 50/50 in Air, (g) SiC/BN = 90/10 in N_2, (h) SiC/BN = 70/30 in N_2, (i) SiC/BN = 50/50 in N_2.

intensity of BN compared with Air atmosphere. This influence of the atmospheres on the crystallographic properties of BN was also observed by IR spectra analysis.

Figure 7 shows the SEM image of the fracture surface of the sintered samples. MA SPS method is effective method in order to prepare homogeneous and fine microstructure.[13] In this research, the small particles of SiC/BN composite exhibit the diameter of less than several hundred nanometers. The influence of the atmosphere on microstructure of MG BN was significant. When the MG BN was prepared in Air, the sintered sample exhibits large grain size and flake like morphology. On the other hand, the sample prepared in N_2 had small grain size (less than several hundred nanometers) and aggregated particle. Selecting N_2 atmosphere led the BN to have not only t BN structure, but also small particle size due to the lack of H_2O and O_2 contamination. However, the influence of the atmosphere on

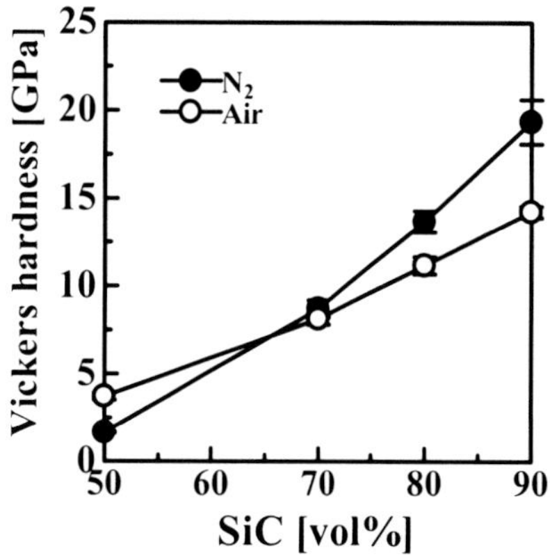

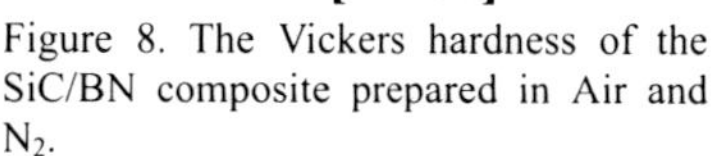

Figure 8. The Vickers hardness of the SiC/BN composite prepared in Air and N_2.

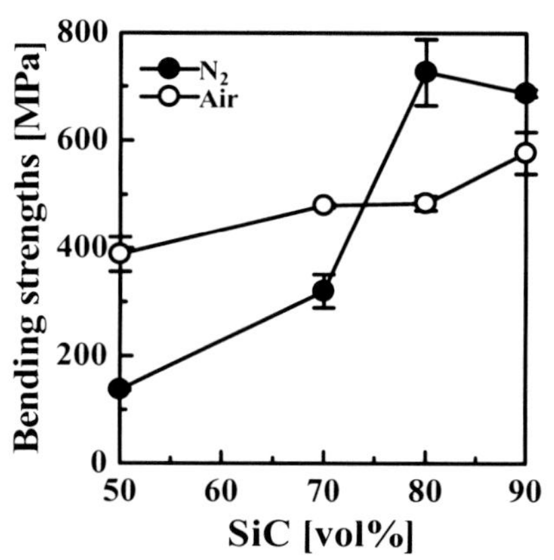

Figure 9. The bending strength of the SiC/BN composite prepared in Air and N_2.

microstructure of SiC/BN was not clear compared with that of MG BN. This is probably caused by the presence of second phases (BN for SiC and vice versa), which inhibits grain growth.

Figure 8 shows the Vickers hardness of the composite with different SiC/BN ratios. The hardness decreased with BN concentration due to the low hardness of h BN. The sample with SiC/BN ratio of 90/10 prepared in N_2 atmosphere exhibited approximately 36% harder than that in Air, because the hardness of t BN is much higher than that of h BN.[13] When the conventional and in situ reaction method was used without sintering aid, Vickers hardness measurement failed for the SiC/BN=46/54 composite consolidated at 2000◦C, due to the high porosity of the products.[23] In contrast, the both of the sample with SiC/BN = 50/50, which were prepared by MA SPS method with Air and N_2 atmosphere, had Vickers hardness of 1.7 and 3.7 GPa, respectively. Additionally, when sintering additive was used in the in situ reaction system, the Vickers hardness of the samples was 2.7 GPa for SiC/BN=46/54 and 8.7 GPa for SiC/BN=75/25.[29] Both values resemble the sample prepared in this research. These facts suggested that MA SPS method is a prommising method for the preparation of SiC/BN composite.

Figure 9 shows the bending strengths of the SiC/BN composite. The strength decreased with BN concentration, because BN has lower bending strength than SiC. In the sample with high BN concentration, Air atmosphere led the SiC/BN to have higher bending strength. This is caused by the existence of H_2O and O_2 contamination, which might affect the strength of grain boundary. The composite prepared in N_2 atmosphere with high BN concentration range (e.g. over 30% of BN), remained t BN structure (see Fig. 5 and 6) and exhibited relatively weak bending strength. The t BN is easy to react with water, which we used during a sample preparation and machining. The reaction was confirmed by pH measurement suggesting the ammonia emission, and caused the deterioration in the bending strength. As without the sintering additive system in this study at SiC/BN=50/50, bending strength of approximately 388 and 137 MPa was obtained in the sample prepared in Air and N_2, respectively. These values were relatively high compared with SiC/BN=46/54 composite consolidated at 2000◦C by conventional and in situ reaction methods with approximately 61 and 56 MPa, respectively.[29]

CONCLUSION

Two types of SiC/BN powder were prepared in Air and N_2 atmosphere with various SiC/BN ratios ranging from 100/0 to 0/100. Both the X ray diffraction and IR spectra analysis indicated that the mixture powder consists of the coexistence of stacking disordered SiC and t BN. The successful consolidation of SiC/BN composite was achieved without sintering additives. This consolidation was based on the mass transfer and the rearrangement of nano size grain, which was probably caused by the accelerated diffusion of defects and atoms in stacking disordered SiC and t BN. The influence of atmosphere during powder preparation on density was not clear. The remarkable change of crystal structure was obtained between the milled and consolidated samples. The powder preparation atmosphere indicated negligible influence for the SiC phase of final product and the structural ordering from the stacking disorder to β SiC. In contrast, the atmosphere clearly affected BN phase. When the Air atmosphere was selected, sintered SiC/BN consisted of h BN as the result of the structural ordering from t BN. In N_2 atmosphere, sintered SiC/BN remained t BN due to the lack of H_2O and O_2 contamination. This difference was also observed by IR spectra analysis.

Homogeneous and small particles with the diameter of less than several hundred nanometers exhibited in SiC/BN composite. Selecting N_2 atmosphere led the SiC/BN to have not only t BN structure, but also relatively small particle size due to the lack of H_2O and O_2 contamination. Without sintering aid, the MA SPS method was able to prepare the composite with higher Vickers hardness compared with referenced sample prepared by conventional process. The bending strength of approximately 388 and 137 MPa was obtained in the sample with SiC/BN=50/50 prepared in Air and N_2, respectively. These values were relatively high compared with referenced sample prepared by conventional process. These facts suggest that MA SPS method is a promising method for the preparation of SiC/BN composite.

References

[1] K. Yamada and M. Mohri, *Silicon Carbide Ceramics 1*, (ed) S. Somiya and Y. Inomata, New York , (1991).

[2] A. Tavassoli, Present Limits and Improvements of Structural Materials for Fusion Reactors /A Review, *Journal of Nuclear Materials*, **302**, 73 88 (2002).

[3] T. Yano, M. Akiyoshi, K. Ichikawa, Y. Tachi, and T. Iseki, Physical Property Change of Heavily Neutron Irradiated Si_3N_4 and SiC by Thermal Annealing, *Journal of Nuclear Materials,* **289**, 102 109 (2001).

[4] H. Heinisch, L. Greenwood, W. Weber, and R. Williford, Displacement Damage Cross Sections for Neutron Irradiated Silicon Carbide, *Journal of Nuclear Materials*, **307**, 895 889 (2002).

[5] A. Lipp, K. Schwetz, and K. Hunold, Hexagonal Boron Nitride: Fabrication, Properties and Applications, *Journal of the European Ceramic Society*, **5**, 3 9 (1989).

[6] R. Vaßen, A. Kaiser, J. Forster, H. Buchkremer, and D. Stover, Densification of Ultrafine SiC Powders, *Journal of Materials Science*, **31**, 3623 3637 (1996).

[7] T Yamamoto, H. Kitaura, Y. Kodera, T. Ishii, M. Ohyanagi, and Z. A. Munir, Consolidation of Nanostructured β SiC by Spark Plasma Sintering. *Journal of the American Ceramic Society*, **87**, 1436 1441 (2004).

[8] M. Ohyanagi, T. Yamamoto, H. Kitaura, Y. Kodera, T. Ishii, and Z. A. Munir, Consolidation of Nanostructured β SiC with Disorder Order Transformation, *Scripta Materialia,* **50**, 111 114 (2004).

[9] Yamamoto T, Kitaura H, Kodera Y, Ishii T, Ohyanagi M, and Munir ZA, "Effect of Input Energy on Si C Reaction Milling and Sintering Process," *Journal of the Ceramic Society of Japan* ,**112**, 940 945 (2004).

[10] Y. Kodera, T Yamamoto, H. Kitaura, T. Ishii, M. Ohyanagi, and Z. A. Munir, "Role of Disorder Order Transformation in Consolidation of Ceramics," *Journal of Materials Science*, **41**, 727 732 (2006).

[11] T. Yamamoto, N. Isibasi, N. Toyofuku, Y. Kodera, M. Ohyanagi and Z. A. Munir, Consolidation of h BN with Disorder Order Transformation, *Innovative Processing and Synthesis of Ceramics, Glasses and Composites, Materials Science and Technology 2006*, 531 538 (2006)
[12] Y. Kodera, N. Toyofuku, H. Yamasaki, M. Ohyanagi, and Z. A. Munir, Consolidation of SiC/BN Composite through MA SPS Method, *Journal of Materials Science*, **43**, 6422 6428 (2008).
[13] N. Toyofuku, N. Yamasaki, Y. Kodera, M. Ohyanagi, and Z. A. Munir, Turbostratic Boron Nitride Consolidated by SPS, *Journal of the Ceramic Society of Japan* ,**117**, 189 193 (2009).
[14] K. Szulzewsky, Ch. Olschewski, I. Kosche, H. D. Klotz and R. Mach, Nanocrystalline Si C N composites, *Nanostructured Materials*, **6**, 325 328 (1995)
[15] V. V. Pujar and J. D. Cawley, Computer Simulations of. Diffraction Effects due to Stacking Faults in SiC: I, Simulation Results, *Journal of the American Ceramic Society*. Soc., **80**, 1653 1662 (1997).
[16] B. Palosz, S. Gierlotka, S Stelmakh, R. Pielaszek, P. Zinn, M. Winzenick,U. Bismayer, and H. Boysen, High pressure High temperature in situ Diffraction Studies of Nanocrystalline Ceramic Materials at HASYLAB. *Journal of Alloys and Compounds*. **286**. 184 194 (1999).
[17] W. G. Spitzer, D. Kleinman, and D. Walsh, Infrared Properties of Hexagonal Silicon Carbide, *Physical Review*, 113, 127 132 (1959).
[18] H. Abderrazak, and M, Abdellaoui, Synthesis and Characterization of Nanostructured Silicon Carbide, *Materials Letters*, **62**, 3839 3841 (2008),
[19] M. Sherif El Eskandarany, K. Sumiyama, K. Suzuki, Mechanical solid state reaction for synthesis of ß SiC powders, *Journal of Materials Research*, **10**, 659 667 (1995).
[20] N. I. Cho, Y. M. Kim, J. S. Lim, C. Hong, Y. Sul, and C. K. Kim, Laser Annealing Effect of SiC Films Prepared by PECVD, *Thin Solid Films*, **409**, 1 7 (2002)
[21] R. Geick and C. H. Perry, Normal Modes in Hexagonal Boron Nitride, *Physical Review,* **146**, 543 547 (1996).
[22] A. S. Rozenberg, Y. U. A. Sinenko, and N. V. Chukano, I.R. Spectroscopy Characterization of Various Types of Structural Irregularities in Pyrolytic Boron Nitride, *Journal of Materials Science*, **28**, 5675 5678 (1993).
[23] G. Zhang and T. Ohji, Effect of BN Content on Elastic Modulus and Bending Strength of SiC BN in situ Composites, *Journal of Materials research Society* ,**15**, 1876 1886 (2000).
[24] G. Zhang, J. Yang, Z. Deng and T. Ohji, Effect of Y_2O_3 Al_2O_3 additive on the Phase Formation and Densification Process of in situ SiC BN Composite, *Journal of the Ceramic Society of Japan*, **109**, 45 48 (2001).
[25] G. Zhang, Y. Beppu and T. Ohji, Reaction Mechanism and Microstructure Development of Strain Tolerant in situ SiC/BN Composites, *Acta materialia* , **49**, 77 82 (2001).
[26] X. Wang, G. Qiao and Z. Jin, Fabrication of Machinable Silicon Carbide Boron Nitride Ceramic Nanocomposites, *Journal of the American Ceramic Society,* **87**,565 570 (2004)._
[27] T. Kusunose, Fabrication of Boron Nitride Dispersed Nanocomposites by Chemical Processing and Their Mechanical Properties, *Journal of the Ceramic Society of Japan*, **114**, 167 173 (2006).
[28] T. Taniguchi, K. Kimoto, M. Tansho, S. Horiuchi and S. Yamaoka, Phase Transformation of Amorphous Boron Nitride under High Pressure, *Chemistry of Materials*, **15**, 2744 2751 (2003).
[29] G. Zhang and T. Ohji, In Situ Reaction Synthesis of Silicon Carbide Boron Nitride Composites" *Journal of the American Ceramic Society*, **84**, 1475 1479 (2001).

FABRICATION OF DENSE Zr , Hf AND Ta BASED ULTRA HIGH TEMPERATURE CERAMICS BY COMBINING SELF PROPAGATING HIGH TEMPERATURE SYNTHESIS AND SPARK PLASMA SINTERING

Roberta Licheri, Roberto Orrù, Clara Musa, Antonio Mario Locci, Giacomo Cao

Dipartimento di Ingegneria Chimica e Materiali, Centro Studi sulle Reazioni Autopropaganti (CESRA), Unità di Ricerca del Consorzio Interuniversitario Nazionale per la Scienza e Tecnologia dei Materiali (INSTM), Unità di Ricerca del Consiglio Nazionale delle Ricerche (CNR) Dipartimento di Energia e Trasporti, Università degli Studi di Cagliari,
Piazza d'Armi
Cagliari, Italy, 09123

ABSTRACT

The combination of the Self propagating High temperature Synthesis (SHS) technique and the Spark Plasma Sintering (SPS) technology is adopted in this work for the fabrication of fully dense MB_2 SiC and MB_2 MC SiC (M=Zr, Hf, Ta) Ultra High Temperature Ceramics (UHTCs). Specifically, Zr, Hf or Ta, B_4C, Si, and graphite powders are first reacted by SHS to successfully form the desired composites. For the case of the Ta based composites, a 20 min ball milling treatment of the starting reactants is required to mechanochemically activate the corresponding synthesis reactions.

The resulting powders are then subjected to consolidation by SPS. In particular, by setting a dwell temperature level equal to 1800 °C, a mechanical pressure P=20 MPa, and a non isothermal heating time t_h= 10 min, products with relative densities greater than 96% can be obtained for all systems investigated within 30 min of total processing time.

The characteristics of the resulting dense UHTCs, i.e. hardness, fracture toughness, and oxidation resistance, are similar to, and in some cases superior than, those related to analogous products synthesized by alternative, less rapid, methods.

Moreover, it is found that the ternary composites display relatively low resistance to oxidation as a consequence of the lower SiC content in the composite, in comparison with the binary systems, as well as to the presence of MC. In fact, although the latter ones are potentially able to increase the resistance to ablation of the composites, they oxidize rapidly to form MO_2 and carbon oxides which lead to sample porosity increase thus enhancing product oxidation.

INTRODUCTION

Transition metal borides and carbides like MB_2 and MC (M=Zr, Hf, Ta) belong to the so called Ultra High Temperature Ceramics (UHTCs) that exhibit several interesting properties such as melting temperatures higher than 3000 °C, high hardness, high electrical and thermal conductivity, chemical stability, good thermal shock resistance and high resistance to ablation in oxidizing environments[1,2]. These characteristics make UHTCs suitable in different application fields where thermal, electrical, chemical, and wear resistance are required, like cutting tools, high temperature crucibles, microelectronics as well as in aerospace industry for the fabrication of thermal protection components[2 5]. In this context, it is also well established the beneficial effect of SiC, used as additive, in terms of oxidation resistance at high temperatures[6 8].

Bulk UHTCs are typically obtained in dense form by Hot Pressing (HP), through which the commercial ceramic constituents in powder form are sintered[7,9 10], or by synthesizing and densifying in a single step appropriate reaction promoters[11 13].

The critical point encountered following this procedure is represented by the fact that in both cases HP requires not only high sintering temperatures and mechanical loads, but expecially prolonged processing times, generally on the order of hours, to achieve acceptable relative density levels. Moreover, under these processing conditions, materials with residual porosity and rather coarse microstructure are typically obtained. The use of some suitable sintering aids like ZrN[14],

HfN[15], Si_3N_4[7], or $MoSi_2$[16] was beneficial in terms of sintering conditions although the total sintering time still remains high.

Along these lines, Spark Plasma Sintering (SPS), a relatively novel technology where the starting powders to be only consolidated or also simultaneously reacted are crossed by an electric pulsed current[17 18], offers a possible convenient tool to overcome the drawback above. In fact, various dense advanced materials with rather uniform and fine microstructure are obtained relatively faster and at lower temperature levels by SPS, with respect to HP. Among these materials several UHTCs are included[20 25].

In this context, the Self propagating High temperature Synthesis (SHS), a well known combustion synthesis method based on the occurrence of strongly exothermic reactions that, once ignited, propagate in the form of a combustion wave through the reacting mixture without requiring any other energy supply,[26 27] represents a convenient complementary method able to synthesize the powders to be subsequently densified by SPS. This approach was for instance successfully applied for the obtainment of several intermetallic compounds[28].

In this paper, this processing route is investigated for the fabrication of highly dense $2MB_2$ SiC and $4MB_2$ 4MC 1.5SiC (M=Zr, Hf, Ta) products starting from Zr/Hf/Ta, B_4C, Si, and graphite powders.

The SPSed optimal products are characterized in terms of microstructure, oxidation resistance, hardness, fracture toughness, and the obtained results are compared with those reported in the literature relatively to analogous composites prepared using alternative fabricating routes.

It should be noted that the compositions of the ternary systems examined in this work fall well within the ranges of volume percentage of ceramic components, i.e. 20 64 vol% MB_2, 20 64 vol% MC, and 10 16 vol% SiC (M=Zr or Hf), corresponding to the higher resistance to ablation[29].

EXPERIMENTAL MATERIALS AND METHODS

The different UHTC composites were prepared by SHS starting from reactants whose characteristics and sources are reported in Table 1. The initial mixtures were obtained by blending reactants according to the following reactions:

$$2M + B_4C + Si \rightarrow 2MB_2 + SiC \quad (1)$$
$$8M + 2B_4C + 1.5Si + 3.5C \rightarrow 4MB_2 + 4MC + 1.5SiC \quad (2)$$

which correspond approximately to ZrB_2 25 vol% SiC, HfB_2 26.5 vol% SiC, TaB_2 27.9 vol% SiC, ZrB_2 40 vol% ZrC 12 vol% SiC, HfB_2 40.6 vol% HfC 11.2 vol% SiC, and TaB_2 39.1 vol% TaC 13.7 vol% SiC respectively. For the sake of simplicity, these systems will be indicated by ZS, HS, TS, ZZS, HHS, and TTS respectively, in what follows.

Table 1. Characteristics of the starting reactants used in the present investigation.

Powders	Vendor	Particle Size	Purity
Hf	Alfa Aesar	< 44 μm	> 99.6 %
Zr	Alfa Aesar	< 44 μm	> 98.5 %
Ta	Alfa Aesar	< 44 μm	99.9 %
B_4C	Alfa Aesar	1 7 μm	> 99.4 %
Si	Aldrich	< 44 μm	> 99 %
Graphite	Aldrich	1 2 μm	

Powders mixing was performed in a SPEX 8000 (SPEX CertiPrep, USA) shaker mill for 30 min using a plastic vial and alumina balls. Mechanochemical activation of Ta based mixtures was carried out using the same mill apparatus with two steel balls (13 mm diameter, 8 grams weight) for 20 min milling time interval and ball to powders or charge ratio (CR) equal to 1. Details on the experimental set up used in this work for SHS and SPS are described elsewhere[27,30]. Depending upon the system investigated, a suitable amount (8 15 g) of the starting powders either only blended or mechanochemically activated were uniaxially pressed to form cylindrical pellets with a diameter of 10 mm, height of 30 mm and a green density of ~50 % of the theoretical value. The combustion front was generated at one sample end by using an electrically heated tungsten coil, which was immediately turned off as soon as the synthesis reaction was initiated. Then, the reactive process self propagates until it reaches the opposite end of the pellet. The temperature during reaction evolution was measured using C type thermocouples (W Re, 127 μm diameter, Omega Engineering Inc., USA) as well as by a two color pyrometer (IRCON, Mirage, USA). To convert the obtained SHS product to powder form, about 4 g of it were milled by means of the shaker mill apparatus mentioned above, using a stainless steel vial with two steel balls (13 mm diameter, 8 g weight) for 20 min. Particle size distribution of the obtained powders was determined using a laser light scattering analyser (CILAS 1180, France).

An SPS 515 apparatus (Sumitomo Coal Mining Co. Ltd, Japan) was used in the temperature controlled mode for powder densification. This machine combines a 50 kN uniaxial press with a DC pulsed current generator (10 V, 1500 A, 300 Hz) to simultaneously provide a pulsed electric current through the sample and the graphite die containing it, and a mechanical load through the die plungers.

A certain amount (3 5 g) of the SHS powders was first cold compacted inside the die (outside diameter, 35 mm; inside diameter, 15 mm; height, 40 mm). To protect the die and facilitate sample release after synthesis, a 99.8 % pure graphite foil (0.13 mm thick, Alfa Aesar, Karlsruhe, Germany) was inserted between the internal surfaces of the die and the top and the bottom surface of the sample and the graphite plungers (14.7 mm diameter, 20 mm height). Both the die and the plungers were composed of AT101 graphite and provided by Atal s.r.l., Italy. In addition, with the aim of minimizing heat losses by thermal radiation, the die was covered with a layer of graphite felt (3 mm thick, Atal s.r.l., Italy). Afterwards, it was placed inside the reaction chamber of the SPS apparatus and the system was evacuated down to about 10 Pa. This step was followed by the application of 20 MPa mechanical pressure through the plungers.

During the process, temperature, applied current and voltage, mechanical load and the vertical displacement of the lower electrode were recorded in real time. In particular, temperature was measured by both a C type thermocouple (Omega Engineering Inc., USA), which was inserted inside a small hole in one side of the graphite die, and a two color pyrometer (IRCON, Mirage, USA). The measured displacement can be regarded as the degree of powdered compact densification, although thermal expansion of the sample as well as that of both electrodes, graphite blocks, spacers and plungers, also contribute to this parameter. Thus, as described in detail in a previous paper[30], a specific procedure was followed to evaluate the sample shrinkage (δ), which will be considered in the following discussion. In any case, the final degree of consolidation was determined by measuring the density of the sample at the end of the process. For the sake of reproducibility, each experiment was repeated at least twice. After the synthesis process, the sample was allowed to cool and then removed from the die.

The relative densities of dense products were determined by the Archimedes' method. The theoretical density of the ZS, HS, TS, ZZS, HHS and TTS composites, i.e. 5.37, 9.17, 9.98, 6.02, 10.92, 12.05 g/cm^3, respectively, were calculated through a rule of mixture[31], by considering the density values of ZrB_2, HfB_2, TaB_2, ZrC, HfC, TaC, and SiC as 6.1, 11.18, 12.6, 6.4, 12.69, 14.48, and 3.2 g/cm^3, respectively. Phase identification was performed by a Philips (The Netherlands) PW 1830 X rays diffractometer using a Ni filtered Cu K_α radiation (λ=1.5405 Å). The microstructure and local phase composition of end products were examined by scanning electron microscopy (SEM) (mod. S4000, Hitachi, Japan) and energy dispersive X rays spectroscopy (EDS) (Kevex Sigma 32 Probe, Noran Instruments, USA), respectively.

Indentation method using a Zwick 3212 Hardness tester machine (Zwick & Co. GmbH, Germany) was used to determine Vickers hardness and fracture toughness (K_{IC}) of the obtained products. The applied loads used for all measurements were in the range 1 10 kg while the dwell time was 18 s.

The oxidation resistance of UHTCs was determined by performing thermogravimetric analysis (TGA) using a NETZSCH (Germany) STA 409PC Simultaneous DTA TGA Instrument under 100 cm^3/min air flow. Specifically, oxidation tests were conducted either under non isothermal conditions by heating slowly (2 °C/min) the specimen from room temperature to 1450 °C or isothermally at 1450 °C for about 4 h, during which the mass sample variation with temperature and time, respectively, was monitored. For the sake of comparison, the obtained results have been normalized by dividing sample mass gain by the external surface of the UHTC material exposed to an oxidizing environment.

RESULTS AND DISCUSSION

Powders synthesis and characterization

While pellets prepared starting from regularly blended powder mixtures according to reactions (1) (2) exhibited a self propagating character when M=Zr and Hf, the Ta based systems do not displayed an analogous behaviour. This fact is consistent with the corresponding enthalpies of reaction shown in Table 2, which are relatively lower for TS and TTS. Thus, in the latter cases the mechanochemical activation of starting mixture under the conditions reported in the Experimental Materials and Methods section was required to promote the self propagating behaviour to the synthesis reactions.

The maximum combustion temperatures measured during the SHS process evolution are reported in Table 3 along with the corresponding average front velocities. Consistently, both parameters are relatively lower for the less exothermic Ta based systems.

Figures 1(a) 1(f) show the diffraction patterns of the obtained products along with those of the corresponding reactant mixtures and, for the case of the TS and TTS systems (cf. Figures 1(e) 1(f)), also those related to the co milled reactants. Other than a slight peaks broadening, as an indication of crystal size refinement and internal strain increase in the processing powders, no additional effects induced by the ball milling treatment can be evidenced from the XRD results. It is likely that the mechanical treatment of the starting mixture favours interfaces formation among reactants thus overcoming the diffusion limitation and enhancing chemical reactivity.

As far as the composition of SHSed products are concerned, the presence of all the major peaks related to the phases constituent the expected composites is revealed from this analysis. In conclusion, it is possible to state that a complete conversion of reactants into desired products is achieved by SHS for all systems investigated. It should be also noted that the secondary phases present as impurities in the starting reactants when synthesizing ZrB_2 based composites (cf. Figures 1(a) 1(b)), were eliminated during the synthesis process.

Table 2. Enthalpy of formation of UHTC composites by reactions (1) (2)[32].

System	ΔH_r^0 [kJ]
ZS	647.266
HS	674.042
TS	348.364
ZZS	2044.51
HHS	2315.634
TTS	1380.762

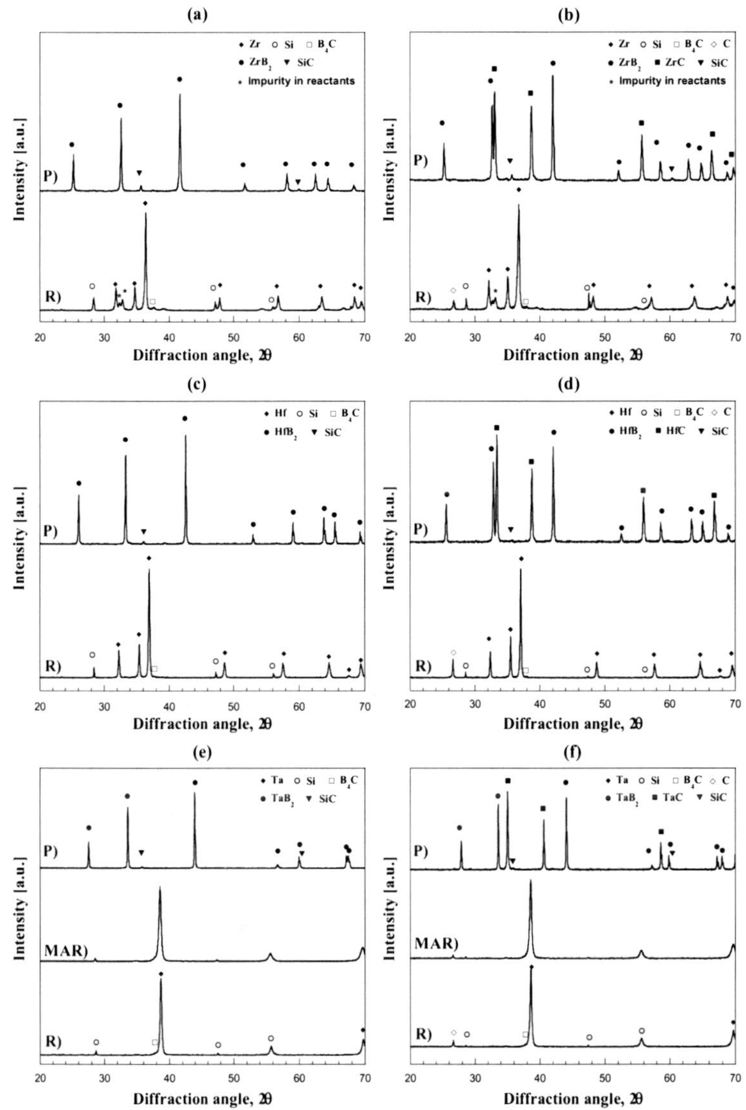

Figure 1. Comparison of XRD patterns of R) original reactants, MAR) mechanochemically activated reactants and P) products obtained by self propagating high temperature synthesis according to reactions (1 2): (a) ZS, (b) ZZS, (c) HS, (d) HHS, (e) TS and (f) TTS.

Table 3. Maximum combustion temperatures and average wave velocities measured during SHS of UHTC composites (*after mechanochemical activation of the starting reactants by ball milling).

System	Combustion temperature [°C]	Average wave velocity [mm/s]
ZS	2200±20	11±1
HS	2150±50	7±1
TS*	1850±50	4.5±0.5
ZZS	2200±50	8±1
HHS	2250±50	10±1
TTS*	2050±50	5.8±0.2

Following the procedure described in the previous section, the consolidation stage by SPS is preceded by a ball milling step required to convert the obtained SHS porous products to powder form.

The resulting powders have been characterized in terms of particle size distribution and microstructure. The cumulative curve obtained for the case of the ZZS product is reported as an example in Figure 2(a), which shows that particle size was less than 70 μm. In addition, from this analysis it was also found that d_{50} = 7.23 ± 0.13 μm.

SEM investigations conducted on the same powders samples are consistent with laser light scattering results. From the SEM back scattered micrograph shown in Figure 2(b), it is observed that each SHS powder particle is a mixture of different phases consisting of ZrC (brighter phase), ZrB_2 (medium bright), and SiC (darker) grains, each of them being typically less than 3 4 μm in size. Similar results are obtained when examining the other composite powders synthesized by SHS. Thus, analogous consideration can be made.

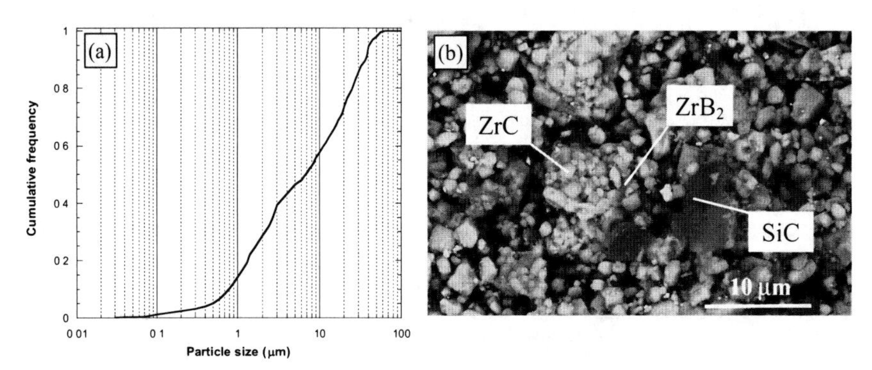

Figure 2. Size distribution (a) and SEM back scattered micrograph (b) of ZZS ball milled SHS powders that are to be densified by SPS.

Products densification and characterization

The influence of the total sintering time and dwell temperature during SPS was systematically investigated for the ZS and HS systems in a previous study[25]. It was found that by setting a dwell temperature level (T_D) equal to 1800 °C, a mechanical pressure P=20 MPa, and a non isothermal heating time t_h= 10 min, near fully dense materials were obtained within 30 min of total processing time. This feature holds also true when consolidating by SPS the remaining composites taken into account in the present paper. Specifically, all SPSed specimens obtained under the conditions above exhibited a relative density ≥ 96% of the theoretical value.

Typical outputs of sample shrinkage (δ), expressed as a percentage relative to its final value, and temperature recorded during the consolidation process by SPS of HS powders are reported in Figure 3(a). Specifically, they refer to the conditions of T_D=1800 °C, t_H=10 min, t_T=30 min, and P=20 MPa. The corresponding current and voltage behavior is shown in Figure 3(b) where the electrical mean values are reported.

No peculiar changes in the sample shrinkage are manifested during the first 4 min of the SPS process (cf. Figure 3(a)). Subsequently, δ increases at an approximately constant rate until the T_D value is reached. During the isothermal stage, densification continues to occur albeit at a minor rate to achieve final density (> 99.9 %) at t_T=30 min. Regarding the electrical behavior of the system (cf. Figure 3(b)), it may be seen that the current and voltage are augmented during the non isothermal heating, to satisfy the chosen thermal program. Afterwards, both parameters rapidly decrease down to the corresponding stationary mean values, i.e. about 920 A and 4.6 V, respectively. Sample shrinkage, temperature, mean current and voltage time profiles recorded during the SPS process of the other UHTC powders investigated in this work were qualitatively similar to those obtained for the HS system and shown in Figure 3(a) 3(b). Analogous behaviour was also displayed when setting different T_D and t_T values.

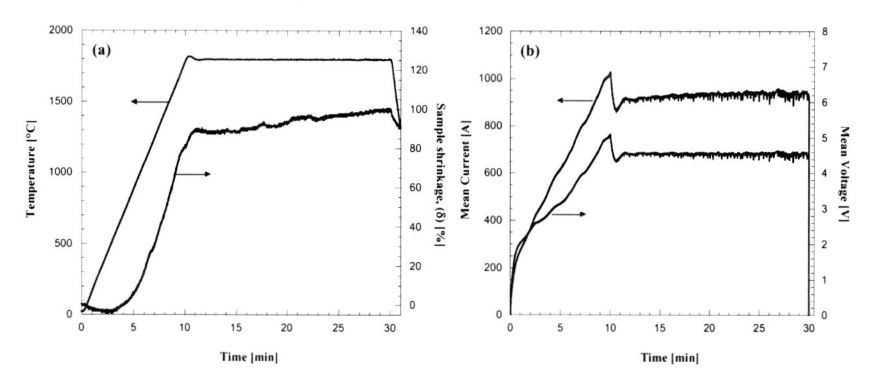

Figure 3. Temporal profiles of SPS outputs during the preparation of dense HS products starting from powders obtained by SHS: (a) temperature and sample shrinkage, (b) mean current intensity and mean voltage (T_D=1800 °C, t_H=10 min, t_T=30 min, P=20 MPa).

Table 4. Properties of dense SPSed UHTC composites.

System	Relative density [%]	Hardness [GPa] (Applied load)	K_{IC} [MPa m$^{1/2}$]
ZS	99.6	16.7±0.4 (1 kg)	5.0±0.3
HS	>99.9	20.55±0.8 (3 kg)	
		19.2±0.6 (10 kg)	7.0±0.7
ZZS	98.7	16.9±0.2 (10 kg)	5.9±0.5
HHS	98.5	17.7±1.5 (3 kg)	
		18.3±1.1 (10 kg)	6.2±0.7

Three back scattered SEM micrographs of the binary dense composites obtained in this work after consolidation by SPS under the conditions T_D=1800 °C, P=20 MPa, t_H=10 min, and t_T=30 min, are shown in Figures 4(a) 4(c). It is seen that two different phases, distributed quite uniformly all over the sample, are easily distinguishable. Specifically, the brighter and darker zones correspond to MB_2 (M=Zr, Hf, Ta) and SiC, respectively. As expected, the presence of a third phase related to ZrC, HfC or TaC was evidenced when investigating by SEM the ternary systems.

The mechanical properties of the SPSed UHTCs are summarized in Table 4 along with the corresponding relative density. Work is in progress as far as the results related to the Ta based ceramics are concerned.
The best result in terms of both Vickers hardness and fracture toughness (K_{IC}) is obtained for the case of the HS system. Anyway, all values are comparable to, and in some cases better than, those reported in the literature for similar systems[15,33 34]. For instance, the ZrB_2 20 vol% SiC and ZrB_2 6.4 vol% ZrC 20 vol% SiC materials fabricated by Wu et al.[34] using the RHP (Reactive Hot Pressing) method displayed Vickers hardness of 13.6 16.7 GPa and K_{IC} of 4.5 5.1 MPa $m^{1/2}$, respectively. Moreover, as far as the Hf based UHTC is concerned, Gasch et al.[33] obtained a completely dense HfB_2 20 vol% SiC product using the HP technique that exhibited a Vickers hardness of 19 21 GPa and K_{IC} of 4.1 4.2 MPa $m^{1/2}$.

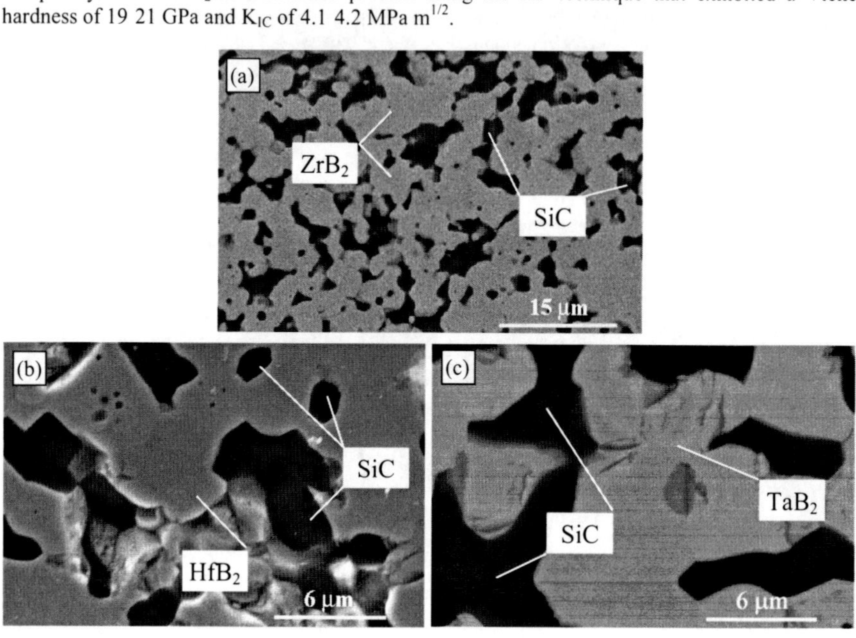

Figure 4. SEM back scattered micrographs of dense SPSed ZS (a), HS (b), and TS (c) products.

The oxidation resistance of the composites prepared in this work has been measured using TGA by monitoring the mass change of the sample subjected to an oxidizing environment (air) at high temperature. The results obtained during dynamic (non isothermal) oxidation tests are reported in Figure 5. It is apparent that the binary systems exhibit higher resistance to oxidation as compared to the ternary composites. Among them, the HS ceramic displays the relatively lower oxidation rate up to 1450 °C. The same indication is provided by the oxidation tests conducted under isothermal conditions at 1450 °C.

The observed oxidative behavior can be interpreted on the basis of several studies reported in the literature on this subject[7,14,35 36]. Briefly, the very volatile B_2O_3, obtained from the oxidation of MB_2, combines with SiO_2 formed from SiC oxidation to give a silica rich borosilicate glass layer. The latter one reduces boria evaporation, other than acting as an oxygen diffusion barrier, thus providing improvement of oxidation resistance of the UHTC material.
The fact that ternary composites display relatively low resistance to oxidation, can be related to the lower SiC content in the composite, in comparison with the case of binary systems, as well as to the presence of MC. In fact, although these compounds are potentially able to increase the resistance to

ablation of the composites, they oxidize rapidly to form MO_2 and carbon oxides. The formation of a porous product is then favoured, thus permitting the oxygen to diffuse through the bulk of the UHTC material.

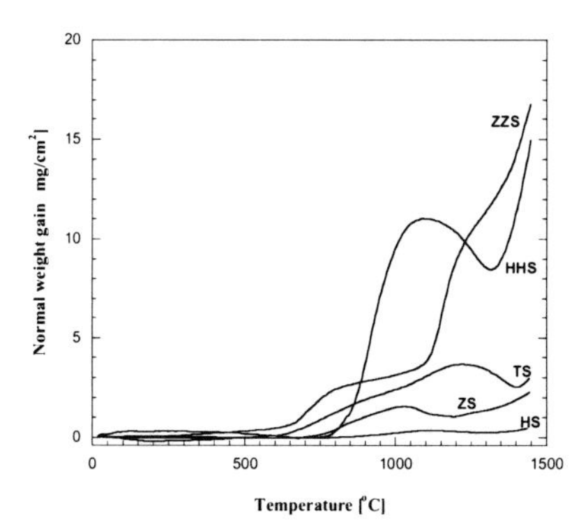

Figure 5. Specific weight change during TGA oxidation in air of the sintered ZS, HS,TS, ZZS and HHS dense samples as a function of temperature (heating rate equal to 2 °C/min).

CONCLUSIONS

Near fully dense MB_2 SiC and MB_2 MC SiC (M=Zr, Hf, Ta) composites were obtained in this work through a processing route resulting from the combination of the SHS and SPS techniques. Taking advantage of the highly exothermic reactions (1) (2), Zr (or Hf), B_4C, Si, and, for the case of the ternary systems, graphite, were firstly reacted by SHS to completion in the appropriate proportions to form the desired UHTCs. For the case of the Ta based composites, the mechanochemical activation of the starting reactants by ball milling for 20 min was required to induce the self propagating behaviour to the corresponding synthesis reactions. The complete conversion of reactants into the desired ceramic phases was achieved also in this case.

The porous SHS products were then ball milled and the resulting powders subsequently consolidated without the addition of any sintering aid using an SPS apparatus. It was found that, by setting a dwell temperature level equal to 1800 °C, a mechanical pressure P=20 MPa, and a non isothermal heating time t_h= 10 min, products with relative densities ≥96% were obtained for all systems investigated within 30 min of total processing time.

The characteristics of the resulting dense UHTCs, i.e. hardness, fracture toughness, and oxidation resistance, are similar to, and in some cases superior than, those related to analogous products synthesized by competitive methods. However, as a relevant difference, the processing route adopted in this work is characterized by shorter processing times and/or lower sintering temperature, when compared to the others fabrication methods proposed in the literature. For instance, when commercially available ZrB_2 and SiC powders were sintered by HP at T=1900 °C and P=32 MPa[9,37], the total time required to obtain a near full dense ZrB_2/20 30 vol% SiC was 450 min. Analogously, the fabrication of dense HfB_2 20 vol% SiC was carried out by HP in 1 h dwell time, to which the non isothermal heating time has to be added, under the conditions of T=2200 °C and P=25 MPa[33]. If the comparison is extended to other methods reported in the literature that make use of the SPS technique for the fabrication of UHTCs[21 22], the required total processing time is analogous, i.e. t_T≤30 min. However, the corresponding dwell temperature (2100 °C) was much higher than that adopted in our study. This finding reveals the improved sintering behavior of MB_2 SiC and MB_2 MC SiC composite powders synthesized by SHS. This result is consistent with the better sinterability displayed by SHSed ZrB_2 powders as compared to the analogous product prepared by the other techniques, i.e. replacement reaction and carbothermic reduction process[38].

Such behaviour is justified by Mishra et al.[38] who interpreted their results on the basis of the defect concentration level reached in the obtained powders. Specifically, SHS products showed higher defect concentration as compared to the powders synthesized by other methods, thus facilitating sintering phenomena.

In another recent paper[24], higher densification levels were achieved using SHSed ZrB_2 ZrC SiC powders instead of starting from relatively finer commercial ZrB_2, ZrC, and SiC powders, while maintaining the same SPS conditions. In this context, SEM investigations revealed that each individual powder particle consisted of a mixture of boride and carbide grains, whose size was finer than the commercial powders. Thus, composites sinterability was in this case likely improved by the reduction of diffusion distances among the three ceramic phases produced in situ by SHS.

ACKNOWLEDGMENTS

IM (Innovative Materials) S.r.l., Italy, is gratefully acknowledged for granting the use of SPS apparatus. The authors thank Eng. Leonardo Esposito (Centro Ceramico di Bologna, Italy) for performing hardness and fracture toughness measurements.

REFERENCES

[1]K.Upadhya, J.M. Yang, W.P. Hoffmann, Materials for ultrahigh temperature structural applications, *Am. Ceram. Soc. Bull.*, **58**, 51 56 (1997).

[2]W.G. Fahrenholtz, G.E. Hilmas, I.G. Talmy, J.A. Zaykoski, Refractory diborides of Zirconium and Hafnium, *J. Am. Ceram. Soc.*, **90**, 1347 1364 (2007).

[3]S.R. Levine, E.J. Opila, M.C. Halbig, J.D. Kiser, M. Singh, J.A. Salem, Evaluation of ultra high temperature ceramics for aeropropulsion use, *J. Europ. Ceram. Soc.*, **22**, 2757 2767 (2003).

[4]R. Rapp, Materials for Extreme environments, *Materials Today*, **9(5)**, 6 (2006).

[5]X. Zhang, G.E. Hilmas, W.G. Fahrenholtz, Synthesis, densification, and mechanical properties of TaB_2, *Mater. Letters*, **62**, 4251 4255 (2008).

[6]W.C. Tripp, H.H. Davis, H.C. Graham, Effect of a SiC Addition on the oxidation of ZrB_2, *Am. Ceram. Soc. Bull.*, **52(8)**, 612 616 (1973).

[7]F. Monteverde, A. Bellosi, The resistance to oxidation of an HfB_2 SiC composite, *J. Eur. Ceram. Soc.*, **25(7)**, 1025 1031 (2005).

[8]F. Monteverde, L. Scatteia, Resistance to Thermal Shock and to Oxidation of Metal Diborides SiC Ceramics for Aerospace Application, *J. Am. Ceram. Soc.*, **90(4)**, 1130 1138 (2007).

[9]W.G. Fahrenholtz, G.E. Hilmas, A.L. Chamberlain, J.W. Zimmermann, B. Fahrenholtz, Processing and characterization of ZrB_2 based ultra high temperature monolithic and fibrous monolithic ceramics, *J. Mater. Sci.*, **39**, 5951 5957 (2004).

[10]J. Marschall, D.C Erlich., H. Manning, W. Duppler, D. Ellerby, M. Gasch, Microhardness and High velocity impact resistance of HfB_2/SiC and ZrB_2/SiC composites, *J. Mater. Sci.*, **39**, 5959 5968 (2004).

[11]M.M.Opeka, I.G. Talmy, E.J. Wuchina, J.A. Zaykoski, S.J. Causey, Mechanical, thermal, and oxidation properties of refractory hafnium and zirconium compounds, *J. Eur. Ceram. Soc.*, **19**, 2404 2414 (1999).

[12]G. J.Zhang, Z. Y. Deng, N. Kondo, J. F.,Yang T. Ohji, Reactive hot pressing of ZrB_2 SiC composites, *J. Am. Ceram. Soc.*, **83(9)**, 2330 2332 (2000).

[13]F. Monteverde, Progress in the fabrication of ultra high temperature ceramics: In situ synthesis, microstructure and properties of a reactive hot pressed HfB_2 SiC composite, *Compos. Sci, & Technol.*, **65(11-12),** 1869 1879 (2005).

[14]F.Monteverde, A. Bellosi, Oxidation of ZrB_2 based ceramics in dry air, *J. Electrochem. Soc.*, **150(11)**, B552 B559(2003).

[15]F.Monteverde, A. Bellosi, Efficacy of HfN as sintering aid in the manufacture of ultrahigh temperature metal diborides matrix ceramics, *J. Mater. Res.*, **19(12)**, 3576 3585 (2004).

[16]A. Balbo, D. Sciti, "Spark Plasma Sintering and Hot Pressing of ZrB_2 $MoSi_2$ under high temperature ceramics, *Mater. Sci. Eng. A*, **475**, 108 112 (2008).

[17]Z.A.Munir, U. Anselmi Tamburini, M. Ohyanagi, The effect of electric field and pressure on the synthesis and consolidation of materials: A review of the spark plasma sintering method, *J. Mater. Sci.*, **41(3)**, 763 777 (2006).
[18]R. Orrù, R. Licheri, A.M. Locci, A. Cincotti, G. Cao, Consolidation/Synthesis of Materials by Electric Current Activated/Assisted Sintering, *Mater. Sci. Eng. R*, **63(4-6)**, 127 287 (2009).
[19]V. Medri, F. Monteverde, A. Balbo, A. Bellosi, Comparison of ZrB_2 ZrC SiC composites fabricated by spark plasma sintering and hot pressing, *Adv. Eng. Mater.*, **7(3)**, 159 163 (2005).
[20]U. Anselmi Tamburini, Y. Kodera, M. Gasch, C. Unuvar, Z.A. Munir, M. Ohyanagi, S.M. Johnson, Synthesis and characterization of dense ultra high temperature thermal protection materials produced by field activation through spark plasma sintering (SPS): I. Hafnium diboride, *J. Mater. Sci.*, **41(10)**, 3097 3104 (2006).
[21]F. Monteverde, C. Melandri, S. Guicciardi, Microstructure and mechanical properties of an HfB_2 + 30 vol.% SiC composite consolidated by spark plasma sintering, *Mater. Chem. & Phys.*, **100(2-3)**, 513 519 (2006).
[22]F. Monteverde, Ultra high temperature HfB_2 SiC ceramics consolidated by hot pressing and spark plasma sintering, *J. Alloys. Compd*, **428(1-2)**, 197 205 (2007).
[23]R. Licheri, R. Orrù, A.M. Locci, G. Cao, Efficient Synthesis/Sintering Routes to obtain Fully Dense ZrB_2 SiC Ultra High Temperature Ceramics (UHTCs), *Ind. Eng. Chem. Res.*, **46**, 9087 9096, (2007).
[24]R. Licheri, R. Orrù, C. Musa, G. Cao, Combination of SHS and SPS Techniques for Fabrication of Fully Dense ZrB_2 ZrC SiC Composites, *Mater. Letters*, **62**, 432 435 (2008).
[25]R. Licheri, R. Orrù, C. Musa, A.M. Locci, G. Cao, Spark Plasma Sintering of UHTC powders obtained by Self propagating High temperature Synthesis, *J. Mater. Sci.*, **43(19)**, 6406 6413 (2008).
[26]Z.A. Munir, U. Anselmi Tamburini, Self propagating exothermic reactions: the synthesis of high temperature materials by combustion, *Mater. Sci. Rep.*, **3**, 277 365 (1989).
[27]A. Cincotti, R. Licheri, A.M. Locci, R. Orrù, G. Cao, A review on combustion synthesis of novel materials: recent experimental and modeling results, *J. Chem. Technol. Biot.*, **78(2-3)**, 122 127 (2003)
[28]K. Hirota, S. Nakane, M.Yoshinaka, O. Yamaguchi, Spark Plasma Sintering (SPS) of Several Intermetallic Compounds Prepared by Self Propagating High Temperature Synthesis (SHS), *Intern. J. SHS*, **10(3)**, 345 358 (2001).
[29]J. Bull, M.J. White, L. Kaufman, Ablation resistant Zirconium and Hafnium Ceramics, US Patent No. 5,750,450 (1998).
[30]A.M. Locci, R. Orrù, G. Cao, Z.A. Munir, Simultaneous spark plasma synthesis and densification of TiC TiB_2 composites, *J. Am. Ceram. Soc.*, **89(3)**, 848 855 (2006).
[31]F.L. Matthews, R. Rawlings: Composite Materials: Engineering and Science. Chapman & Hall, Great Britain (1994).
[32]I. Barin Thermochemical data of pure substances, VHC (1989).
[33]M. Gasch, D. Ellerby, E. Irby, S. Beckman, M. Gusman, S. Johnson, Processing, properties and arc jet oxidation of hafnium diboride/silicon carbide ultra high temperature ceramics, *J. Mater. Sci.*, **39(19)**, 5925 5937 (2004).
[34]W.W. Wu, G.J. Zhang, Y.M. Kann, P.L. Wang, Reactive Hot Pressing of ZrB_2 SiC ZrC Ultra High Temperature Ceramics at 1800 °C, *J. Am. Ceram. Soc.*, **89(9)**, 2967 2969 (2006).
[35]J.W. Hinze, W.C. Tripp, H.C. Graham, The High Temperature Oxidation Behavior of HfB_2+20 v/oSiC Composite, *J. Electrochem. Soc.*, **122(9)**, 1249 1254 (1975).
[36]F. Peng, R.F.Speyer, Oxidation resistance of fully dense ZrB_2 with SiC, TaB_2, and $TaSi_2$ additives, *J. Am. Ceram. Soc.*, **91(5)**, 1489 1494 (2008).
[37]A.L. Chamberlain, W.G. Fahrenholtz, G.E. Hilmas, D.T. Ellerby, High strength zirconium diboride based ceramics, *J. Am. Ceram. Soc.*, **87(6)**, 1170 1172 (2004).
[38] S. K. Mishra, S. Das, L. C. Pathak, Defect structures in zirconium diboride powder prepared by self propagating high temperature synthesis, *Mater. Sci. Eng. A*, **364(1-2)**, 249 255 (2004).

Novel, Green, and Strategic Processing

MICROWAVE SINTERING OF MULLITE AND MULLITE ZIRCONIA COMPOSITES

Subhadip Bodhak, Susmita Bose and Amit Bandyopadhyay*
W. M. Keck Biomedical Materials Research Laboratory, School of Mechanical and Materials Engineering, Washington State University, Pullman, WA 99164, USA.
*Corresponding author: amitband@wsu.edu

ABSTRACT

The objective of this research is to evaluate microwave sintering as a viable option to process high strength mullite and mullite zirconia composites utilizing reduced time and energy. Mullite samples were sintered using a 3 KW, 2.45 GHz microwave furnace. Sintering temperatures were in the range of 1400 to 1500°C, keeping the sintering time constant at 60 minutes. With an increase in sintering temperature from 1400°C to 1500°C, the porosity decreases from 30 to 13% and the compressive strength increases from 128±18 MPa to 387±21MPa when 1wt% MgO was added as a sintering aid to mullite. Furthermore, 5 wt% yttria stabilized tetragonal zirconia (YTZP) was incorporated to prepare zirconia toughened mullite composites. A maximum compressive strength of ~ 632±29 MPa was obtained for 5 wt% ZrO2 reinforced 1wt% MgO doped mullite composites sintered at 1500°C in a microwave furnace. Microwave sintering data, when compared with those of conventional sintering, revealed that localized heating of microwaves led to faster neck growth between mullite grains by surface diffusion and significantly enhanced compressive strength with a minimal change in density in mullite zirconia composites.

INTRODUCTION

Mullite ($3Al_2O_3.2SiO_2$) is considered as a promising material for use in high temperature structural applications due to its excellent high temperature strength, good chemical and thermal stability along with high creep resistance properties [1, 2]. Over the last few decades, significant amounts of research have been invested on the processing of mullite ceramics exhibiting suitable thermal, electrical and mechanical properties for a variety of high temperature applications, such as high temperature furnace equipments, infrared transparent windows, substrates for microelectronic packages, high performance protective coating, turbine engine components etc [3 5]. However, the processing of mullite compacts with near theoretical density has always been a challenge as the slow diffusion kinetics of Si^{4+} and Al^{3+} ions make the material so difficult to sinter. Literature results suggest that sintering of pure mullite to a relative density of more than 95% often necessitating using temperatures above 1700°C in a conventional electrically heated furnace which is a time consuming as well as energy intensive procedure [6]. To address this limitation, several researchers have employed different low temperature mullite synthesis techniques such as, mullite synthesis from organic precursors [7], sol gel techniques [8], colloidal mixing techniques [9] etc. However, these methods were found to only lower the mullatization temperature since the densification has still taken place by solid state reaction at high temperatures. In addition, presence of impurities and mineralizers in those solutions based synthesis techniques can be a major problem concerning the high temperature mechanical properties of dense mullites. Therefore, mullite synthesis and sintering has been continuously developed for many years.

Recently the use microwave heating is found of interest to sinter ceramics at relatively low temperature [10, 11]. During microwave heating the materials first volumetrically absorb the electromagnetic energy by coupling with microwaves and then subsequently convert the energy into heat [12]. This is fundamentally different from conventionally heating in which heating occurs through surface heating by the mechanisms of conduction, radiation and convection. In conventional heating, heat is first transferred onto the material surface from the heating elements and then moves inward. Therefore, for poor thermal conducting materials such as mullite (thermal conductivity ~ 6 W/mK), conventional heating can cause a large thermal gradient from the surface to the center of the material and consequently weakens the mechanical properties. In contrast, volumetric heating through microwave offers several potential advantages like sintering rate enhancement, uniform heating, selective energy absorption, high efficiency and reduced costs [9, 13, 14]. However, it has been observed that as an alternative sintering technique microwave heating has become more popular for oxide ceramics such as ZrO_2, Al_2O_3, but only few researches have investigated the microwave sintering of mullite ceramics.

In our current research, a comprehensive study has been done to investigate the feasibility of fabrication of dense mullite ceramics by microwave sintering technique focusing on understanding the efficacy of microwave heating on densification and mechanical performance of sintered mullite and mullite zirconia compacts. We have investigated the hypothesis that the addition of a dopant and/or other good microwave absorber can stimulate the microwave mullite coupling and therefore enhance densification as well as mechanical properties of mullite ceramics. To study this hypothesis, we have used 1 wt% MgO as an additive and 5 wt% yttria stabilized tetragonal zirconia polycrystalline (YTZP) has been incorporated to improve sintering and/or densification kinetics and thus mechanical performances. The influence of microwave heating has been evaluated by the bulk density calculation, scanning electron microscopy (SEM) microstructural observation, compressive strength measurement and results are compared with conventional heating to appreciate the economic advantage for microwave sintering of mullite ceramics.

MATERIALS AND METHODS

In our experiments commercial grade high purity mullite powders were procured from CE Minerals (King of Prussia, PA, USA) and used as a starting powder. High purity MgO powders were obtained from Fisher Scientific (Fair Lawn, NJ, USA) and used as an additive to aid in sintering kinetics. 1 wt% MgO doped mullite powders were weighed and ball milled in 250 ml polypropylene bottles using 5 mm diameter zirconia milling media. Wet ball milling was done in ethanol for 24 h. After ball milling the mixtures were dried at 100°C for overnight in an oven. After drying, the measured amount of powder mixtures were uniaxially pressed under a pressure of 150 MPa to prepare the specimens. Two different steel molds were used to prepare disc and cylindrical shaped specimens. Disc compacts [12 mm (Φ) X 1 mm (h)] were used for microstructural analysis and bulk density calculation, and cylindrical specimens [6 mm (Φ) X 12 mm (h)] were used for measuring compressive strength. Pure mullite powders without any dopant were also compacted under identical conditions. Finally, all the samples were sintered in a 2.45 GHz, 3 KW fully automated commercial microwave sintering furnace (MW L0136V, Changsha Longtech Co. Ltd., China). Samples were placed on a SiC baseplate and surrounded by a hollow SiC cylinder, which was used as the susceptor to enhance microwave heating through efficient coupling during initial period of heating. The entire assembly was covered with ceramic fiber insulation. Sample temperature was measured continuously with the help of an optical pyrometer from the top. Both pure mullite and mullite 1wt% doped MgO samples were sintered in a microwave furnace at three different temperatures i.e. 1400°C, 1450°C, and 1500°C for 1 h. For comparison, pure mullite and 1wt% MgO doped mullite samples were also sintered in an electrically heated conventional muffle furnace at identical sintering temperatures i.e. 1400°C, 1450°C, and 1500°C for 2 h.

To prepare zirconia reinforced mullite composites, commercially grade high purity yttria stabilized tetragonal zirconia polycrystalline powders (YTZP) with an average particle size (d_{50}) of 50 μm were procured from Sulzer Metco (NV, USA). Measured amount of YTZP powders were mixed with mullite powders to prepare 5 wt% zirconia reinforced composites and 1 wt% MgO was added as an additive to the composite mixtures. After wet ball milling, the composite mixtures were dried in an oven and then green discs and cylindrical specimens were prepared using a uniaxial press. Finally the composites specimens were sintered in a microwave furnace at 1450°C and 1500°C for 1 h. Composite

samples were also sintered in a conventional furnace at 1500°C and 1600°C for 2 h in order to compare the properties of microwave processed samples.
Bulk densities of the sintered samples were measured from the known volumes and weights of the compacts. To compare the densification of microwave sintered samples with conventional sintering, relative densities (%) were calculated from the theoretical densities of the starting powders. The surface morphologies and microstructures of as processed sintered samples were observed using a field emission scanning electron microscope (FEI Inc., OR, USA) and a scanning electron microscope (SEM, Hitachi's 570, Japan). The ultimate compressive strength of all sintered samples was measured using a screw driven Instron machine (Norwood, MA, USA) with a constant crosshead speed of 0.5mm/min. Ultimate compressive strength was calculated from the maximum load applied immediately prior to failure.

RESULTS AND DISCUSSIONS

The relative densities of pure and 1 wt% MgO doped mullite compacts sintered in both microwave and conventional furnaces are shown in Fig. 1. It can be seen that the relative density of microwave sintered pure mullite was increased from 69.38 ± 1.24 % to only 74.35 ± 0.76% by increasing the sintering temperature from 1400°C to 1500°C. Not much difference in densification behavior was observed in conventionally sintered mullite compacts which exhibited little lower relative density of 72.33 ± 1.01% at 1500°C. This clearly suggests the difficulties in sintering the mullite powders in both microwave and conventional heating. This indicates that mullite is not a good microwave energy absorber since it has very low dielectric loss property. It can be recalled that microwave radiation can efficiently heat a material with high dielectric loss factor [15]. But the addition of 1 wt% MgO as a sintering aid produced significant improvement in densification behavior of mullite powders. From Fig. 1, it can be clearly seen that the relative density was increased from 70.29 ± 0.53% to as high as 87.29±0.38% for 1wt% MgO doped mullite samples as the sintering temperature increased from 1400°C to 1500°C in a microwave furnace. Moreover, the mullite grain morphology was also observed to change with the incorporation of 1wt% MgO additive. Fig. 2a and b presents the SEM microstructural images of as processed pure mullite and 1wt% MgO doped mullite samples sintered at 1500°C in a microwave furnace. From SEM microstructural observation the mullite grains were found as equiaxed in pure mullite (Fig. 2a), but in presence of 1 wt% MgO, morphology of the mullite grains was changed into acicular or needlelike shape (Fig. 2b). It can be concluded that at high temperature, MgO formed a glassy layer along the mullite grain boundary with the exsolution of Al_2O_3 and SiO_2 from mullite grain. Because of this, the morphology of the grains was changed at this stage to acicular or needle shape from the original equiaxed shaped grain structure of mullite. The formation of glassy phase eventually increased the density of the doped mullite samples. This observation is consistent with previously reported results. It has been reported that the sintered density of mullite increased significantly with increasing the amount of MgO [16, 17]. Montanaro et al. [18] showed that more than 95% densification can be achieved by sintering mullite at 1550°C when more than 2 wt% MgO was added. However, it can also be recalled that increasing the liquid phase can also decrease the high temperature mechanical properties of mullite. Because of this in the present research the MgO additive content has been restricted at only 1wt%.

Zirconia ceramics are known as efficient microwave absorbers and can be preferentially heated in a microwave field [19]. Therefore, in our research, 5 wt% yttria stabilized zirconia polycrystalline (YTZP) powders were incorporated into 1 wt% MgO doped mullite matrix with an aim to enhance microwave absorption efficiency of the system, and thus to improve the composite's sinterability as well as mechanical properties. Table 1 summarizes the relative densities of zirconia reinforced mullite composites sintered in microwave and conventional furnaces at different temperatures. For comparison, relative density data for pure and doped mullite are also presented. From Table 1, it can be seen that with incorporation of 5wt% ZrO_2, the relative density was increased from 87.29±0.38% to

89.26 ±0.947% for 1500°C microwave sintered samples. The effect of zirconia addition on microwave heating was more prominent at low temperature, i.e. at 1450°C, when 7% decrease in porosity was observed in 5wt% ZrO_2 mullite composites. This increase in density can be solely attributed to the efficient microwave coupling of mullite zirconia compacts which led to localized heating and thereby enhanced the densification kinetics. It is believed that at this low sintering temperature, i.e. 1450°C, the possible influence of glassy phase on mullite densification is minimal. However, at higher temperature, i.e. at 1500°C, both the ZrO_2 and MgO induced glassy phase contributed to the higher sintered density. From the conventional sintering densification data, a similar trend has also been observed. Apparently, in mullite zirconia systems no significant difference in sintered density was found after sintering at 1500°C in both microwave and conventional furnaces. However, the rapid heating rate, volumetric heating, uniform crosslinked acicular shaped grains arrangement, and low amount of glassy phase in composite microstructure have been expected to enhance the mechanical properties of microwave sintered mullite zirconia composites. Fig. 2c presents the representative SEM surface microstructures of as processed 5wt% zirconia reinforced mullite composites sintered in microwave furnace at 1500°C. Overall, the mullite zirconia composites were characterized by a crosslinked acicular or needle shape mullite grains arrangement where zirconia grains were mainly located at the mullite grain boundaries. However, in conventional sintering, with an increase in sintering temperature from 1500°C to 1600°C, relative density was observed to decrease for 5 wt% zirconia reinforced mullite composites. This observation can be explained from XRD analysis which indicated that with increasing the sintering temperature due to exsolution of yttria from YTZP grains, tetragonal zirconia became partly unstabilized. This led to the transformation of tetragonal zirconia into the monoclinic phase and the associated volume change induced residual pores within 1600°C conventionally heated samples.

Fig. 3 shows the compressive strength results for pure mullite, 1wt% doped mullite and mullite ZrO_2 composites sintered in microwave and conventional furnaces different sintering temperatures. It can be seen that pure mullite exhibited significantly low strength of 84±16 MPa at 1400°C which increased only to 162±27 MPa when the processing temperature increased to 1500°C in the microwave furnace. This low strength number was due to the porous microstructure of pure mullite which reduced its mechanical property. However, as the porosity decreased from 30 to 13% when 1wt% MgO was added as a sintering aid to mullite, the compressive strength also increased to a maximum of 387±21 MPa for 1500°C microwave sintered mullite samples. It has been observed that rapid and volumetric heating phenomena in microwave processed samples led to uniform and homogeneous microstructure in MgO doped mullite compacts, which in turn produced better mechanical properties in comparison to conventionally sintered samples. From the compressive strength results as obtained for microwave sintered mullite zirconia composites, it has been observed that incorporation of 5 wt% zirconia into mullite matrix significantly improved the microwave coupling efficiency and thus resulted in 39 % enhancement of mechanical strength for 1wt% MgO doped mullite composites sintered under identical microwave heating conditions. Interestingly, a maximum compressive strength of ~ 632±29 MPa was obtained for 5wt% ZrO_2 reinforced 1wt% MgO doped composite sintered at 1500°C in a microwave furnace. However, mullite zirconia composites sintered in a conventional furnace at 1500°C exhibited lower strength in spite of having the comparable density in comparison with microwave sintered composites. This difference can be explained by the good microwave energy absorbance ability of ZrO_2 particles which led to localized heating and helped in faster neck growth between mullite particles by enhancing surface diffusion kinetics with a minimal increase in density [19]. This improvement in sintering kinetics resulted in considerable improvement (~ 18.5 %) of mechanical strength in microwave sintered mullite zirconia composites compared to conventionally sintered samples even though both composites exhibited comparable density.

CONCLUSIONS

In this report, the feasibility of fabrication of dense mullite ceramics by microwave heating technique was investigated and parallel conventional heating studies were also carried out to compare the results. Experimental results showed that pure mullite exhibited poor densification behavior even at 1500°C sintering temperature irrespective of the heating techniques used. However, the addition of 1 wt% MgO as a sintering aid produced significant improvement in densification behavior of mullite powders in a microwave furnace. A maximum sintered density of 87.29± 0.38% was achieved for 1wt% MgO doped mullite which produced a significant increase in compressive strength of (~ 387±21MPa) when sintered in a microwave furnace for 1 h. Interestingly, incorporation of 5 wt% zirconia into the mullite matrix significantly improved the microwave coupling efficiency and thus resulted in 39% enhancement of mechanical strength for microwave sintered 5 wt% ZrO_2 1wt% MgO doped mullite composites. Our research findings indicate that microwave heating could be a suitable alternative of conventional heating to commercially produce high strength mullite and mullite zirconia composites utilizing reduced time and energy.

ACKNOWLEDGEMENT

Authors like to acknowledge financial support from the Office of Naval Research under the grant no. N00014 01 05 0583.

REFERENCES

[1]R. Atisivan, S. Bose, and A. Bandyopadhyay, Porous mullite preforms via fused deposition, *J. Am. Ceram. Soc.*,**84 [1]**, 221 23 (2001).

[2]M. G. M. Ismail, Z. Nakai, and S. Somiya, Microstructure and mechanical properties of mullite prepared by the sol gel method, J. Am. Ceram. Soc., **70 [l]**, C 7 C 8 (1987).

[3]A. Bandyopadhyay, Functionally designed 3 3 mullite aluminum composites, *Adv. Eng. Mater.*, **1 [3-4]**, 199 201 (1999).

[4]R. Soundararajan, R Atisivan, G. Kuhn, S. Bose, and Amit Bandyopadhyay, Processing of mullite Al composites, *J. Am. Ceram. Soc.*, **84 [3]**, 509 13 (2001).

[5]S. Kanzaki, H. Tabata, T. Kumazawa, and S. Ohta, Sintering and mechanical properties of stoichiometric mullite, *J. Am. Ceram. Soc.*,**68 [1]**, C 6 C 7 (1985.

[6]B.G. Ravi, V. Praveen, M. P. Selvam, and K.J. Rao, Microwave assisted preparation and sintering of mullite and mullite zirconia composites from metal organics, *Mater. Res. Bul.*, **33 [10]**, 1527 36 (1998).

[7]H. Ivankovic, E. Tkalcec, R. Nass, and H. Schmidt, Correlation of the precursor type with densification behavior and microstructure of sintered mullite ceramics, *J. Eur. Ceram. Soc.*, **23,** 283 92 (2003).

[8]D. Amutharani, and F.D. Gnanam, Low temperature pressureless sintering of sol gel derived mullite, *Mater. Sci. Eng. A*, **264,** 252 261 (1999).

[9]S. Maitra, A. Rahaman, A. Sarkar, and A. Tarafdar, Zirconia mullite materials prepared from semi colloidal route derived precursors, *Ceram.Int.*, **32,** 201 206 (2006).

[10]M. Mizuno, S. Obata, S. Takayama, S. Ito, N. Kato, T. Hirai, and M. Sato, Sintering of alumina by 2.45GHz microwave heating, *J. Eur. Ceram. Soc.*, **24**, 387 91 (2004).

[11]A. Chanda, S. Dasgupta, S. Bose, an A. Bandyopadhyay, Microwave sintering of calcium phosphate ceramics, *Mater. Sci. Eng. C.*,**29**, 1144 49 (2009).

[12]Z. Huang, M. Gotoh, and Y. Hirose, Improving sinterability of ceramics using hybrid microwave heating, *J. Mater. Proc. Tech.*, **209 [5]**, 2446 52 (2009).

[13]D. Clark, W. Sutton, and D. Lewis, Microwave: theory and application in materials processing IV, *Am. Ceram. Soc Bul.*, 61 96 (1997).

[14]E. David, C. F. Diane, and K. W. Jon, Processing materials with microwave energy, *Mater. Sci. Eng.*, **A287**, 153 58 (2000).
[15] R. W. Chun Chan, and B. B. Krieger, Kinetics of dielectric loss microwave degradation of polymers: lignin, *J. Appl. Pol. Sci.*, **26 [5]**, 1533 53.
[16]L. Montanaro, J. M. Tulliani, C. Perrot, and A. Negro, Sintering of industrial mullites, *J. Eur. Ceram. Soc.*, **17**, 1715 23 (1997).
[17]P. M. Souto, R. R. Menezes, and R.H.G.A. Kiminami, Sintering of commercial mulite powder: effect of MgO dopant, *J. Mater. Proc. Tech.*, **209 [1]**, 548 53 (2009).
[18]L. Montanaro, C. Perrot, C. Esnouf, G. Thollet, G. Fantozzi, and A. Negro, Sintering of industrial mullites in the presence of magnesia as a sintering aid, *J. Am. Ceram. Soc.*, **83**, 189 96 (2000).
[19]Y. I. Fang, J. Cheng, R. Roy, D. M. Roy, D. K. Agrawal, Enhancing densification of zirconia containing ceramic matrix composites by microwave processing, *J. Mater. Sci.*, **32** 4925 30 (1997).

Table I. Relative densities (%) of pure mullite, 1 wt% MgO doped mullite samples and mullite 1 wt% MgO 5 wt% ZrO_2 composites sintered in microwave and conventional furnaces at different sintering temperatures.

Samples	**Microwave Sintering**		**Conventional Sintering**	
	1450°C	**1500 °C**	**1500 °C**	**1600 °C**
Pure mullite	72.61±1.10	74.35±0.76	72.33±1.01	
Mullite 1wt% MgO	76.96±0.34	87.29±0.38	82.76±0.43	
Mullite 1wt% MgO 5wt% ZrO_2	83.59±0.723	89.26 ±0.947	89.92±0.9236	86.23±1.19

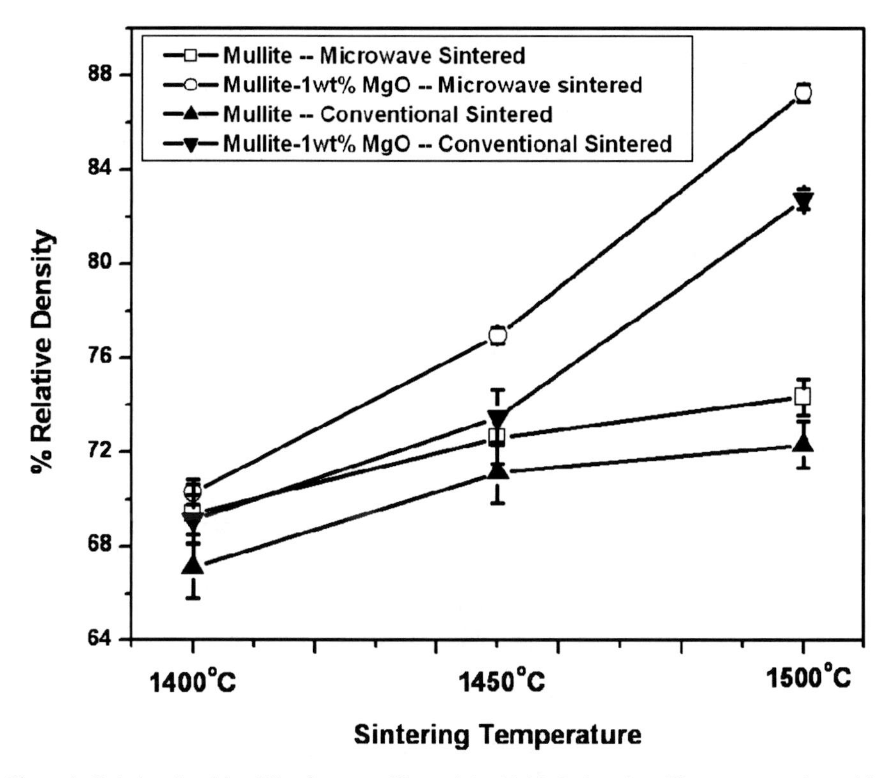

Figure 1. Relative densities (%) of pure mullite and 1wt% MgO doped mullite compacts sintered in microwave and conventional furnaces at different sintering temperatures.

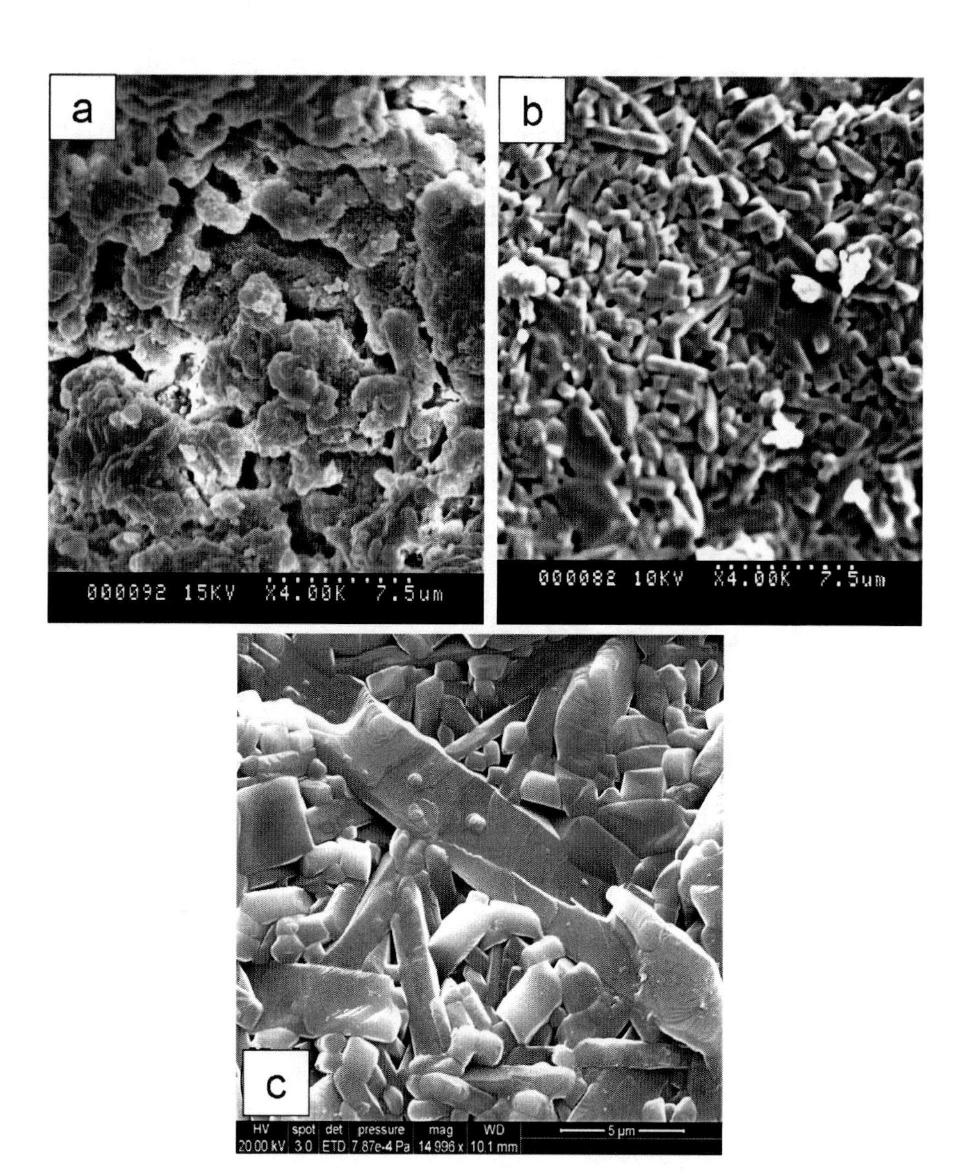

Figure 2. SEM images illustrating the microstructures of (a) pure mullite and (b) 1 wt% MgO doped mullite and (c) mullite 1 wt% MgO 5 wt% ZrO_2 composite sintered in a microwave furnace at 1500°C for 1 h.

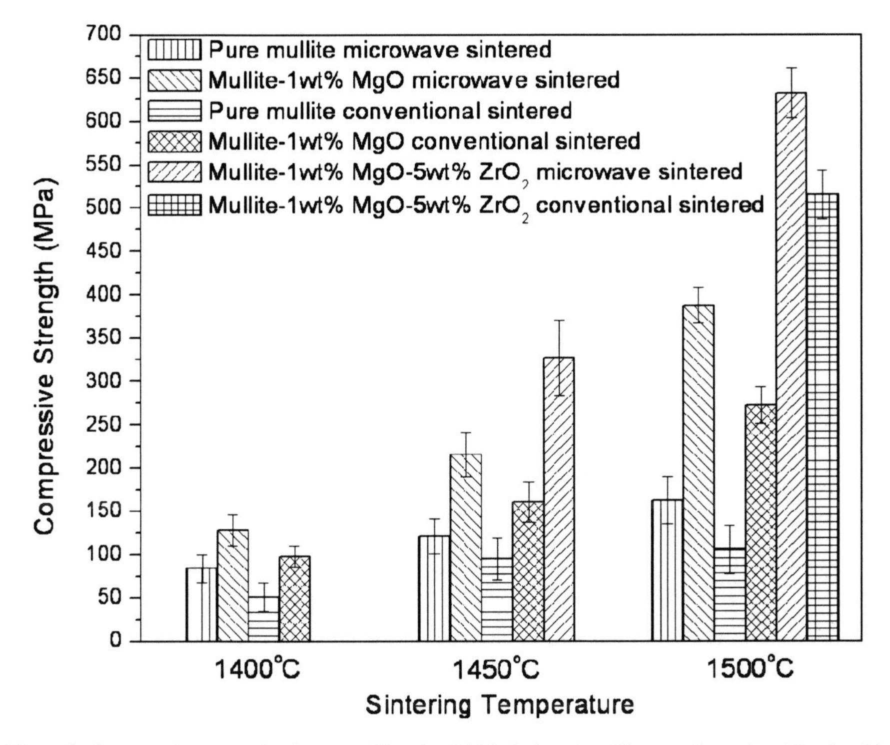

Figure 3. Compressive strength of pure mullite, 1 wt% MgO doped mullite samples and mullite 1 wt% MgO 5 wt% ZrO_2 composites sintered in microwave and conventional furnaces at different sintering temperatures.

IDLE TIME AND GELATION BEHAVIOR IN GELCASTING PROCESS OF PSZ IN ACRYLAMIDE SYSTEM

Nasim Sahraei Khanghah, Mohammad Ali Faghihi Sani*
Department of Materials Science and Engineering, Sharif University of Technology, Azadi St., Tehran, Iran.

ABSTRACT

Gelcasting is a novel forming method in fabricating complex three dimensional ceramic parts, and has many parameters and characteristics required to be specified. Up to now, few articles have been published on determination of the idle time of gelation precisely. In this work the chemorheology of gelation in aqueous solution of acrylamide and N, N' methylenebisacrylamide monomer, and zirconia (PSZ) suspensions of this solution was investigated. As the viscosity of gel system increases abruptly in gelation point, idle time can be determined precisely by measurement of viscosity against time. Idle time can also be determined through temperature measurement against time since the reaction of gelation is exothermic. This work compared these two methods of measurement of idle time. It was investigated in this work that idle time as a critical parameter in gelcasting, decreases with increasing of variables, such as acrylamide, N,N' methylenebisacrylamide and initiator concentrations and also temperature. In addition, presence of ceramic powder also enhances the rate of gelation intensely. In comparison of the two methods, it can be concluded that the rheological method is more reliable because the idle time can be perceived more rapidly after the beginning of the gelation.

Keywords: gelcasting, idle time, rheological properties, organic precursors.

INTRODUCTION

Gelcasting is a facile near net shape forming technique for fabricating complex, three dimensional bodies[1]. Controllable casting and solidification are possible using polymerizable organic monomers in high solid volume fraction colloidal suspension of ceramic powders. It uses systems of low organic content, which undergo gelation, via in situ polymerization of monomer species[1 3] or cross linking of existing polymeric species in solution[4]. During gelation, a macromolecular network forms to hold the ceramic particles together.

This process is suitable for complex shape fabrication by offering short molding times (on the order of several minutes), high green strength (~30 MPa)[5], and low cost machining[6]. This generic technique has been successfully applied to ceramics such as alumina, silicon nitride, and sialon[1].

In spite of abundant literature on various gel systems in gelcasting technology as well as various ceramic materials manufactured by this technique, few papers have been published on determining the effects of components concentration on the idle time of gelation process. In gels with acrylamide as monofunctional monomer, when polymer chains concentration reaches to critical overlap concentration (C_c), the chains become close enough to crosslink. In this stage, cross linking and gelation occurs, resulting in increase of viscosity[7, 8]. Therefore, idle time can be determined through viscosity measurement against time. Generally, radical polymerization of AM is directly proportional to the amount and ratio of the monomers and the initiator, as well as temperature[9].

In their original work on gelcasting, Young et al.[1] pointed out that the idle time of acrylamide N,N' methylenebisacrylamide system was reduced by presence of alumina powder. The chemorheological study of this gelling system, performed by Babaluo et al.[10] showed that the presence of alumina powder caused reduction in activation energy of polymerization, resulting in shorter idle times. Potoczec[11] has also stated the catalytic effect of alumina powder on gelation of methacrylamide. Morissete et. al.[4] investigated chemorheology of a gelcasting system based on aqueous alumina poly vinyl alcohol (PVA) suspensions, which was cross linked by an organometallic coupling agent.

In this work, viscosity profile has been investigated during gelation to recognize the effect of precursor composition on idle time in acrylamide N,N' methylenebisacrylamide (AM MBAM) system. This study highlights the importance of composition (i.e. monomer, cross linking agent, initiator, and solids content) and processing temperature on gelation behavior of AM MBAM solutions.

MATERIALS AND METHODS

The essential components of the gelcasting process, used in this work, were monofunctional acrylamide, $C_2H_3CONH_2$ (AM) (Merck), and di functional N, N' methylenebisacrylamide $(C_2H_3CONH)_2CH_2$ (MBAM) (Merck). These monomers were dissolved in deionized water to give premix solution. To form the gel structure, this premix solution undergoes free radical initiated vinyl polymerization in presence of ammonium per sulfate, $(NH_4)_2S_2O_8$ (APS) (Merck) as initiator. The reaction was accelerated by heating. Partially stabilized zirconia powder (PSZ) was used as ceramic powder. The average particle size and the density of the powder were 0.2 μm and 5.65 gr/cm^3, respectively. Dolapix CE64 was also used as dispersants.

The procedure flowchart is presented in Fig. 1. In this regard, premix solution with AM, MBAM, and APS concentration of 15, 0.6, and 0.1 wt %, respectively was selected as reference. Effect of AM (3 18 wt %), MBAM (0.2 3.1 wt %) and APS (0.3 1.5 wt %) concentration, all based on the premix solution, were investigated on rheological behavior of the solution during gelation, while other parameters were constant. In addition, effect of temperature (45, 50, and 55 °C) was studied on idle time of gelation in reference premix solution. At last, the suspension was prepared with various volume fractions of PSZ (0.15 0.45) to investigate the effect of solid volume fraction on idle time. The dispersant amount was 0.6 wt % (by total weight of suspension). All measurements were started immediately after adding initiator.

A controllable stress rheometer (Physica MCR300, Anton Paar, Australia) was used to characterize the rheological behavior of pre mix solutions and suspensions. All measurements were carried out with cylindrical chamber (No. C PTD 200) at 50°C unless otherwise noted.

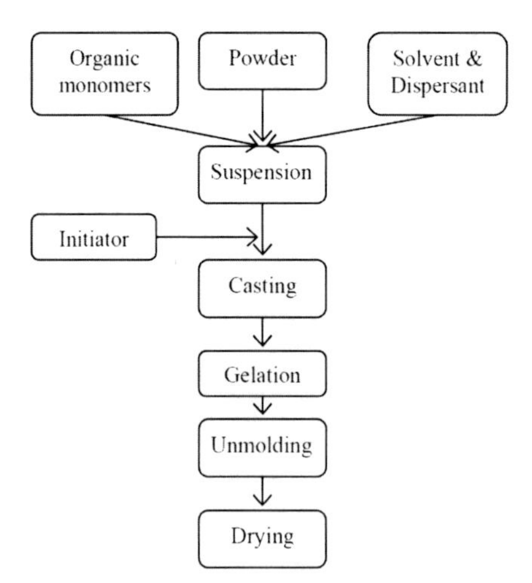

Fig. 1. Flowchart of the gelcasting process.

RESULTS AND DISCUSSION

1) Effect of AM concentration

Changes of apparent viscosity (η) as a function of time with constant crosslinking agent and APS (0.6 and 0.1 wt %, respectively) for various concentration of acrylamide at 50 °C, are shown in Fig. 2(a). In each curve, abrupt increase of viscosity shows gelation moment. According to this diagram, gelation was retarded with decrease in AM concentration. When AM concentration increases, probability of polymerization reaction increases. As a result, concentration of polymer chains reaches to critical overlap concentration (C_c) faster. During this measurement, no significant gelation was observed at 2.9 wt % according to Fig 2(a). It seems that the concentration of AM was not enough to yield a gel in this case.

The idle time is usually estimated from the x axis intercept of the tangent to the viscosity curve at elevated viscosity in viscosity time diagram. The most important problem of idle time measurement in this method is that the stress of spindle can easily breaks the gel structure. Therefore, the increase of viscosity happens later, resulting in longer idle time.

In this work, viscosity time curves have also been plotted in semi logarithmic scale to study the viscosity increment at the beginning of gelation precisely. Fig. 2(b) shows diagram 2(a) in semi logarithmic scale. As it can be easily seen, in AM concentration of 2.9 wt %, viscosity increment can be distinguished clearly in semi logarithmic diagram (Fig. 2(b)). This increment corresponds to the beginning of gelation, but the chain concentration is not high enough to yield a three dimensional gel structure under the shear stress.

In AM concentration of 6 wt%, also some alteration in slope of the curve can be distinguished in Fig 2(b), resulting in significant delay in the idle time measured at elevated η from intercept of the tangent to the viscosity curve (t_{idle}) in Fig. 2(a). This premix solution undergoes gelation even under the shear stress, because AM concentration is high enough to reach the critical overlap concentration. It can be obviously distinguished that t_{idle} is recognizably higher than the idle time calculated from the first step of elevating viscosity in semi logarithmic diagram in Fig. 2(b) ($t_{idle}^{semi-log}$). The gel structure becomes more tight and strong in higher concentration of AM (15 and 18 wt %), resulting in less delay in t_{idle} in comparison with $t_{idle}^{semi-log}$.

Figure 2(c) shows the two measured idle times against AM concentration. As anticipated, t_{idle} and $t_{idle}^{semi-log}$ difference is larger in lower AM concentration, due to weakness of gel structure. These results show that it is more precise to use semi logarithmic scale to measure idle time.

2) Effect of MBAM concentration

Figure 3 shows the effect of MBAM concentration on gelation kinetics of solution with constant AM and APS concentration (15 and 0.1 wt%, respectively) in 50 °C. Figure 3(a) cannot be interpreted because the curves have intercepted each other, due to the applied shear stress that ruptures the gel formation. In semi logarithmic diagram (Fig. 3(b)), the sequence of $t_{idle}^{semi-log}$ based on MBAM concentration can be clearly distinguished. Comparing Figs. 2 and 3 shows that AM concentration affects the idle time more intensely than MBAM concentration. In fact, MBAM concentration affects the gelation and crosslinking reaction, while AM concentration affects polymerization reaction and critical overlap concentration. Under the stress, physical properties and behavior of gel change in various MBAM concentrations (Fig 3(b)).

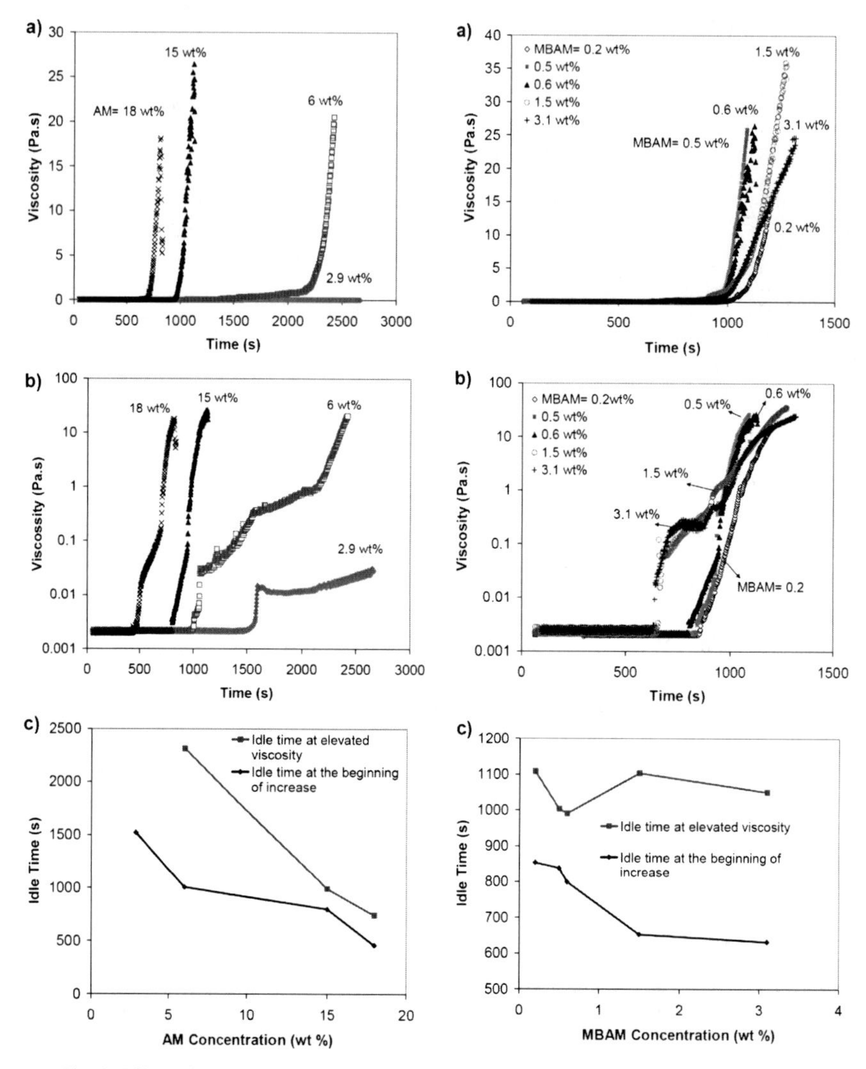

Fig. 2. Effect of AM concentration on idle time with constant MBAM and APS concentrations at 50 °C, apparent viscosity of premix solution versus time in nonlogarithmic (a) and semi logarithmic (b) scales, idle time versus AM concentration (c).

Fig. 3. Effect of MBAM concentration on idle time with constant AM and APS concentrations at 50 °C, apparent viscosity of premix solution versus time in nonlogarithmic (a) and semi logarithmic (b) scales, idle time versus MBAM concentration (c).

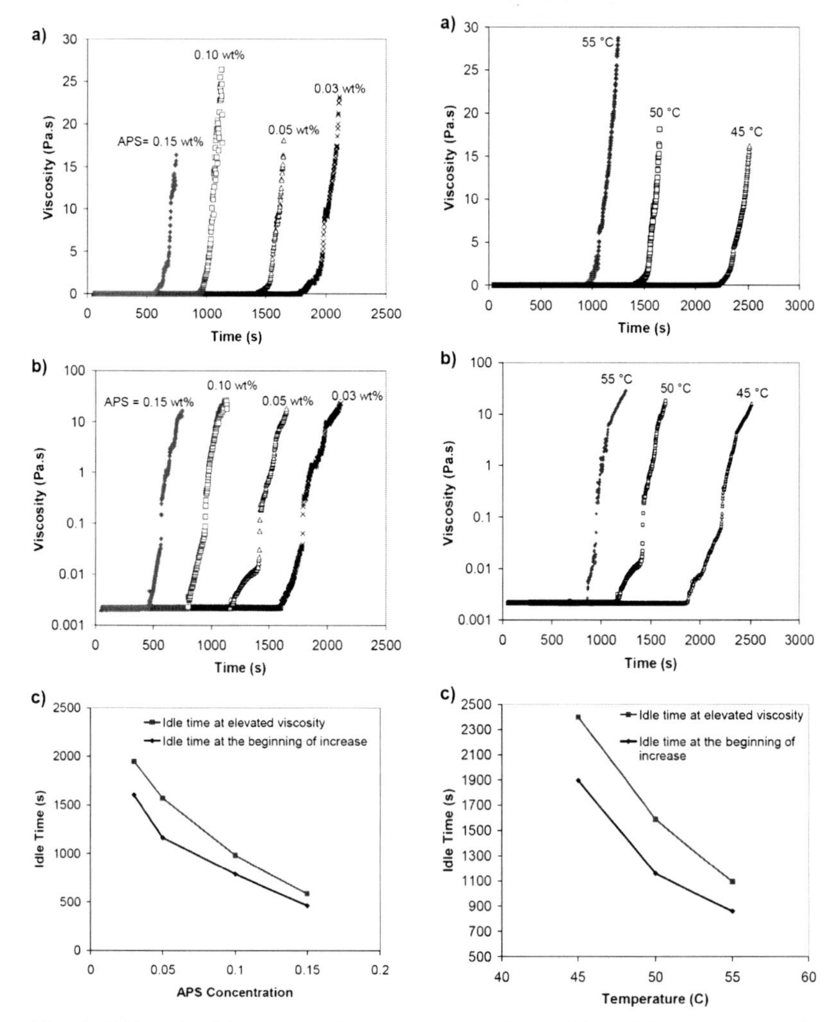

Fig. 4. Effect of APS concentration on idle time with constant AM and MBAM concentrations at 50 °C, apparent viscosity of premix solution versus time in nonlogarithmic (a) and semi logarithmic (b) scales, idle time versus APS concentration (c).

Fig. 5. Effect of Temperature on idle time with constant AM, MBAM and APS concentrations, apparent viscosity of premix solution versus time in nonlogarithmic (a) and semi logarithmic (b) scales, idle time versus temperature (c).

Fig 3(c) shows t_{idle} and $t_{idle}^{semi-log}$ against MBAM concentration. In this figure $t_{idle}^{semi-log}$ decreases with MBAM concentration, but t_{idle} has unexpected rise in larger MBAM concentration (1.5 and 3.1 wt %) which is related to frequent changes in slope of η t curve (Fig 2(b)) due to gel's physical properties under stress.

3) Effect of APS concentration

Figs. 4 (a), (b) and (c) show the effect of APS concentration on gelation kinetics when the AM and MBAM concentrations are constant (15 and 0.6 wt%, respectively). Fig. 4(a) shows that the idle time is decreased with increase in APS concentration. Higher APS concentration increases probability of polymerization reaction, resulting in lower idle time. According to Fig. 4(b) the physical properties of gel structure is not affected by APS concentration, due to constant behavior of solution during gelation. Small difference between t_{idle} and $t_{idle}^{semi-log}$ in Fig. 4(c), is due to the fact that viscosity curves slope in Fig. 4(b) does not have remarkable rupture.

4) Effect of temperature

Apparent viscosity as a function of gelation time for constant AM, MBAM and APS concentration (15, 0.6 and 0.1 wt %, respectively) at various temperatures is shown in Figs. 5 (a) and (b). These figures show that the idle time decreases dramatically with increase in temperature. The strong temperature dependence of the gelation kinetic can be easily seen in Fig 5(c). This dependency is considered advantageous, because one can overcome the accelerating effect of solid addition on gelation rates by handling such systems at low temperatures prior to casting.

The idle time is inversely proportional to the rate of production of free radicals, which is related to the solution temperature, T. This relationship can be expressed by Arrhenius type equation (1):

$$t_{idle} = A\ \exp(\frac{E_a}{RT}) \tag{1}$$

where R is gas constant, and E_a is the activation energy of gelation. Arrhenius plot of the measured idle time for the reference solution is presented in Fig.6, showing linear changes of Ln($t_{idle}^{semi-log}$) with 1/T, and an activation energy of 68.74 KJ/mole.

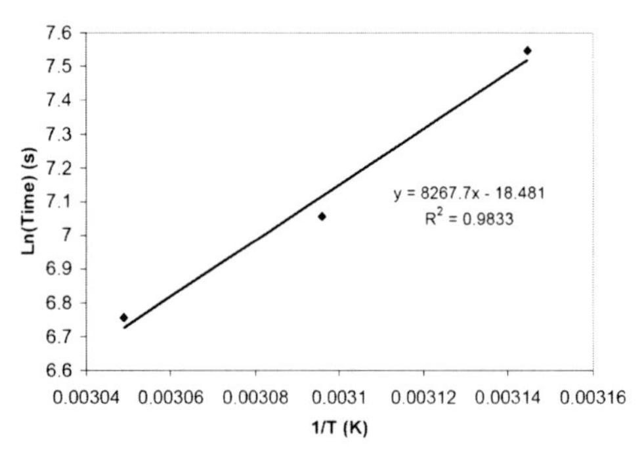

Fig. 6. Arrhenius plot of Idle time ($t_{idle}^{begining}$) for reference solution.

5) Sensitivity of idle time

Figure 7 shows sensitivity of idle time to the above mentioned parameters. In this figure, y axis shows the sensitivity of idle time, which means percentage of difference in $t_{idle}^{semi-log}$ with respect to the idle time of reference specimen. X axis also represents sensitivity of each parameter, which means percentage of difference in each mentioned parameter with respect to the same parameter of the reference specimen. In this figure, the slope of temperature curve is the highest comparing with other parameters (AM, MBAM and APS concentration), which means that idle time is more sensitive to temperature. In addition, it is clear that MBAM concentration has less effect on idle time than AM and APS concentration.

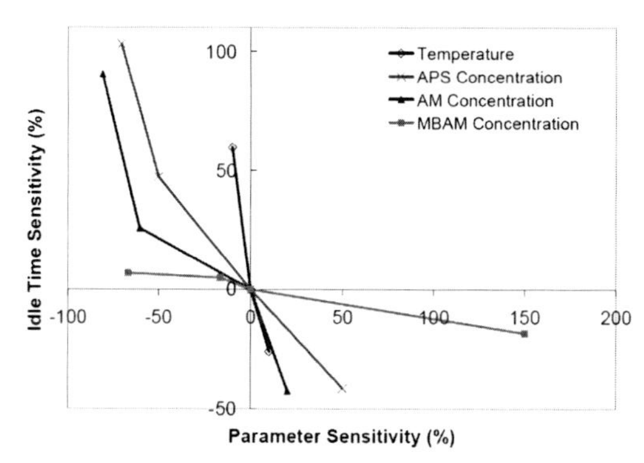

Fig. 7. Sensitivity of idle time as a function of various gelation parameters.

6) Effect of solid volume fraction

Effect of solid volume fraction on gelation behavior is also studied. The apparent viscosity of suspension as a function of time for various solid volume fractions at 50 °C is plotted in Fig. 8 (a) and (b) in constant AM, MBAM and APS concentration (15, 0.6 and 0.05 wt% in solution, respectively). Behavior of the corresponding premix solution (without ceramic powder) is also included in Fig 8(a) for comparison. Accordingly, the gelation kinetics was enhanced with increase in PSZ volume fraction. These data suggest that gelation is induced by solid particles. Confining the analysis only to suspension behavior, it is clear that idle time decreases significantly with increase in PSZ volume fraction, suggesting that the solid particles can not be considered inert.

According to Fig 8(b), initial suspension viscosity increases with increase of powder volume fraction. In addition, slope of increment during gelation increases intensively (according to distance between curve's points) with increase in powder volume fraction, suggesting that the kinetic of gelation is enhanced.

Fig 8(c) shows excerpted idle times versus solid volume fraction. Increase of initial viscosity of suspensions and enhancement of gelation rate cause reduction of difference between t_{idle} and $t_{idle}^{semi-log}$. In conclusion, stress induced by spindle can delay t_{idle}, however presence of solid particles can reduce this delay.

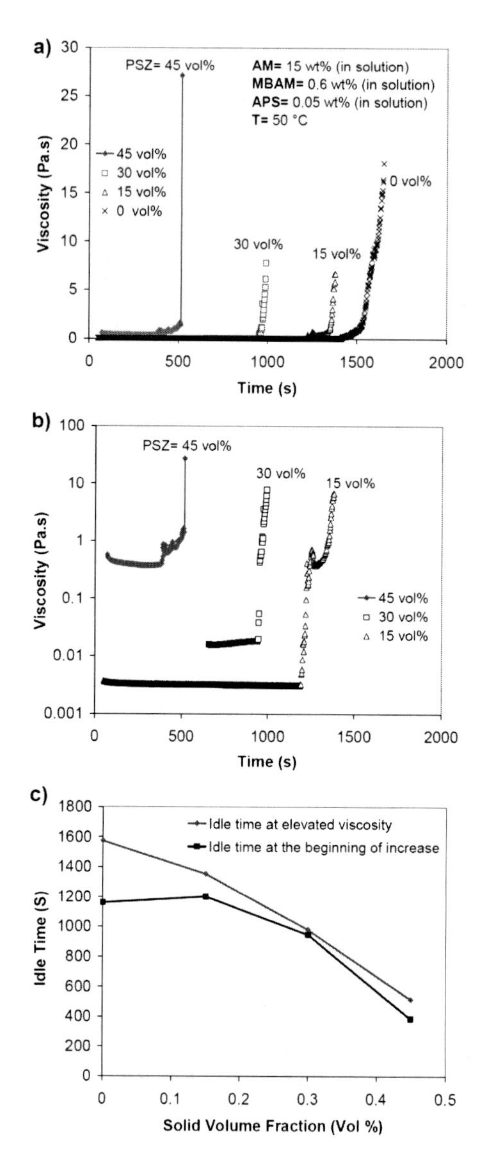

Fig. 8. Effect of solids volume fraction on idle time in reference solution at 50 °C, apparent viscosity versus time in nonlogarithmic (a) and semi logarithmic (b) scales, idle time versus solids volume fraction (c).

CONCLUSION

The chemorheological properties of AM MBAM system were investigated in this work. Accordingly, dependence of idle time to various components concentration as well as temperature was determined. Based on these observations, increase of AM, MBAM and APS concentration, as well as temperature and solid volume fraction decreases the idle time.

The results suggest that semi logarithmic plot is more appropriate to determine the idle time, by eliminating the effect of shear stress applied by measurement equipment. In this semi logarithmic diagram, in which changes of viscosity less than 1 (Pa.s) is detectable more accurately, the increase of viscosity can be observed on the first propagation instant of gel network.

Comparison of the idle time sensitivity with various parameters shows that the most and least effective parameters are temperature and MBAM concentration, respectively.

ACKNOWLEDGMENTS

This work was supported by the Advanced Materials and Technology Research Center.

REFERENCES

[1] Young, A. C., Omatete, O. O., Janney, M. A., & Menchhofe, P. A., Gelcasting of Alumina, *J. Am. Ceram. Soc.*, **74** [3], 612 18, (1991).

[2] Janney, M. A., Omatete, O. O., Walls, C. A., Nunn, S. D., Ogle, R. J., & Westmoreland, G., Development of Low Toxicity Gelcasting Systems, *J. Am. Ceram. Soc.*, **81** [3], 581 91, (1998).

[3] Kokabi, M., Babaluo, A. A., & Barati, A., Gelation process in low toxic gelcasting systems, *J. Eur. Ceram. Soc.*, **26**, 3083 90, (2006).

[4] Morissette, Sh. L., & Lewis, J. A., Chemorheology of Aqueous Based Alumina Poly (vinyl alcohol) Gelcasting Suspensions, *J. Am. Ceram. Soc.*, **82** [3], 521 28, (1999).

[5] Vlajic, M. D., Krstic, V. D., Strength and machining of gelcast SIC ceramics, *J. Mater. Sci.*, **37,** 2943 47, (2002).

[6] Tan, Q. Q., Gao, M., Zhang, Zh. T., & Tang, Z. L., Polymerizing mechanism and technical factors optimization of nanometer tetragonal polycrystalline zirconia slurries for the aqueous gel tape casting process, *Mater. Sci. Eng. A*, **382,** 1 7, (2004).

[7] Orakdogen, N., Kizilay, M. Y., & Okay, O., Suppression of inhomogeneities in hydrogels formed by free radical crosslinking copolymerization, *Polymer,* **46**, 11407 15, (2005).

[8] Boyko, V. B., N Vinylcaprolactam based Bulk and Microgels: Synthesis, Structural Formation and Characterization by Dynamic Light Scattering, PhD thesis, Faculty of Mathematic and Natural Sciences, Dresden University of Technology, Germany, (2004).

[9] Ha, C. G., Jung, Y. G., Kim, J. W., Jo, C. Y., & Paik, U., Effect of particle size on gelcasting process and green properties in alumina, *Mater. Sci. Eng. A,* **337**, 212_221, (2002).

[10] Babaluo, A. A., Kokabi, M., & Barati, A., Chemorheology of alumina aqueous acrylamide gelcasting systems, *J. Eur. Ceram. Soc.*, **24**, 635 644, (2004).

[11] Potoczek, M., A catalytic effect of alumina grains onto polymerization rate of methacrylamide based gelcasting system, *Ceram. Int.,* **32,** 739 744, (2006).

CHARACTERIZATION OF THE MESOPOROUS AMORPHOUS SILICA IN THE FRESH WATER SPONGE CAUXI

Ralf Keding, Martin Jensen and Yuanzheng Yue
Section of Chemistry, Aalborg University, DK-9000 Aalborg, Denmark

ABSTRACT

The freshwater sponge Cauxi has been collected at the river bank of the Rio Negro at Praia Grande in the Amazon basin 60 km west of Manaus, Brazil. The skeleton of the sponge consists of spicules (average diameter: 15.6 μm, average length 305 μm) that are cemented through organic junctions. The spicules have been characterized using different techniques such as vacuum hot extraction (VHE),optical microscopy, scanning electron microscopy, X ray diffraction as well as thermogravimetry. The results reveal that the spicules themselves consist of a glassy phase and mesopores (average diameter: 23 nm). During the VHE scan, the water residing inside the mesopores is released in a temperature range between 250 and 1000 °C, indicating a tight bonding of water in the mesopores. Heating of the skeleton at 10 K/min up to 1450°C removes the organic binder and allows amorphous silica to act as a binder that maintains the main structure of the skeleton. The glassy phase is mostly converted into the cristoballite phase. A faster heating up to 2000°C converts the mesoporous spicules into a foam like amorphous silica. This paper describes the macro and microstructural features and the properties of the Cauxi skeleton before and after heat treatments.

INTRODUCTION

Amorphous silica has a wide variety of applications such as membranes [1], columns [2], heat proof materials [3], nanoglue [4], and optical communication fibres [5]. Some of these applications rely on the high temperature stability of the amorphous silica. The application field of amorphous silica can be expanded further if the silica is mesoporous as the mesoporosity adds new functionalities [6 8]. However, mesoporosity increases the surface area of the silica, and as the crystallization of silica is known to emanate from the surface, the thermal stability might be reduced by the mesoporosity. Very recently, we have discovered that the spicules constituting the skeleton of the freshwater sponge Cauxi are made of highly pure, mesoporous, amorphous silica [9]. The spicules are cemented by organic junctions to form the sponge. The surface of the spicules is partially covered with iron oxide from the river. After removal of the surface layer and the organic material by bleaching, the amorphous phase of the spicules is found to consist of 99.7 wt% SiO_2 [9]. Based on these findings, a growth mechanism for biological formation process has been suggested which opens the possibility for a biomimetic production of the mesoporous amorphous silica. However, the high temperature stability of these mesoporous silica fibers is still unknown. Due to the importance of thermal stability and the effect of mesoporosity on this stability, the current paper focuses on the characterization of mesoporous amorphous silica at different temperatures.

EXPERIMENTAL

Extraction:

The freshwater sponge Cauxi was collected at the river bank of the Rio Negro at Praia Grande in the Amazon basin 60 km west of Manaus, Brazil (3°03'22.5S and 60°30'33.56W) in the dry time (November). The river bank is inundated during the rain time due to the increase of the level in the Rio Negro. It is believed that the sponge was dried in the tropical sun for longer time, at least 1 month. There were several hundreds of Cauxi sponges on the branches of trees that are flooded in the rain season.

Bleaching:

A selected amount of the sponge was added to a Teflon container and mixed 35 % H_2O_2, 69 % HNO_3, and water in the volume ratio 8:7:1. The Teflon container was placed in a 90 °C water bath for 20 min to accelerate the removal of organic material from the sponge. The solvents were then removed and fresh solvents added. The container was replaced in the water bath for 20 min t. The procedure was repeated until a white material without any colorants appeared in the container. The solvents were then removed and the material rinsed once with water. The material was heated at 160 °C for 2h to remove any remaining bleaching agents and, to keep the bulk properties of the sponge skeleton unaffected.

Heat treatments:

A sample was prepared by heating untreated sponge in an alumina crucible in air up to 1450°C for 17h in a electric furnace (Ängelholm, Sweden). Other treatments were done in a simultaneous thermal analyzer (NETZSCH STA 449C Jupiter (Selb, Germany)) in platinum crucibles. Initially the crucible was held 5 minutes at an initial temperature of 60°C. Thereafter an upscan to 1270 °C and a subsequent downscan were performed with 10 K/min. The purge gas was air with a flow of 40 ml/min.

Scanning electron microscopy (SEM):

The untreated sponge was cut with a diamond wire, glued on an Al sample holder, and subsequently coated with gold using an evaporation process. The pictures were recorded using a Zeiss DSM 940 A (Oberkochen, Germany) microscope. The SEM measurements on heat treated samples were done on a Zeiss 1540 XB scanning electron microscope (Oberkochen, Germany) using uncoated samples. The secondary electrons were recorded.

Optical microscope:

The pictures from the structure were obtained from a digital camera attached to a stereomicroscope Carl Zeiss Technival (Jena, Germany). The phase contrast pictures was obtained from a digital camera attached to a Carl Zeiss JENAPOL interphako μ map microscope (Jena, Germany).

Transmission electron miroscopy (TEM):

After the untreated sponge was heat treated in an electric furnace at 1720 K for 17 h it was finely ground using an agate mortar. The obtained power was dispersed in isopropanol and after thorough dispersing, a droplet of the dispersion was allowed to dry on a carbon coated copper grid. The ground, heat treated sponge fragments kept on the grid were imaged in a high resolution transmission electron microscope (JEOL JEM 4010, acceleration voltage 400 keV, point to point resolution: 0,155 nm).

X ray diffraction (XRD):

The samples were ground to a fine powder and were measured with a Seifert FPM HZG4 diffractometer (Jena. Germany) with Fe Kα radiation. The scan was conducted in the range $5° < 2\theta < 65°$.

Vacuum hot extraction (VHE):

The measurement was done on bleached fibers. Prior to measurement, the sample was kept at 200 °C to remove adsorbed water. The sample was cooled to room temperature in the device. Subsequently, the degassing rate was recorded when heating the sample with a rate of 10 K/min to 1500 °C by a mass spectrometer (MS). The MS signal was calibrated with a Gypsum ($CaSO_4*2H_2O$) sample. This allows to calculate the release of water from the MS current at 18 g/mol.

Thermogravimetry:

Measurements were conducted in the STA 449C Jupiter at a heating rate of 10 K/min up to 1270 °C. During the measurement, the furnace of the STA was purged with argon at a flow of 40 mL/min.

RESULTS AND DISCUSSION

Fig 1 shows a photography of the sponge after the transportation to the lab in a plastic bag. The color of the specimen is brown with gray tones and it is spherical with a diameter of 3 18cm. The brownish color of the skeleton is mainly due to the organic parts of the sponges and partially due to the precipitated humic acids from the Rio Negro. The structure of the sponge is rather durable against mechanical compression as it is barely impossible to break it by a compression with the bare hands. The density of Cauxi as shown in Fig 1 is approx. 15g/l. The sponge is not ignitable by a lighter. The odor of the gaseous pyrolysis products is like burned proteins and the remaining solid material has a yellowish, slightly sintered structure.

Fig 1: Photography of Cauxi as extracted.

Detailed investigations of the skeleton of Cauxi are given in Fig 2 through Fig 4. Fig 2 shows a selected typical area of the sponge skeleton. The structure of interconnected rings with a size of around 500 μm is composed of a large number of oriented spicules. Some of the spicules are seen close to the center of the picture. The structure is arranged in such a manner that hydraulic resistance to water is minimized..

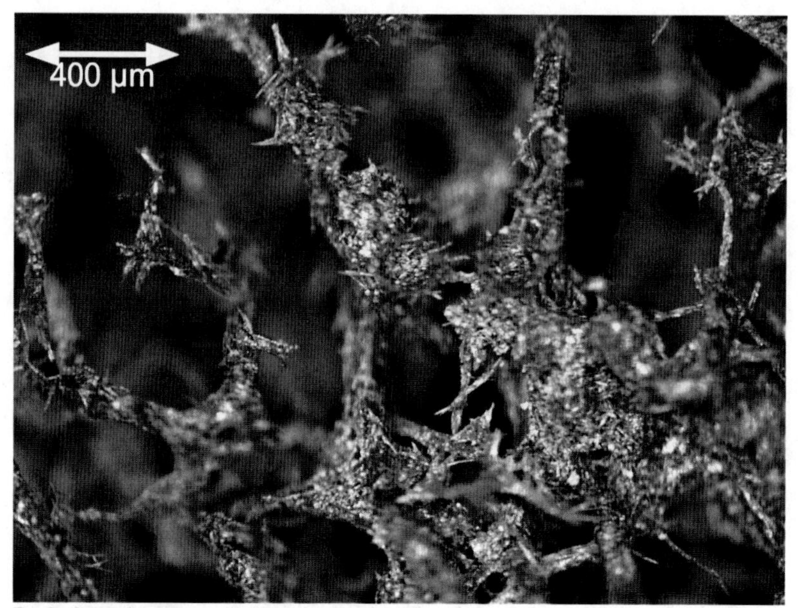

Fig 2: Optical micrograph of a typical area in the sponge

A back scattering electron (BSE) micrograph of the untreated sponge reveals the structure of the skeleton (Fig 3). The spicules are bundled and aligned according to the primary skeleton structure. The orientation of the fibers is mixed in the interconnection to support the structure of the skeleton. Besides the spicules, some smaller structures are seen on the surface of the bundled fibers. These structures are mainly organics of sponge and precipitations from the river.

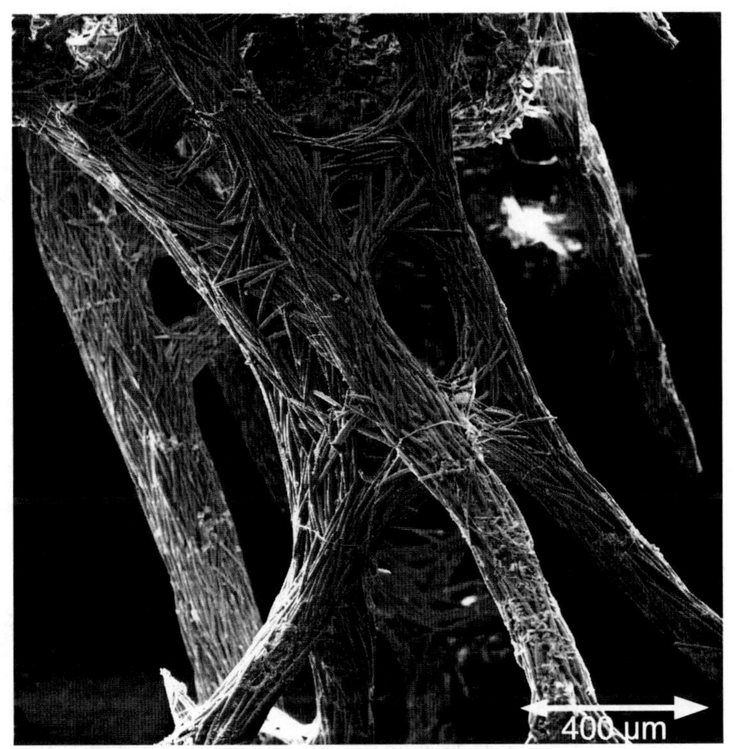

Fig 3: BSE Micrograph of the structure of the alignment of the spicule in the skeleton structure. The sample is coated with gold.

The junction between spicules is illustrated in a higher magnification in Fig 4. The junction in the lower middle of the picture shows some small holes at the right spicule that is the interface between the inorganic spicule and the binder cementing the junctions. The binder in Cauxi is organic [9] whereas an inorganic silica binder has been found in other sponges [10].

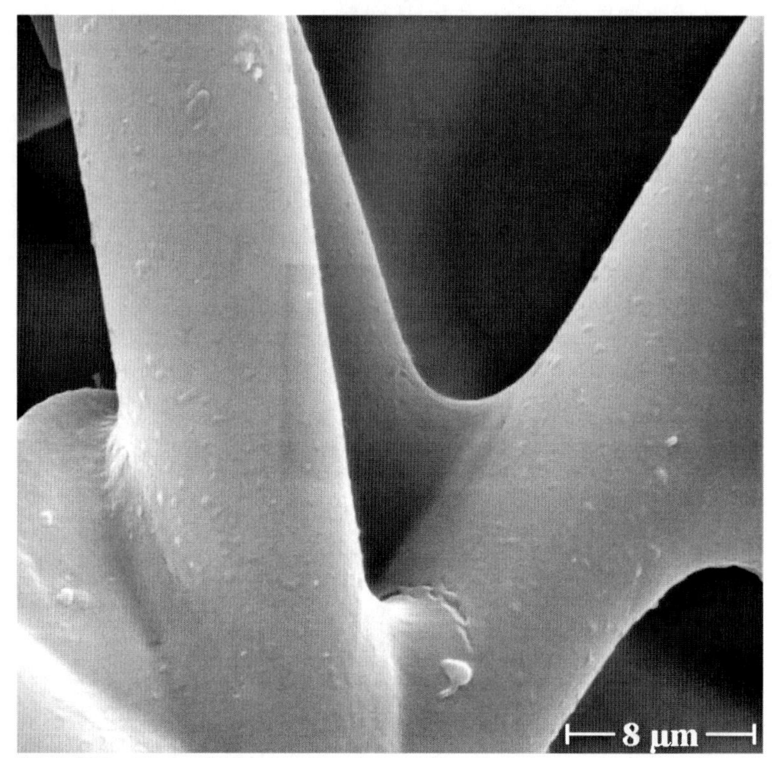

Fig 4: BSE micrograph of the junctions between the spicule in untreated cauxi. The sample is gold coated.

Fig 5 shows an optical micrograph of a single spicule that was collected from the dust that was produced during the sample preparation for SEM imaging. This spicule was selected from a large number of objects with almost the same size and embedded in immersion oil. The spicule possesses a channel in the middle as described in [9] harboring a protein filament known to catalyze the polymerization of silica [11,12]. The spicule growth originates from this filament by protein catalyzed polymerization of silicic acids [9]. The surface of the spicule is partially contaminated with smaller (approx. 1 3 µm) strings that appear brownish. The strings probably contain the dried residuals from the organic structure of the sponge as well as attached components from the river. The optical microscope in transmission is not able to detect any structure in the bulk of the spicule beside the central channel. Pictures obtained with crossed polarization filter did not give any hint for birefringence. This exclude mechanical stress in the spicule and all crystalline phases that exhibits birefringence like quartz.

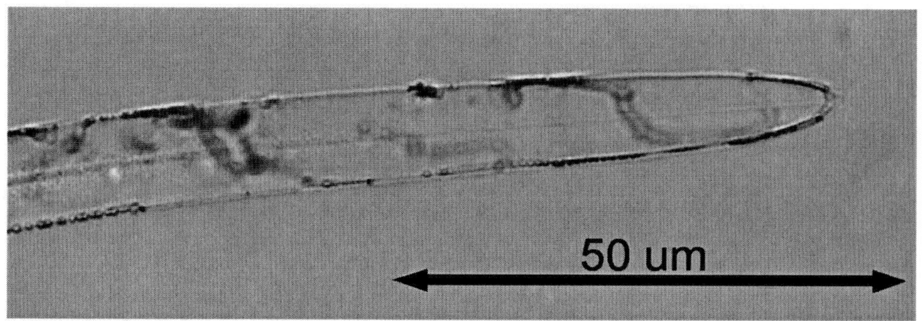

Fig 5: Transmission optical microscopy of a untreated spicule.

By means of optical microscopy, the length and width of the spicules have been determined to be 305 ± 18 and 15.6 ± 1.5 μm, respectively.

The sponge consists of amorphous silica fibers cemented by an organic binder [9]. The organic binder can influence the thermal properties of the inorganic phase, e.g., as a heat treatment will cause a combustion of the organic material, but pyrolysis products can serve as nuclei for crystal growth and hence reduce the stability of the glassy silica. To elucidate this effect some of the sponge has been bleached to remove the organic material. The bleaching destroys the bundled structure and therefore results in unconnected spicules. Both bleached and untreated sponge have been heated to 1270 °C in air at a rate of 10 K/min. After the heat treatment, the cemented network of the untreated spicules has been lost due to combustion of the organic binder. In fact, the bundled structure is already lost at 550 °C [9]. The XRD pattern and SEM micrographs of the two samples after the treatment are shown in Fig 6 and Fig 7, respectively.

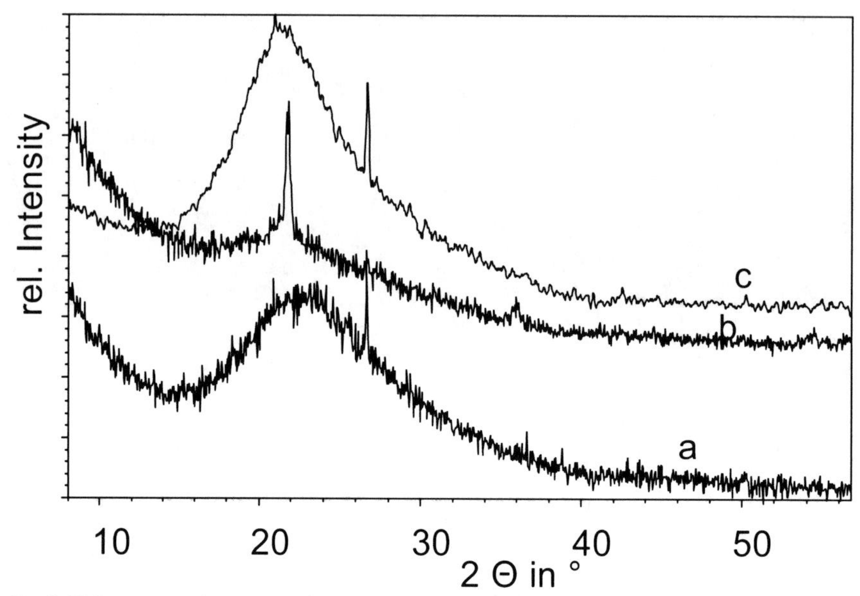

Fig 6: XRD patterns of a: untreated sponge prior to heat treatment, b: untreated sponge after heat treatment to 1270 °C, c: bleached sponge after heat treatment to 1270 °C.

The XRD pattern of the untreated sponge prior to heat treatment (Fig 6 a) demonstrates that the spicules are amorphous due to the broad peak from 15 to 30°. However, a narrow peak around 26.5° that must arise from a crystalline structure is present. The crystalline phase cannot be identified from a single peak, though. Pattern c has been recorded from a sample bleached prior to the DSC measurement, i.e., the organic compounds have been removed from the surface of the spicule prior to the heat treatment. The small, crystalline peak is seen in both pattern a and c. This demonstrates that the crystalline phase is insoluble in the acidic and oxidizing bleaching agent and that the phase is stable to a heat treatment in air to 1270°C. After heat treatment of the untreated spicules, the peak at 26.5° has vanished, but instead a peak at 21° occurs (pattern b). Hence, the crystalline phase at 26.5° can be reduced by the organic material during combustion. These features are normally exhibited by the Fe_2O_3 phase. This inference is also supported by a magnetic test as the untreated spicules as well as the heat treated ones are not attracted by a 1T magnet. This excludes the existence of magnetite and metallic iron phases.

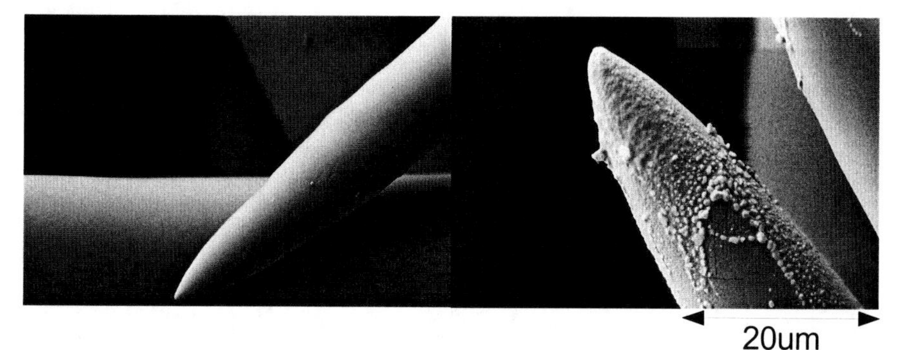

Fig 7: SEM micrograph of spicules after heat treatment to to 1270 °C a: (left) bleached spicules b: (right) as extracted spicules.

The SEM micrograph (Fig 7 a) of the bleached spicules after heat treatment to 1270 °C shows that they have a smooth, particle free surface. This observation agrees with the XRD results. On the surface of the non bleached spicules (Fig 7 b), crystal like particles can be seen. These structures most likely cause the crystalline peak at 21°. The single peak is insufficient to identify the phase.

As the bleached material has been proven to be the most thermal resistant and does not contain any combustible organic material at the surface, the bleached spicules are selected for the determination of mass loss during heat treatment. The mass loss upon heating is determined by both mass spectrometry (MS) coupled with vacuum hot extraction (VHE) and thermogravimetry (TG) in Ar (Fig 8). In order not to measure the water adsorbed on the surface of the spicules, the mass of the samples is indexed to 100 % at 200 °C as adsorbed water is evaporated below this temperature [13].

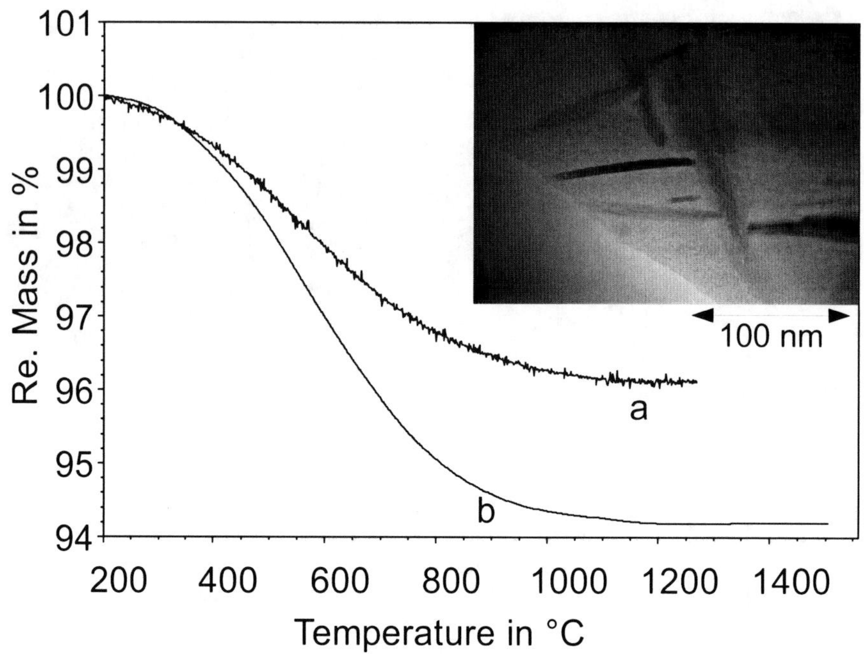

Fig 8: Thermogravimetry (a), vacuum hot extraction with 10K/min (b) of the bleached spicule. The mass was normalized to 100% at 200°C. Upper right: TEM micrograph of the spicule after heat treatment at 1450°C for 17h

The release rate of water from the bleached spicules depends on both the temperature and the atmosphere conditions. The main release of water in the VHE experiment occurs in the temperature range between 290 and 850 °C with a second increase in the water release around 1100 °C that is close to the glass transition temperature (1136 °C) of the SiO_2 in Cauxi [9]. The water content can be calculated to be 5.5 % when averaging two VHE measurements. The glassy silica that composes the spicules has been found to contain approximately 400 ppm hydroxyl species, i.e., structural water [9].

The main water release in the TG measurement starts at 250°C and ends around 1000°C. In contrast to the VHE, the TG has been recorded in Ar. The water content calculated from the weight loss determined by this method is only 3.8%, i.e., a discrepancy between TG and VHE exists. Since the equilibrium between liquid water and gaseous water is shifted to lower temperatures with decreasing gas pressure, the change in the main release temperature range can be explained as follows. The Ar with 1 bar pressure decreases the thermodynamic driving force compared to vacuum, and therefore the water content calculated from the TG is smaller. However, it is notable that the temperature 1270 °C is insufficient to vaporize all the water and that the temperature range of the water release is broad and occurs at high temperatures. This phenomenon can be explained by the fact that the spicules contain mesopores filled with the water [9], i.e., the water is entrapped inside these small capillaries. These mesopores can be seen on the TEM image (Fig 8). Due to pronounced decomposition, TEM images of untreated spicules could not be obtained. Although the bleached sample underwent massive radiation damage upon TEM inspection, the amorphous nature of the spicules was doubtlessly disclosed. The samples heat treated at 1450°C, 17h do not undergo a radiation damage in a extended scale anymore, but the channel structure is altered. The channels contain cristobalite instead of water in this case. The capillaries could affect the water release by two phenomena: either by impeding the diffusion of the water inside them (kinetic hindrance) and/or by elevating the boiling point of the entrapped water (thermodynamic hindrance). As the release rate differs between the two methods conducted at the same rate, the kinetic factor can be excluded. The possible remaining factor is the elevation of the boiling temperature and this can be explained as follows. A large fraction of the water molecules are in direct contact with the SiO_2 wall as the average diameter of the capillaries has been determined to approx. 23 nm [9] The number of channels is sufficient to explain a weight loss of 5.5%. It is expected that the chemical reaction between the water and the SiO_2 forms a sort of a gel layer. Since a capillary effect is linked to the interaction between the liquid and the solid of the capillary material, the state of water in 23 nm capillaries should be described by a combination between capillary effects, adsorption, and gel layer effects. The combination of those will allow a qualitative description of the high release temperatures of water. The high release temperatures demonstrates the water in the pores is tightly bound.

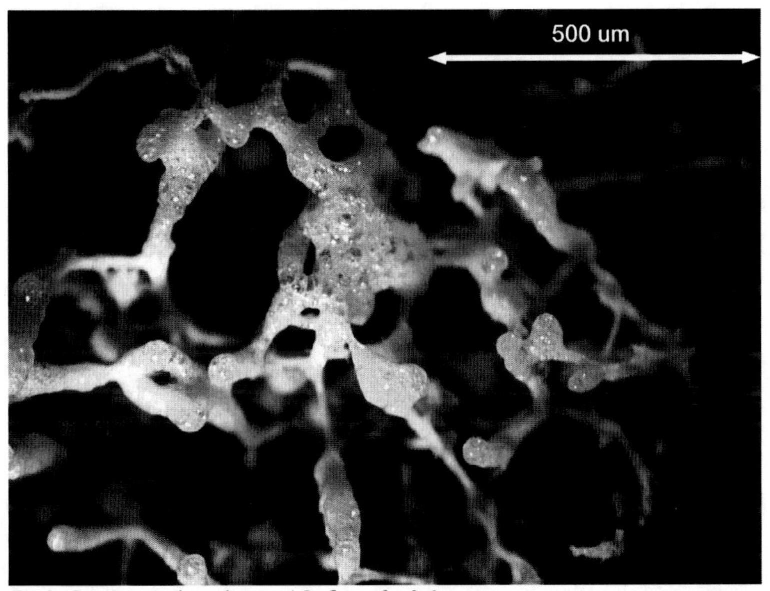

Fig 9: Cauxi treated in a butane / O_2 flame for 3 6 s.

The presence of water in channels of the spicules has been confirmed by heating the untreated sponge with a butane / O_2 burner for 3 6 s (Fig 9). The procedure is associated with a bright light emission and gives an odor of burned leather. Since the temperature of the flame can reach 2000°C, it is sufficient to convert the siliceous spicules into a viscose melt. The viscosity is high enough to maintain the main structure of the skeleton, though. The sudden and intense heating does not allow the water to evaporate from the pores before the SiO_2 is converted into a viscose melt. Therefore, the water remains encapsulated and expands in the viscose melt. This results in the foam like structure shown in

Fig 9. The brownish color of the skeleton has disappeared due to the combustion of the organic components.

Damages that the encapsulated water causes on the spicules during the intense heating will disclose the real changes that are occurring in the skeleton. In order to observe these, a part of the sponge is heated at 1450 °C for 17h. The heat treatment of the Cauxi skeleton with a heating rate of 10K/min to 1450°C and a subsequent dwell time of 17 h results in a pale orange structure. The shape of the skeleton is not notably altered, but is found to much more brittle than that of the untreated sponge. This is in contrast to heat treatments at temperatures of 1270°C or 550°C which result in a break down of the skeleton leaving only the individual spicules.

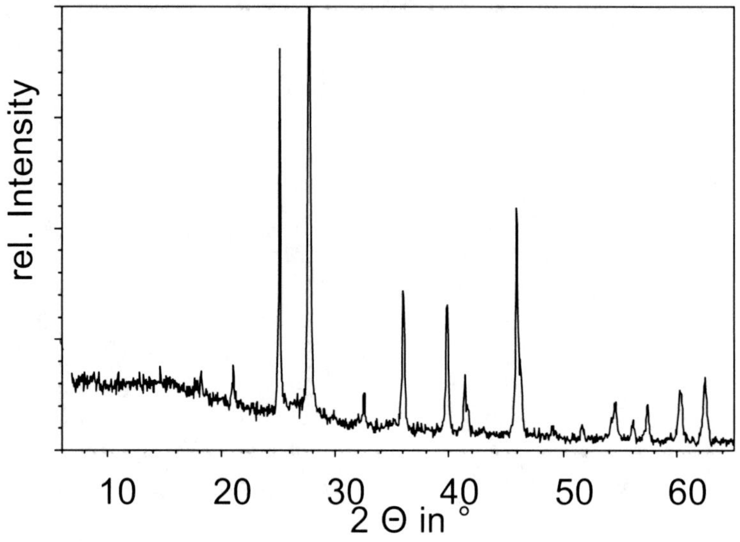

Fig 10: XRD of the spicules heat treated at 1450 °C for17 h

Fig 10 shows the XRD of the spicules after the treatment at 1450 °C. The broad amorphous peak observed in the untreated spicules has vanished and several sharp peaks have arisen. Through comparison with the pattern from the JCPDS, the peaks can be identified as α cristobalite (75 0923). The cristobalite crystals can be seen in Fig 11.

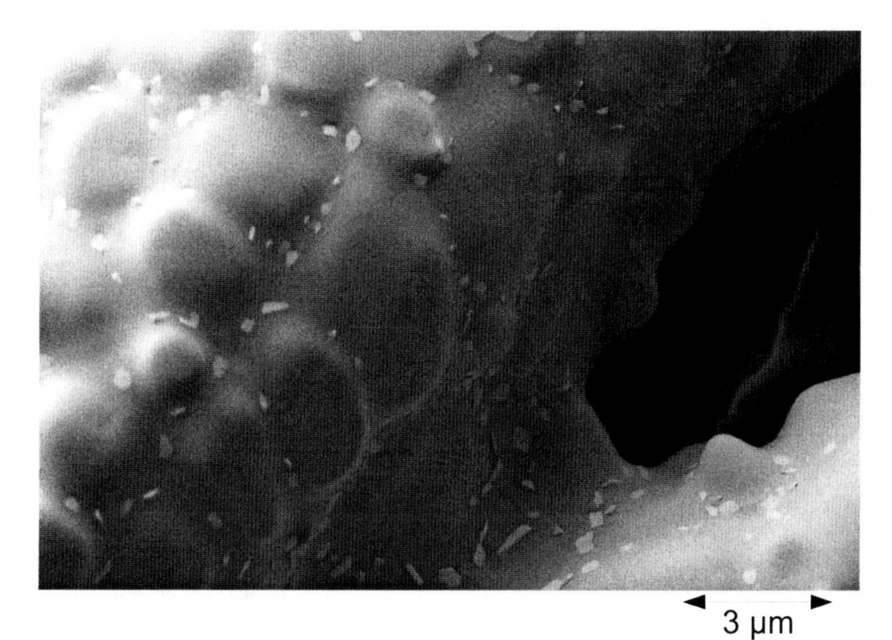

Fig 11: SE micrograph of the sponge skeleton after 17h treatment at 1450°C. The junction of the spicule is shown.

Fig 11 shows a junction between 2 spicules after the treatment. However, the junction is no longer organic as in the untreated skeleton, but now instead a sinter neck between the spicules. The change from flexible organic binder to sinter neck accounts for the increased brittleness of the skeleton as the sinter neck is unable to release the stress by elastic deformation. The spicule surface itself exhibits approx. 1 μm large bumps at the surface that can be assigned to the cristobalite crystals. On the grain boundaries of the cristobalite crystals, bright structures are attached to the surface. They could contain the inorganic impurities of the sponge. The results of the XRD does not reveal the nature of this second crystal phase. However, it is known that the spicules are coated with iron oxide from the river and as the melting point of iron oxide has not been exceeded, the iron coating must still be present. A magnetic test demonstrates that the material is not magnetic thereby excluding the presence of Fe_3O_4 and as the heat treatment has been done in air, FeO is excluded aswell leaving only Fe_2O_3. The bright structures on the cristobalite grain boundaries are to small for an element analysis in the SEM, therefore the presence of Fe_2O_3 cannot be confirmed here. The junction and the grain boundaries show both bright impurity spots. It is suggested that the junction and grain boundaries represent the remaining glassy phase and the bumps represent the cristoballite. This can be attributed to the fact that during the heat treatment the viscosity of the melt was sufficiently low to allow a viscose flow into the sinter neck. Additionally, the temperature is sufficiently high for the crystallisation of the SiO_2 in the spicule. The mechanism of viscous flow and crystallisation explains why the macroscopic shape of the skeleton is retained during the heat treatment to 1450 °C. This high thermal stability at low heating rates explains the addition of Cauxi to pottery produced by the precolumbian natives in the amazon basin [14]. Thereby they utilized the Cauxi spicules for fiber reinforcing their pottery..

CONCLUSIONS

The skeleton of the freshwater sponge Cauxi consists of an organic binder and a skeleton of amorphous, mesoporous silica of high purity. The organic binder links the SiO_2 spicules together and form an open porous skeleton of low density and low hydraulic resistance. Even if the biological part of the sponge is dried out, the remaining proteins still cement the spicules. Heat treatment at 550°C combusts the binder, whereas bleaching dissolves it both causing a collapse of the structure leaving a powder. The spicules remain individually in the powder. The vacuum hot extraction results in a release of 5.5 wt% water at a temperature up to 850°C while the thermogravimetry in argon releases a lower about of water (3.8wt%) at a higher temperature (until 1000°C). Since the structural water in the SiO_2 is only 400 ppm, the released water originates from the mesopores of the spicules. The high release temperatures demonstrate the water in the pores is tightly bound. Heating within a view seconds to 2000°C with a burner let the SiO_2 become a viscose melt before all water is evaporated. This results in an encapsulation of the water by the viscose melt and finally in a foam like structure. A slower heating of 10 K/min up to 1450 °C allows the water to evaporate without encapsulation. The structure of the Cauxi skeleton is maintained after the heat treatment at 1450 °C as the junctions are filled by amorphous SiO_2 through viscous flow. This results in a high brittleness of the entire structure. The organic bonded structure is less brittle because the junctions can release the stress by elastic deformation.

ACKNOWLEDGEMENT

We acknowledge Ralf Müller (VHE, BAM Berlin), Thomas Höche (TEM, MPI for Microstructure Physics in Halle), Christina Apfel (XRD, Universität Jena) and Thomas Kittel (Optical Microscopy, Universität Jena)

---

[1] R. M. de Vos and H. Verweij, High-Selectivity, High-Flux Silica Membranes for Gas Separation, *Science* **279** , 1710–1711 (1998).

[2] J. Dai, X. Yang, P. W. Carr, Comparison of the Chromatography of Octadecyl Silane Bonded Silica and Butadiene-coated Zirconia Phases based on a Diverse set of Cationic Drugs, *J. Chromatogr., A*, **1005**, 63–82 (2003).

[3] K. Saito, N. Ogawa, A. J. Ikushima, Y. Tsurita, K. Yamahara, Effects of Alumina Impurity on the Structural Relaxation in Silica Glass, *J. Non-Cryst. Solids,* **270**, 60–65 (2000).

[4] C. A. Morris, M. L. Anderson, R. M. Stroud, C. I. Merzbacher, D. R. Rolison, Silica Sol as a Nanoglue: Flexible Synthesis of Composite Aerogels, *Science,* **284**, 622-624 (1999).

[5] L. M. Tong, R. R. Gattass, J. B. Ashcom, S. L. He, J. G. Lou, M. Shen, I. Maxwell, E. Mazur, Subwavelength-diameter Silica Wires for Low-loss Optical Wave Guiding, *Nature,* **426**, 816–819 (2003).

[6] T. Asefa, C. Yoshina-Ishii, M. J. MacLachlan, G. A. Ozin, New-Nanocoposites: Putting Organic Function „Inside" the Channel Walls of Periodic Mesoporous Silica, *J. Mater Chem.*, **10**, 1751-1755 (2000).

[7] T. R. Sathe, A. Agrawal, S. Nie, Mesoporous Silica Beads Embedded with Semiconductor Quantum Dots and Iron Oxide Nanocrystals: Dual-Function Microcarriers for Optical Encoding and Magnetic Separation, *Anal. Chem.*, **78**, 5627-5632 (2006).

[8] A. Stein, Advances in Microporous and Mesoporous Solids – Highlights of Recent Progress, *Adv.*

Mater., **15**, 763-775 (2003).

[9] M. Jensen, R. Keding, T. Höche, Y. Z. Yue, Biologically Formed Mesoporous Amorphous Silica, *J. Am. Chem. Soc.,* **131**, 2717 2721(2009).

[10] J. Aizenberg, J. C. Weaver, M. S. Thanawala, V. C. Sundar, and D. E. Morse, P. Fratzl, Skeleton of Euplectella sp: Structural Hierarchy from the Nanoscale to the Macroscale, *Science*, **309**, 275 278 (2005).

[11] J. N. Cha, K. Shimizu, Y. Zhou, S. C. Christiansen, B. F. Chmelka, G. D. Strucky, and D. E. Morse, Silicatein Filaments and Subunits from a Marine Sponge Direct the Polymerization of Silica and Silicones *in Vitro*, *Proc. Natl. Acad. Sci. U.S.A.*, **96**, 361 365 (1999).

[12] K. Shimizu, J. Cha, G. D. Stucky, and D. E. Morse, Silicatein α: Catephsin L like Protein in Sponge Biosilica, *Proc. Natl. Acad. Sci. U.S.A.*, **95**, 6234 6238 (1998).

[13] L. L. Hench & J. K. West, The Sol Gel Process, *Chem. Rev.*, **90**, 33 72 (1990).

[14] M. L. Costa, D. C. Kern, A. H. E. Pinto, J. R. T. Souza, The ceramic artifacts in archeological black earth (terra preta) from lower Amazon region, Brazil, *Acta Amazonica* **34**, 165 178 (2004).

NOVEL CHEMISTRY MODIFICATION APPROACH FOR SYNTHESIS OF SiAlON FROM FLY ASH

J. P. Kelly, J. R. Varner, W. M. Carty, V. R. Amarakoon
Kazuo Inamori School of Engineering, NYS College of Ceramics at Alfred University
Alfred, NY 14802

ABSTRACT

This research focuses on making SiAlON ceramics from fly ash, a by product of burning coal to produce electricity. Fly ash chemistry is complex. If fly ash is to be considered a viable raw material for synthesis of SiAlON ceramics, then adequate means of controlling chemistry to predict the properties and characteristics of the SiAlON are necessary. An approach based on grouping the functionality of fly ash constituents during the carbothermal reduction and nitridation process is proposed. A ternary functionality diagram is presented and is tested using six fly ashes. The average solid solution value of the resulting β SiAlON powders, determined by lattice refinement, was $z = 2.0$, which agrees with theoretical predictions. Phase analyses, microstructure analyses, and the mechanical properties were evaluated after sintering compacts with 4 wt% yttria additions. The average hardness of two phases was determined to be 12.9 and 8.3 GPa respectively. The average crack lengths after Vicker's indents at loads of 5, 10, and 20 kgf held for 15 seconds were determined to be 224, 394, and 663 μm respectively. Statistical analyses of the mechanical properties support the null hypothesis with 99% confidence that the samples are not statistically different.

INTRODUCTION

SiAlON, silicon aluminum oxynitride, is an advanced structural ceramic that exhibits excellent thermal shock resistance, high strength, good fracture toughness, good high temperature strength, low thermal expansion, good oxidation resistance, and excellent resistance to corrosion by non ferrous metals. The SiAlON phases exhibit extensive solid solution ranges that offer the opportunity to customize properties for various applications. Despite the excellent properties of SiAlON materials, they have not seen their full potential and only satisfy a niche market because of their high cost. Raw materials account for approximately 30 50% of the cost of making nitrogen ceramics, with the remaining costs being associated with processing.[1] The use of low cost starting materials subjected to the carbothermal reduction and nitridation (CRN) process is attractive for lowering the cost of SiAlON products and opening new market opportunities.

Fly ash is an attractive raw material for making SiAlON because of its secondary impact on the environment, low cost, and overall characteristics. Fly ash is a byproduct of burning coal. In 2003, 70 million tons were generated. Approximately 39% of the fly ash was beneficially used and the remainder was landfilled.[2] The disposal of large amounts of fly ash is an environmental concern of increasing magnitude, and the beneficial uses of fly ash are sought.[3] The American Coal Ash Association (ACAA) is dedicated to this purpose. Fly ash is an aluminosilicate with small particle size, is primarily amorphous, has carbon necessary for the CRN process, and has alkali and alkali earth oxides which are capable of stabilizing the α SiAlON phase.

Indeed, several researchers have worked on converting fly ash into SiAlON.[4 9] The referenced papers are not an exhaustive list. In much of the reported work, the SiAlON powders that were made are multi phase and often have undesirable phases. Also, the prior work only accounts for part of the chemistry while leaving some constituents out. This ultimately results in lack of predictability for the phases that are present in the SiAlON made from fly ash and the degree of solid solution for those phases. There is also no clear indication that reproducible SiAlON can be synthesized from multiple fly ash sources. The present work offers a novel approach for fly ash chemistry modification that demonstrates predictability and reproducibility.

Fly ash is complex, containing many oxide constituents, and having a chemistry that depends on where the coal was mined, how it was burned in the reactor, and post burning processes. A behavioral diagram that includes each component individually would be exceedingly difficult to interpret. On the other hand, a behavioral diagram based on the primary function of each constituent would allow for grouping of constituents, thereby significantly simplifying the interpretation. Such a diagram can be constructed in the form of a ternary functionality diagram, where the primary function of silica is to be reduced and nitrided, that of alumina is to react to form a solid solution with the resulting silicon nitride, and that of alkali oxide and alkali earth oxide is to serve as glass formers that will form a liquid phase at high temperatures, as well as serve as cations that are capable of stabilizing the α SiAlON phase.

EXPERIMENTAL PROCEDURE

A ternary functionality diagram was constructed with the β SiAlON solid solution range superimposed onto it. In addition, the α SiAlON solid solution region, based on literature data for Ca^{2+} cation stabilization, was superimposed onto the diagram.[3] Six fly ashes from three different sources were obtained. Four of the fly ashes were from the same source, but came from different reactors. Chemical analysis of the fly ashes was done using standard techniques by an independent testing laboratory.*

The fly ash chemistry was plotted on the ternary functionality diagram according to the Unity Molecular Formula (UMF) approach used in glaze development.[10] Iron (III) oxide was considered to belong to the alkali earth oxide group, because it is expected to reduce to iron (II) oxide in a nitrogen atmosphere. However, special attention to the iron (III) oxide content is required. Iron and its compounds are known to be catalysts for the CRN process and are thought to be associated with the low eutectic temperature when combined with silica.[1] This is further supported by the fact that iron silicide forms upon cooling. However, iron silicide forms inclusions after sintering that are detrimental to the mechanical properties; therefore it should be removed before sintering.[11] This shift in chemistry caused by the removal of iron silicide is accounted for in predicting the target chemistry. The iron (III) oxide content was normalized to the fly ash with the highest content and was held constant at 8 wt% of the target chemistry after iron silicide removal to account for its catalytic effects.

Beyond accounting for the shift in chemistry due to removal of iron silicide, two further corrections in chemistry must be made. Carbon additions were made according to stoichiometry given by equation 1. When plotted on the ternary functionality diagram, all of the fly ash chemistries are rich in alkali and alkali earth oxide compounds such that the chemistries were outside of the SiAlON formation regions. Therefore, silica and alumina diluents need to be added to achieve the target chemistry. The target chemistry was arbitrarily chosen. The z value of the β SiAlON and the m and n values of α SiAlON were 2, 3.4, and 0.6 respectively. These values correlate to the degree of solid solution for the SiAlON phases. Kudyba Jansen et al. give the chemical reactions for the SiAlON phases according to the z , m , and n values and are given as equations 1 and 2.[4] According to their work, the m value is out of the range of the single α SiAlON phase region at 1700 °C, and is in the region where an α/β SiAlON product is expected. Equations 3 through 5, used for correcting the chemistry, have been derived from equations 1 and 2, where CaO has been replaced by the summation of alkali and alkali earth oxide compounds.

$$(6\quad z)SiO_2+\left(\frac{z}{2}\right)Al_2O_3+\left(4\quad \frac{z}{2}\right)N_2$$
$$+\left(12\quad \frac{3z}{2}\right)C\rightarrow Si_{6\ z}Al_zO_zN_{8\ z}+\left(12\quad \frac{3z}{2}\right)CO \tag{1}$$

$$\left(\frac{m}{2}\right)CaO + (12 \quad (m+n))SiO_2 + \left(\frac{m+n}{2}\right)Al_2O_3 + \left(\frac{16 \quad n}{2}\right)N_2$$
$$+\left(\frac{24 \quad 3n}{2}\right)C \rightarrow Ca_{\frac{m}{2}}Si_{(12 \quad (m+n))}Al_{m+n}O_nN_{16 \quad n} + \left(24 \quad \frac{3n}{2}\right)CO \tag{2}$$

$$z = \frac{6}{\left(\frac{Si}{Al}+1\right)} \tag{3}$$

$$m = \frac{24\left(\frac{\sum_i R_{ix}}{Si}\right)\left(\frac{\sum_i R_{ix}}{Al}\right)}{\left(\frac{\sum_i R_{ix}}{Si}\right)+\left(\frac{\sum_i R_{ix}}{Al}\right)}, \{i \mid 1,2\} \tag{4}$$

$$n = \frac{m\left(\frac{1}{2} \quad \left(\frac{\sum_i R_{ix}}{Al}\right)\right)}{\left(\frac{\sum_i R_{ix}}{Al}\right)}, \{x \mid 1,2\} \tag{5}$$

For all six fly ashes, a batch consisting of 30 g of fly ash, colloidal silica (Sigma Aldrich, St. Louis, MO), colloidal alumina (WesBond Corporation, Wilmington, DE), iron (III) oxide (<5 μm, Sigma Aldrich, St. Louis, MO), and carbon in the form of Lampblack 101 (Degussa North America, Parsippany, NJ) in the appropriate proportions, 250 g of 1 mm yttria stabilized zirconia grinding media (Advanced Materials, Farmington, CT), and 100 mL of isopropanol was mixed for 30 minutes in a 300 cc capacity porcelain milling jar using a laboratory rapid mill (Ceramic Instruments, S.r.l., Sassuolo, MO, Italy). After mixing, the slurry was drained through a 125 μm sieve into a collection pan. The slurry was dried in a drying oven at 95 °C for 24 hrs. The dry powder was mixed with a mortar and pestle to eliminate the effects of particle size segregation caused by settling effects.

For each of the six batches, 20 g of pre synthesis powder was added to a 64 mL alumina combustion boat. The powder was synthesized in a tube furnace using a linear nitrogen flow rate of approximately 50 cm/min, 20 K/min heating and cooling rates, and a dwell time of 3 hrs at 1400 °C with CO detectors installed for safety purposes. After synthesis, hydrochloric acid washing was performed by stirring in 250 mL of 6N hydrochloric acid for 48 hrs, followed by rinsing with 500 mL of deionized water before drying for 24 hrs in a drying oven at 95 °C.

X ray diffraction was performed by scanning from 20 to 100° 2θ using a step size of 0.04° 2θ and a dwell time of 10 sec. Lattice refinement was performed using Jade 8 software to get the lattice parameters. The z value of the powders was determined by using the lattice parameters according to equations 5 and 6 and averaging the results.

$$a(nm) = 0.7603 + 0.00296z \tag{5}$$

$$c(nm) = 0.2907+0.00255z \tag{6}$$

The density of the powders was measured using an AccuPyc 1330 helium pycnometer (Micromeritics Instrument Corporation, Norcross, GA). The surface area was determined by BET analysis using a TriStar 3000 surface area and porosity analyzer (Micromeritics Instrument Corp., Norcross, GA).

Four of the fly ashes were further evaluated after sintering. Each SiAlON powder batch was mixed with 4 wt% yttria (Nyacol Nano Technologies, Inc., Ashaland, MA) sintering additive by milling 40 g of solids with 250 g of 1 mm yttria stabilized zirconia grinding media and 150 mL of deionized water in a rapid laboratory mill for 30 minutes. Particle size of the suspension was measured with a BI XDC particle analyzer (Brookhaven Instrument Corporation, Holtsville, NY). The suspension was dried in a drying oven, held at 95 °C for 24 hours, and the dry powders were mixed with a mortar and pestle to negate the effects of settling.

Pellets were formed by pressing a mixture of 1 g of powder with 0.5 g of deionized water in a 0.5 inch diameter die at 4000 psi and then isostatically pressed at 20,000 psi. Sintering of these pellets was performed in a boron nitride powder bed enclosed inside two boron nitride crucibles, which were in a boron nitride powder bed that was enclosed graphite crucible. The pellets were sintered at 1625 °C, as measured by an optical pyrometer, and held at temperature for 2 hours. Heating and cooling rates were both 10 K/min.

The sintered pellets were polished to 1 μm using diamond abrasives. X ray diffraction and lattice refinement was performed in a similar manner as was done for the SiAlON powders. Microstructural evaluation was performed for the samples using an Environmental Scanning Electron Microscope (ESEM). Vickers hardness of two observable regions was determined from the average of the two diagonals of ten indentations, using a 200 gf load applied for 15 seconds. Since indentation fracture toughness is not widely accepted in the fracture mechanics community[12], the average crack lengths of the two radial cracks that form after indentations are reported without calculating K_{IC} specifically. Crack length measurements were made for 10 indents each at of 5 , 10 , and 20 kgf loads held for 15 seconds. Analysis of Variance (ANOVA) statistical tests were performed on the five mechanical property metrics with a confidence level of 99%.

RESULTS AND DISCUSSION

The ternary functionality diagram is given in Figure 1. The chemistry of the fly ashes, according to the UMF approach, and the target chemistry are plotted in the figure. All of the fly ash chemistries lie near a line of constant silica to alumina ratio (z = 2) on the diagram and are represented as letters. The letter labels include the contribution as iron (II) oxide, whereas a prime superscript demonstrates the shift in chemistry due to the removal of iron silicide due to inherent iron (III) oxide contents in the fly ash. The target chemistry is given in the diagram, which is just off of the single phase β SiAlON solid solution line and in a region where both α and β SiAlON phases are expected to be formed after heat treatment at 1700 °C, based on the work of Kubdya Jansen et. al.[4]

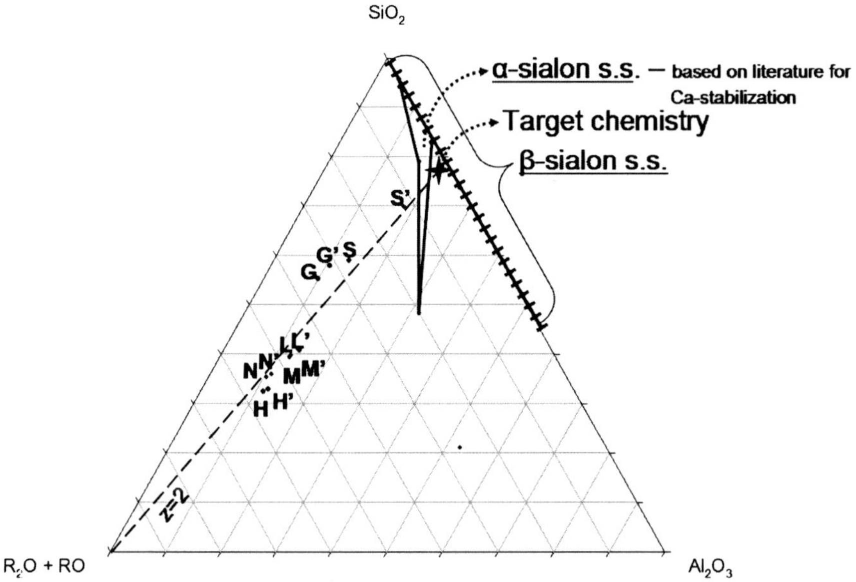

Figure 1. Ternary functionality diagram for the synthesis of SiAlON.

X ray diffraction results of the synthesized SiAlON powders are shown in Figure 2. The same letter notation as in Figure 1 is used to identify the fly ash source. The β SiAlON peaks are also identified. For five of the six SiAlON powders, β SiAlON was the only significant phase. A few unidentifiable peaks exist at the trace level and may be associated with impurity induced secondary phase formation or decomposition phases. For the SiAlON made from S fly ash, a minor phase was also detected and identified as O' SiAlON. The chemistry of the S fly ash was closest to the SiAlON formation regions. Since colloidal additives were being added, a bimodal particle size distribution is expected. With less chemistry modification, the larger particle size peak of the bimodal particle size distribution is more prevalent; therefore, the kinetics of the reaction may be slightly inhibited due to less free energy of the system, associated with the total amount of surface. The O' SiAlON phase is thought to be a transient phase in the formation of the β SiAlON phase and is a result of incomplete reaction. Earlier experiments with less powder allowed the reaction to complete with no O' SiAlON formation. No α SiAlON is present, and this may be because the diagram is constructed for behavior at 1700 °C.

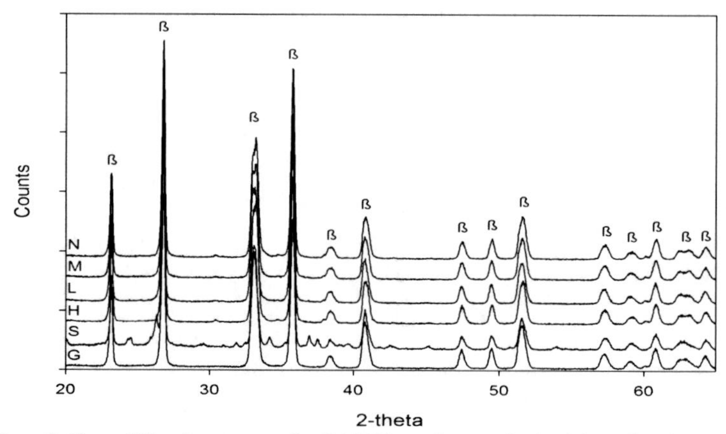

Figure 2. X ray diffraction patterns for SiAlON powders synthesized from fly ashes.

The X ray diffraction results of the sintered samples are shown in Figure 3. It is clear from the presence of minor amounts of zirconium (IV) oxide and zirconium (II) oxide that zirconia contaminate from the milling media was introduced during mixing of the SiAlON powders with yttria sintering additive. There appears to be a trace amount of mullite present and may be a decomposition phase on the surface of the sample. No α SiAlON phase is observed, but sintering was performed at 1625 °C rather than 1700 °C, which the formation regions on the ternary functionality diagram are worked out for, so a direct comparison can not be made.

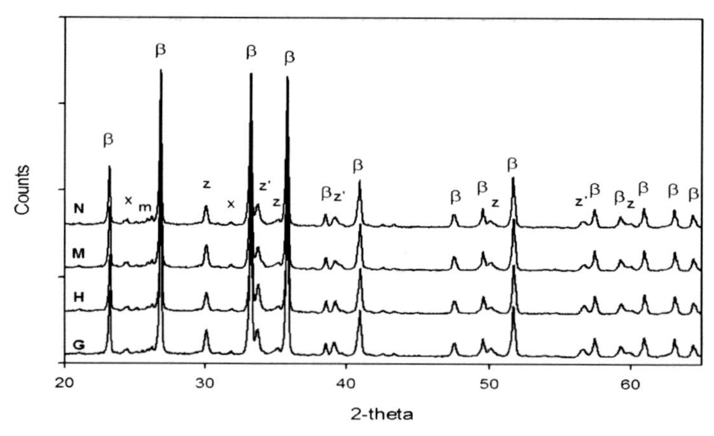

Figure 3. X ray diffraction patterns for sintered SiAlON pellets. (β= β SiAlON, z=zirconium (IV) oxide, z'=zirconium (II) oxide, m=mullite)

Comparison of the X ray diffraction data in Figure 2 and Figure 3 suggests that there is solid solution heterogeneity of the SiAlON powders. In Figure 2, the peaks are broader than in Figure 3. Localized shifts in the degree of solid solution correspond to a slight shift in peaks. These slightly shifted peaks may overlap and form one broadened peak. Possible solid solution heterogeneity of the sample is demonstrated by the fact that multiple peaks are distinguishable for widely broadened peaks such as those around 33 and 63 °2θ. After the sintering heat treatment, distinguishable peaks, such as those around 61 and 63 °2θ, are still noticeable, but much less pronounced.

During lattice refinement of the SiAlON powders, the z value determined by equation 5 was consistently lower than that predicted theoretically, whereas the z value determined by equation 6 was consistently higher. This suggests either less than ideal crystallinity or that alkali and alkali earth cations have stressed the lattice, resulting in elongation of the c axis while shortening the a axis correspondingly. This may be a precursor to α SiAlON formation, which varies in structure from β SiAlON by a c glide plane.[1] To simplify the interpretation, the average of the two z value results is reported for the synthesized powders and the sintered specimens and is given in Figure 4. The average of the reported z values for the SiAlON powders synthesized from the various fly ashes is 1.95. This is in acceptable agreement with the theoretically predicted z value of 2.0. One would expect the z value of the SiAlON powder from S fly ash to be higher, since O' SiAlON is silicon rich, which would make the β SiAlON phase aluminum rich, but this is not the case. This is not fully explainable, but perhaps the presence of a secondary phase affected the lattice refinement results. The z values of the sintered specimens were consistently lowered to an average value of 1.41, significantly less than the theoretical value. This is consistent with what is expected when sintering with yttria. Studies on the grain boundary crystallization in the yttria SiAlON system typically yield yttrium aluminum garnet.[13] This suggests that aluminum is pulled out of the SiAlON structure, therefore lowering the z value. If this is true, a trend should be observed in yttria content and the drop in z value after sintering.

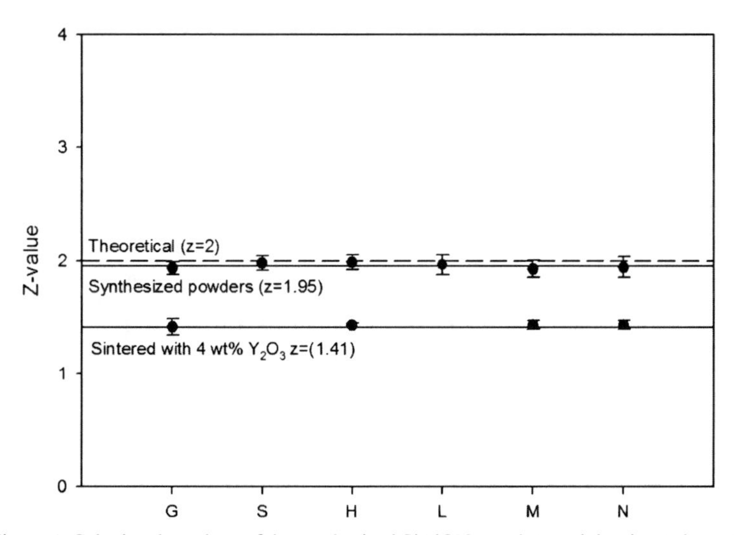

Figure 4. Calculated z values of the synthesized SiAlON powders and the sintered samples starting from different fly ashes.

The densities, surface areas, and particle size information for the SiAlON powders are given in Table I. The theoretical density of SiAlON with z = 2 is 3.12 g/cc. The density of all of the samples is lower than this. The five SiAlON powders with no significant identifiable secondary phase formation had densities of approximately 3.04 g/cc. The low densities are further evidence suggesting less than ideal crystallinity. The SiAlON powder from Seminole fly ash had a density of only 2.95 g/cc, and is likely to be lower than the rest because of the formation of a secondary phase. The significantly lower surface area and larger particle size also supports the idea that the secondary phase is a result of the fact that kinetics for this sample were hindered.

Table I. Density, Surface Area, and Particle Size of the SiAlON Powders.

Fly Ash Source	Density (g/cc)	Surface Area (m^2/g)	Particle Size (μm)		
			d99	d50	Avg.(St.dev.)
G	3.04	12.4	1.218	0.302	0.437(0.257)
S	2.95	4.2	2.310	1.659	1.691(0.213)
H	3.04	12.4	1.517	0.273	0.536(0.379)
L	3.04	14.4	0.948	0.299	0.359(0.167)
M	3.05	10.1	1.145	0.487	0.536(0.265)
N	3.04	9.2	1.155	0.270	0.418(0.281)

The microstructure of the four sintered samples is similar. Figure 5 is an example of a representative microstructure. There are three distinct regions containing the β SiAlON phase. There are large, dark needles of β SiAlON with large aspect ratios. The two other regions have micrometer sized β SiAlON grains with low aspect ratios mixed with another phase, one of which makes up the bulk of the sample and the other forming nodular clusters in the microstructure. There are also zirconium (IV) oxide grains (the brightest phase) dispersed throughout the microsctructure as well as micro porosity. Figure 6 shows a closer look at the microstructure and Energy Dispersive Spectroscopy (EDS) maps for aluminum, silicon, and zirconium. The EDS maps suggest that the bulk mixture region containing β SiAlON is mixed with the liquid phase that was introduced by the addition of the yttria sintering aid. It is apparent that the bulk β SiAlON mixed region is depleted in aluminum content, which is in agreement with the suggestion that the liquid phase, containing yttria, pulls aluminum from the β SiAlON structure, as was determined by the decrease in z value. The EDS maps also indicate that the nodular cluster region containing β SiAlON grains are depleted in silicon. This is not completely understood, but believed to be associated with the effects of zirconium oxides in the structure. Zirconium (IV) oxide appears as individual grains or very small clusters that are not mixed with other phases. Zirconium (II) oxide, which has been reduced during sintering, is present in the larger, nodular clusters with β SiAlON grains, where the depletion of silica occurs.

The hardness was measured in two noticeably distinguishable regions in the microstructure: dense regions and regions containing significant amount of micro porosity. The results are plotted in Figure 7. It was determined that the data points were normally distributed, satisfying the assumptions for ANOVA. Similarly, crack length measurements after indentations at loads of 5 , 10 , and 20 kgf loads are plotted in Figure 8. ANOVA was performed for these data sets as well.

The results from ANOVA are shown in Table II. In all comparisons of the mechanical property metrics, the F statistic is lower than the critical F statistic; therefore, the null hypothesis is accepted (i.e. the specimens were not statistically different with 99% confidence). The average hardness values among the samples for dense regions and micro porous regions were determined to be 12.9 and 8.3 GPa respectively. The average crack lengths after indentations at 5 , 10 , and 20 kgf loads were determined to be 224, 394, and 663 μm respectively.

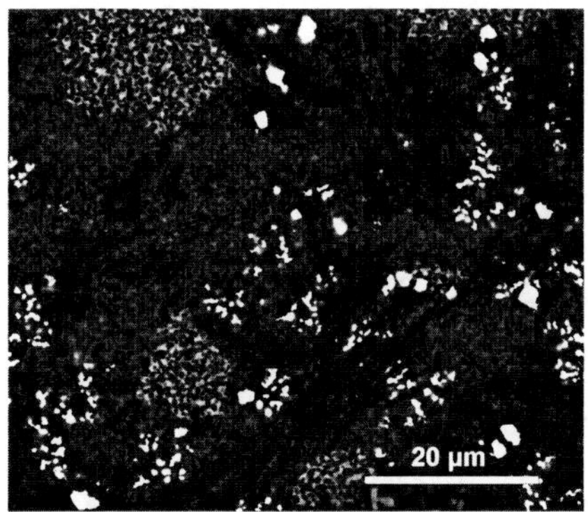

Figure 5. ESEM image, taken at an initial 2,000x magnification, of a sintered SiAlON specimen made from SiAlON that was synthesized from H fly ash.

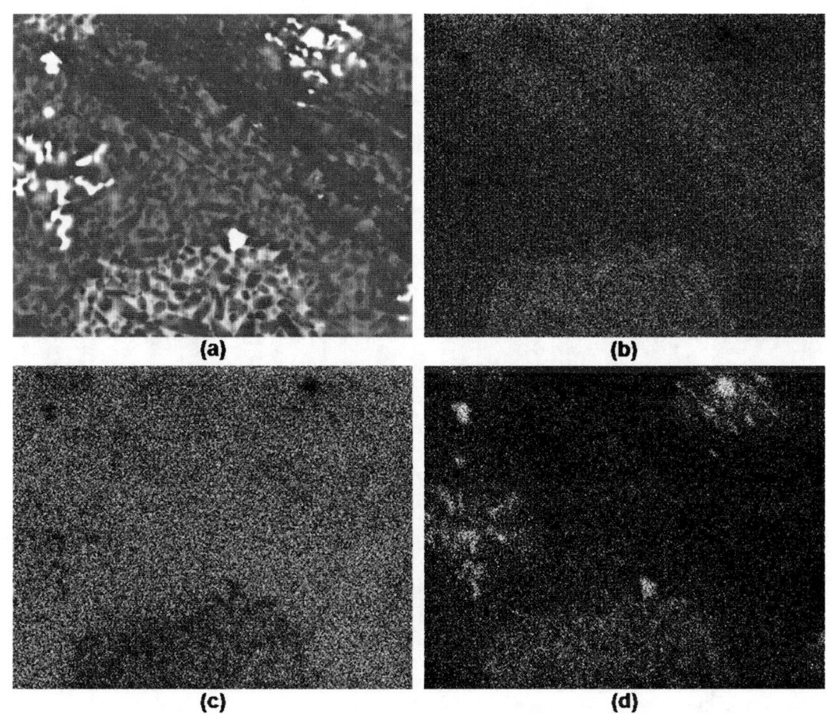

Figure 6. (a) ESEM image of a sintered SiAlON specimen made from N fly ash for comparing Energy Dispersive Spectroscopy (EDS) maps of (b) aluminum, (c) silicon, and (d) zirconium.

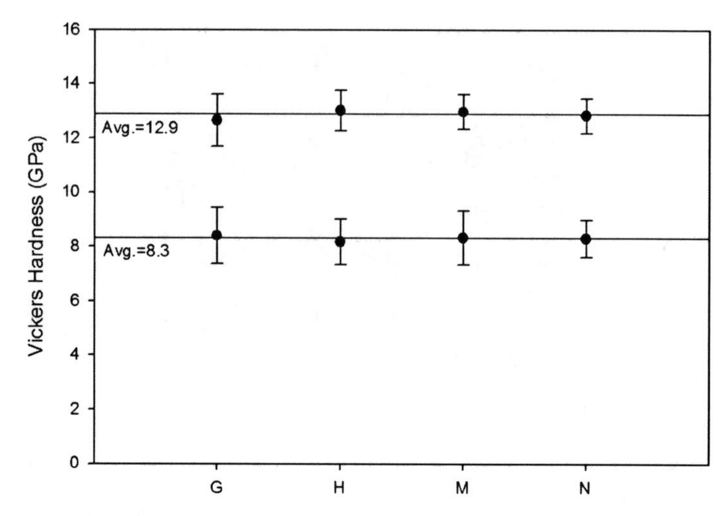

Figure 7. Vickers hardness for sintered specimens for SiAlONs made from different fly ashes. Indents were taken in a hard (dense) region and a soft (micro porous) region.

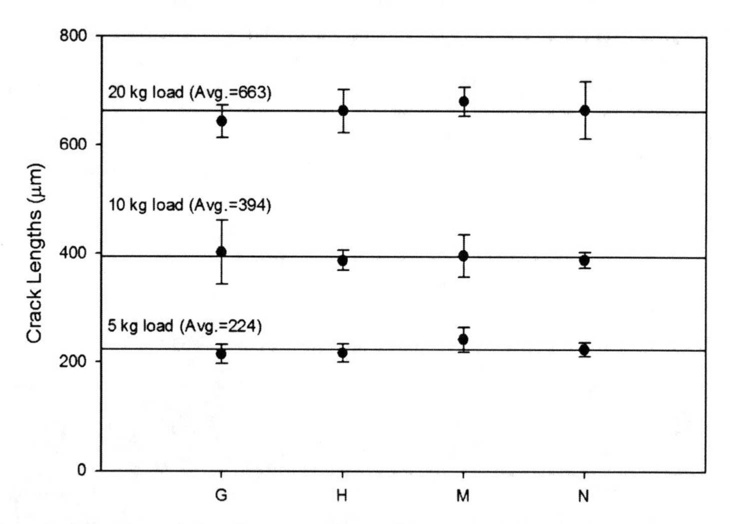

Figure 8. Average crack length measurements of the two radial cracks formed after Vickers indentations at various loads.

Table II. Analysis of Variance for the Mechanical Property Metrics

Mechanical Property Metric	Feature of Test	F-statistic	$F_{critical}$	Null Hypothesis (μ_G μ_H μ_M μ_N)
Hardness	Dense region	0.24	4.51	Accepted
Hardness	Micro-porous regions	0.50	4.42	Accepted
Crack Length	5-kgf load	3.05	4.38	Accepted
Crack Length	10-kgf load	0.36	4.38	Accepted
Crack Length	15-kgf load	1.56	4.40	Accepted

CONCLUSION

A ternary functionality diagram has been constructed that provides a systematic approach for the synthesis of SiAlON powders from fly ash by carbothermal reduction and nitridation. Six fly ashes were tested using the diagram, and the experimental z value for the resulting β SiAlON powders matched theoretical predictions well. No α SiAlON phase was detected, even though the functionality diagram suggests chemistry within a region that it should form. However, the formation regions in the ternary functionality diagram were experimentally determined for a temperature of 1700 °C with no reference to dwell time, so a direct comparison can not be made, even after sintering at 1625 °C. The resulting β SiAlON powders and compacts have an elongated c axis while having a compressed a axis, which may be caused by the alkali and alkali earth cations and could be a precursor to α SiAlON formation, by generating a stress capable of forming the c glide plane in the α SiAlON structure. The synthesized powders also have solid solution heterogeneity, which becomes less apparent after sintering, though not eliminated.

After sintering, the z values of the specimens remained consistent. In addition, ANOVA of the mechanical property metrics were also consistent. Qualitatively, the microstructures are also consistent. The results indicate that the use of a ternary functionality diagram can be used to simplify interpretation of fly ash chemistry and demonstrates reproducibility and predictability of the resulting SiAlON powders synthesized by the CRN process. Reproducibility of the sintered specimens was demonstrated, but predictability could be improved since there was a drop in z value, which is believed to be associated with the amount of yttria sintering aid used in the system. Reproducibility and predictability were demonstrated despite the fact that the fly ash sources and characteristics were different. Therefore, the ternary functionality diagram has been proven to be acceptable for further investigations into the design of SiAlON ceramics from fly ash or other complex, mixed oxide sources. This may help reduce the cost and open up new markets for the useful application of the advanced structural ceramic.

FOOTNOTES

* Performed by ACME Analytical Laboratory, LTD

REFERENCES

[1]F.L. Riley, "Silicon Nitride and Related Materials," *J. Am. Ceram. Soc.*, **83** [2] 245 65 (2000).

[2]EPA, "Using Coal Ash in Highway Construction: A Guide to Benefits and Impacts" (2005) Environmental Protection Agency, Department of Energy, Federal Highway Administration, The American Coal Ash Association, and The Utility Solid Waste Activies Group. Accessed on: March, 2008. Available at <http://www.epa.gov/C2P2/pubs/greenbk508.pdf>.

[3]A.S. Wagh, D. Singh, and W. Subhan, "Cement Based Materials: Present, Future, and Evironmental Aspects," *Ceram. Trans.*, **37**, 139 52 (1993).

[4]A.A. Kudyba Jansen, H.T. Hintzen, and R. Metselaar, "Ca α/β Sialon Ceramics Synthesised from Fly Ash Preparation, Characterization and Properties," *Mater. Res. Bull.*, **36** [7 8] 1215 30 (2001).

[5]Q. Qiu, V. Hlavacek, and S. Prochazka, "Carbonitridation of Fly Ash. I. Synthesis of SiAlON Based Materials," *Ind. Eng. Chem. Res.*, **44** [8] 2469 76 (2005).
[6]Q. Qiu and V. Hlavacek, "Carbonitridation of Fly Ash. II. Effect of Decomposable Additives and Whisker Formation," *Ind. Eng. Chem. Res.*, **44** [8] 2477 83 (2005).
[7]Q. Qiu and V. Hlavacek, "Carbonitridation of Fly Ash. III. Effect of Indecomposable Additives," *Ind. Eng. Chem. Res.*, **44** [19] 7352 8 (2005).
[8]S. Ueno, H. Kita, and T. Ohji, "Role of Mn and Co additives on the Carbothermal Nitridation of Fly Ash," *J. Ceram. Process. Res.*, **6** [4] 290 3 (2005). Also available at <http://jcpr.kbs lab.co.kr/english/ch0301.htm>.
[9]M. Kamiya, R. Sasai, and H Itoh, "Preparation of SiAlON Based Materials from Coal Fly Ash Using Carbothermal Reduction and Nitridation Method," *Ceram. Trans.*, **193**, 1 8 (2006).
[10]W.M. Carty, "Unity Molecular Formula Approach to Glaze Development", *Ceram. Eng. Sci. Proceedings.*, **21** [2] 95 107 (2000).
[11]F.K. Van Dijen, "The Carbothermal Production of $Si_3Al_3O_3N_5$ from Kaolin, Its Sintering and its Properties"; Ph.D. Thesis. Eindhoven University of Technology, Eindhoven, The Netherlands, 1986.
[12]G.D. Quinn and R.C. Bradt, "On the Vickers Indentation Fracture Toughness Test," *J. Am. Ceram. Soc.*, **90** [3] 673 80 (2007).
[13]T. Ekström, "SiAlON Cermics Sintered with Yttria and Rare Earth Oxides," pp. 121 32 in *Silicon Nitride Ceramics: Scientific and Technological Advances*, Materials Research Society Symposium Proceedings, Vol. 287. Edited by I. W. Chen, P.F. Becher, M. Mitomo, G. Petzoq, and T. S. Yen. Materials Research Society, Pittsburgh, PA, 1993.

PATTERNING OF CLOSED PORES UTILIZING THE SUPERPLASTICALLY FOAMING METHOD

A. Kishimoto, Y. Nishino and H. Hayashi

Graduate School of Natural Science and Technology, Okayama University
Division of Chemistry and Biochemistry, Graduate School of Natural Science and Technology, Okayama University, 3 1 1 Tsushima naka, Kita ku, Okayama 700 8530, Japan

ABSTRACT

We have already innovated superplastically foaming method in which pore expands after densification of the matrix utilizing the superplastic deformation. Based on this superplastically foaming method using 3YSZ (3 mol% yttria stabilized zirconia) and α SiC as matrix and foam agent, respectively, we have successfully fabricated several dots closed pore pattern and *C* shaped as well as *S* shaped closed tube patterns by optimizing the amount of matrix 3YSZ. In addition to the shape of the closed pore, the protuberance based on the expansion of the pore changed with the location of the foam agent in the matrix. The shape of the closed pores and the interaction between them can be controlled by the concentration of the foam agent and the amount of the matrix.

INTRODUCTION

Porous ceramics incorporates pores to improve several properties including thermal insulation maintaining inherenet ceramic properties such as corrosion resistance and large mechanical strength. It is widely used as thermal insulator, filtration membrance and catalysis support, so on[1 2)]. Conventional porous ceramics is usually fabricated through an insufficient sintering[1)]. Since the sintering accompanies the exclusion of pores [3)], it must be terminated at the early stage to maintain the high porosity, leading to degraded strength and durability. Especially, so called ceramic foams are fabricated also through the insufficient sintering of green powder network derived from solution precursor containing air bubbles. As a result high level of porosity would be expected while the inter grain bonding results in extremely small. Contrary to this, we have innovated superplastically foaming method to make ceramic foam only in the solid state[4 12)]. In this method, the previously inserted foam agent evaporates after the full densification of matrix at around the sintering temperature. Closed pores expand utilizing the superplastic

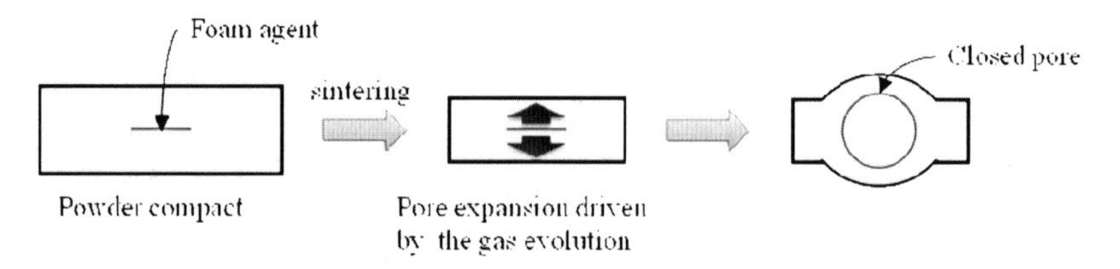

Figure 1. Schematic illustration of forming a closed pore through the superplastisically foaming method

deformation driven by the evolved gas pressure. In other word the pores are introduced into full densified matrix. The superplasticity is a phenomenon exhibiting large strain in polycrystalline materials far below the melting point [13)14)]. The main mechanism is said to be "grain boundary sliding" or "atomic rearrangement", however the detailed mechanism has not been settled yet.

Figure 1 illustrates the schematic procedure of the formation of pores in the superplastically foaming method. When using silicon carbide as the foam agent, the evolved gasses are thought to be SiO and CO formed during the active oxidation of SiC as follows,

$$SiC(s)+O_2(g)\rightarrow SiO(g)+CO(g)$$

The typical features of this superplastically foaming method are listed as follows,

1. The pores are introduced after sintering the solid polycrystal.
2. Only closed pores are introduced, improving the insulation of gas and sound in addition to heat.
3. The pore walls are fully densified expecting a large mechanical strength.
4. Compared with the melt foaming method, the superplastically foaming method is practical because the fabrication temperature (around the sintering temperature) is far below the melting point and it does not need mold.
5. The size and the location pores can be controlled by the amount and position of the foam agent [5)].

In the present study we focused on the last feature to make several patterned closed pores utilizing 3 mol% yttria stabilized zirconia (3YSZ) and a silicon carbide (α SiC) as matrix and foam agent, respectively. First, *C* character, *S* character and dots shaped controlled closed pores have been fabricated. Next, the location dependence of pore figure was examined in the single dot closed pore. Finally, effect of the amount of the matrix and the concentration of foam agent on the relationship between the neighboring pores were examined.

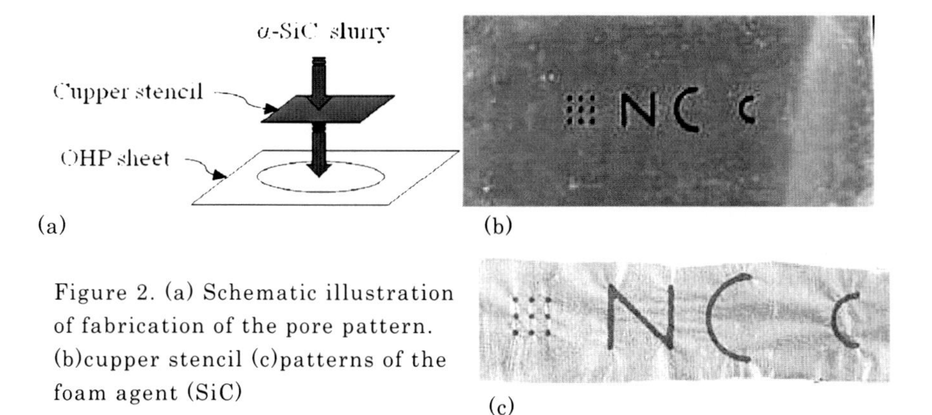

Figure 2. (a) Schematic illustration of fabrication of the pore pattern. (b)cupper stencil (c)patterns of the foam agent (SiC)

EXPERIMENTAL

Patterning of foam agent

In order to form patterns of foam agent α SiC on a compressed 3YSZ matrix powder, dots or line drawing of foam agent were first fabricated on 3YSZ thick film previously prepared by a tape casting. To make dots or line drawing reproducibly, cupper sheet with hole through of desired pattern was used as a stencil (Fig.2(b)), to make slurry patterns dispersed with foam agent.

A 1 wt. % methylcellulose aqueous solution was prepared to make slurry dispersed with matrix or foam agent powder. The latter was prepared by mixing α SiC powder (OY 15, Yakushima denko, Japan) into the methylcellulose aqueous solution with several predetermined ratio[15]. From the matrix slurry with solution/powder mixing ratio of 6 g, 0.12 g, 3YSZ (TZ3P, Toso, Japan) powder based green sheet was prepared by spreading it followed by drying. After fixing the 3YSZ sheet on a substrate polyethylene sheet the cupper stencil was put on it as shown in Fig. 2(b). Then the α SiC slurry was spreaded with a glass bar to make slurry patterns after the hole through (Fig.2(c)). After dried, α SiC dots or line patterns were formed on the matrix 3YSZ green sheet.

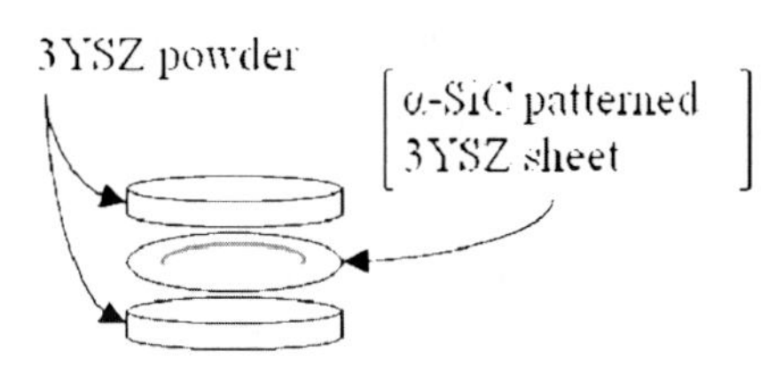

Figure 3. Schematic illustration of uniaxial compression. Loading of matrix powder and patterned sheet (SiC pattern on 3YSZ sheet).

Fabrication of sintered body

The resultant α SiC patterned matrix green sheet was inserted into matrix 3YSZ powder in a steel die (Fig.3) followed by uniaxial pressing under 14 MPa for 1 min.[3]. In the present study, the patterned sheet was inserted in the middle of the matrix powder compact, the amount of the matrix was represented by half of the weight of total matrix powder.

After decomposing the methylcellurose binder at 500°C for 10 min., CIP treatment was conducted again under 200 MPa for 1 min. The resultant powder compact containing α SiC pattern was heat treated at 1600°C for 8 h to sinter following pore expansion.

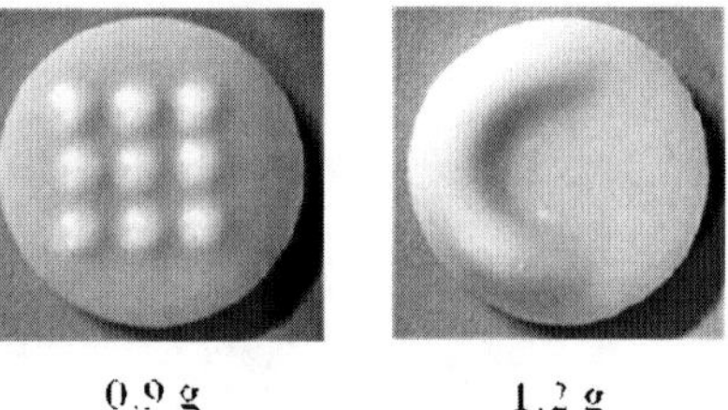

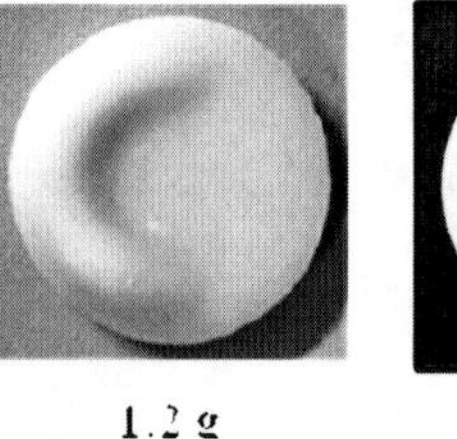

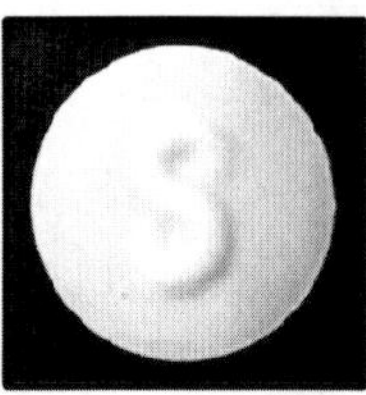

Figure 4. Clearly patterned protuberances through the superplastically foaming method.

RESULTS AND DISCUSSION

First, fabrication condition to make clear protuberance after patterned closed pores was optimized by changing the fabrication parameters such as matrix amount. As a result, three kinds of pore/protuberance patterns have been successfully fabricated with different amount of matrix shown in Fig.4. According to these results, complex shaped closed pore or tubular path could be fabricated by changing the patterns of foam agent, leading to a high temperature reaction field made of ceramics.

Then the necessity of CIP as well as binder decomposition process was examined to make clear and discrete dot patterns. As mentioned in the experimental section, preparation of dot pore patterns includes fabrication of patterned foam agent, powder compaction, binder decomposition, CIP treatment followed by sinter/foaming. Influence of the omission either of binder decomposition or CIP treatment on the figure of pore patterns was examined. As shown in Fig.5, both decomposition omission and CIP omission lead to the pore connection making a united large pore not discrete patterned pores. This result indicates that both binder decomposition and CIP treatment are necessary to make clear discrete pore patterns. The space left on decomposition of the methylcellulose would become void where the foaming gas was easy to be flown, leading to the pore connection.

Next we focused on the single

Figure 5. Connected pores accompanied by the omission of processes.

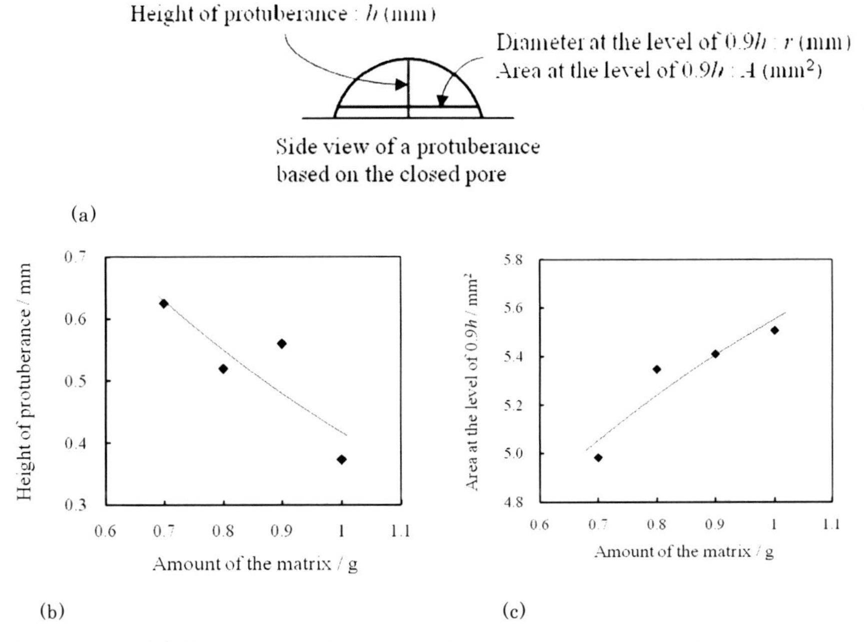

Figure 6. (a)Side view of a protuberance based on the closed pore. (b)Relation between the height of protuberance and the amount of matrix. (c)Relation between the area of protuberance and the amount of matrix.

dot pore, protuberance figure dependence on the location of foam agent was evaluated. In the present study, patterning of foam agent was conducted using the cupper stencil, then the amount of the α SiC as well as the quality of the evolved gas were assumed to be constant in a single pore.

Then the matrix amount dependence of foam height h (mm), radius r (mm) of $0.9h$ from the top, and the correspondence area A (mm^2) of the protuberance were measured and calculated. From the h dependence of the matrix amount as shown in Fig.6(b), the foam height decreased with the amount of the matrix.

Also from the area dependence of the amount of matrix shown in Fig.6(c), the area increases with the amount of matrix, indicating a gradual increase of trail with a decreased height.

It can be concluded that the figure of the protuberance varies with the location of the foam agent even with the same amount. In detail, height as well as the slope of the protuberance decreases with the distance of the foam agent from the surface.

Above examination dealt with the protuberance according to the pores, then the figure of the pore itself was examined from the cross section of the pores by cutting the plane crossing the top. Since the shape of the pores cut across the top was estimated as oval, longitudinal axis vertical to the pellet plane as well as the horizontal axis parallel to pellet were measured as shown in Fig.7(a). The longitudinal /horizontal axis ratio was plotted against the amount of matrix as Fig.7(b).

With increasing the amount of matrix, the figure of the pore changes from tall to broad, revealing the change of the pore figure in addition to the figure change of protubarance.

From these results, change in the figure of the closed pores as well as the protuberance was schematically illustrated in Fig.8. When the foam agent is inserted near the matrix surface, the resultant

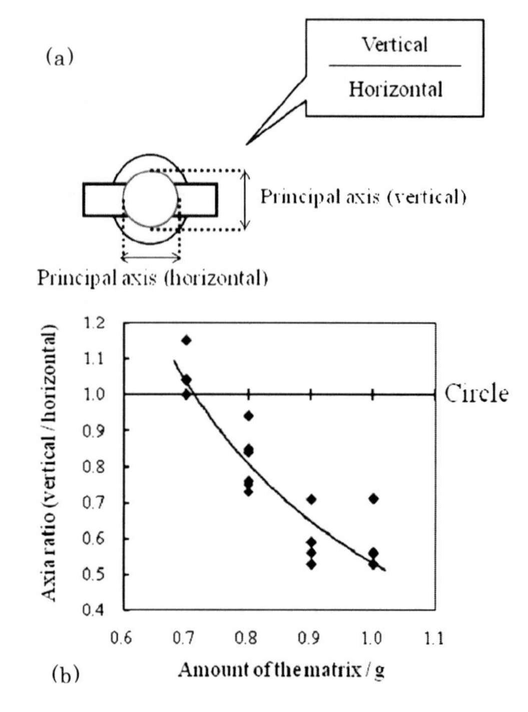

Figure 7. (a)Schematic illustration of cross section of the closed pore, principal axis (vertical) and principal axis (horizontal). (b)Relation between the axis ratio and the amount of the matrix in formed closed pore.

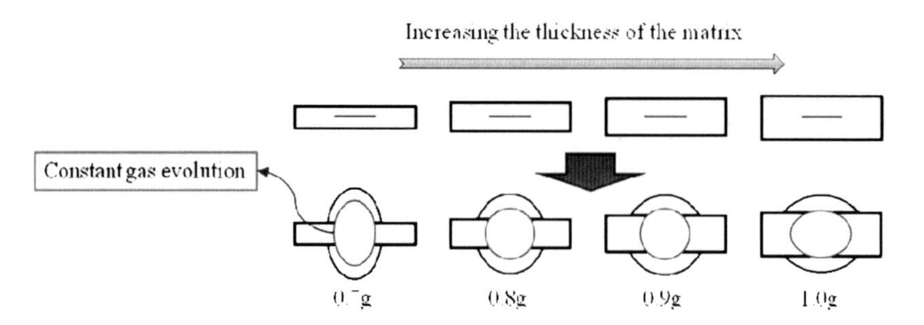

Figure 8. Change in the figure of superplastically foamed closed pores accompanied by the thickness of the matrix.

protuberance after the pore is high and clear. From the foam agent far from the matrix surface, low and gentle sloped protuberance is formed. The shape of the pore itself also changes from tall to wide when the foam agent departs from the matrix surface.

In the present study, disk like foam agent was introduced into matrix pellet. Then the distance from the foam agent to the matrix surface was different between horizontal and vertical. As a result, only longitudinal pore expansion displacement is thought to become small with increasing the amount of matrix in this direction while the pore expansion pressure should be eqidirectional.

Next, in the case that some interactions between the neighboring dot pores are assumed to occur, change in the figure of pores depending on the amount of matrix and concentration of foam agent were examined. Cross sections were observed on patterned dots fabricated by changing the amount of matrix and concentration of foam agent. Resultant pore figures can be classified into three groups and illustrated in Fig.9(a) against the two parameters examined.

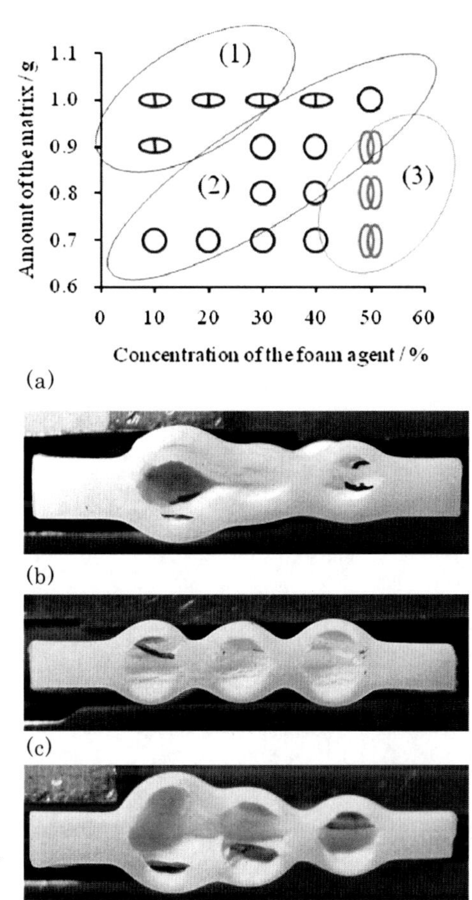

Figure 9. (a) Mutual relationship between adjacent pores with various amount of the matrix and concentration of the foam agent. Cross sectional view belong to group (1) (b), group (2) (c) , and group (3)

Features of these pore connection types, (1), (2) and (3) can be described as follows.

The group (1) shown in Fig.9(b) is characterized by the connected pores with preferentially horizontal growth. This is probably ascribed to the excess amount of matrix compared with the evolved gas. In such case, vertical pore growth would be supressed to make preferential pore growth paralell to the pellet surface, resulting in pore connection with relatively small channel.

In another group (2), cross sectional figure of the pores are almost independent circle as shown in Fig.9(c). In this case, the gas evolution and the amount of the matrix are well balanced to give a spherical pore. Finally, in the group (3), the resultant pores are well grown especially to the vertical direction probably due to the large gas evolution against relatively small amount of matrix. The growth of pores are large in the vertical direction, however, such vertical growth would drag the horizontal pore growth leading to pore connection with relatively large channel compared with those of the group (1).

Depending on this chart, we can maintain the figure of formed pore by changing the concentration of foam agent to accord with the amount of matrix. For example, Fig.9(a) shows that the pore figure is approximately circle with matrix amount of 0.7 g using 10% of α SiC foam agent. When only the matrix is increased from this condition, pore growth becomes anisotropic or horizontal oriented resulting in pore connection. Even with this amount of matrix, the pore figure returns to circle with larger concentration of foam agent facilitating the vertical pore growth.

By collecting such empirical data, quantitative relationship between concentration of foam agent and the figure of pores in a given matrix would be obtained also taking into acount of the amount of evolved gas, inner pressure change and the amount of the superplastic deformation.

CONCLUSIONS

Three mol percent of yttria stabilized zirconia (3YSZ) was used as foam matrix and α SiC as foam agent. Two kinds of powder slurries based on methyl cellulose aqueous solution were first prepared. Using the 3YSZ slurry, green sheet was fabricated using a glass bar. After dried, patterning of SiC slurry was conducted on the 3YSZ sheet through a copper stencil. Three patterned sheets were cut and buried in a 3YSZ powder compacts. Through the heat treatment at the sintering temperature, a ceramic foam containing several isolated closed pores derived from the SiC pattern was successfully fabricated. In this procedure binder decomposition followed by CIP treatment is necessary to make clear patterns. In addition to the shape of the closed pore, the protuberance based on the expansion of the pore changed with the location of the foam agent in the matrix. The shape of the closed pores and the interaction between them can be controlled by the concentration of the foam agent and the amount of the matrix.

REFERENCES

[1]Susumu Mizuta, Kunihito Koumoto, Ceramic Zairyou, Tokyo Daigaku Shuppann sha 211(1986)

2Masatsugu Satou, Advanced Ceramics Binnrann, Ohumu sha 334(1992)

[3]Shingo Iizumi, Ceramics no Kagaku, Maruzenn Kabushiki Gaisha 127(1982)

[4]A. Kishimoto, T. Higashiwada, H. Asaoka and H. Hayashi,The exploitation of superplasticity in the successful foaming of ceramics following sintering, *Adv. Eng. Mater.*, **8(8)** 708 711(2006).

[5]A. Kishimoto, T. Higashiwada, M. Takahara and H. Hayashi, Solid state foaming and free forming of closed pore utilizing the superplasticity of zirconia ceramics, *Mater. Sci. Forum* **544-545**641 644(2007).

[6]T. Higashiwada, H. Asaoka, H. Hayashi and A. Kishimoto, Effect of additives on the pore evolution of zirconia based ceramic foams after sintering, *J. Eur. Ceram. Soc.* **27** 2217 2222(2007).

[7]A. Kishimoto, M. Obata and H. Hayashi, Fabrication of alumina based ceramic foams utilizing superplasticity, *J. Eur. Ceram. Soc.* ,**27**, 41 45(2007).

[8]Y. Hashida, H. Hayashi, and A. Kishimoto, Fabrication of Solid State Foams Based on Full Stabilized Zirconia Ceramics Facilitating the Superplasticity with Dispersoids, *J. Jpn. Soc. Powder Powder Metallurgy*, **54,** 732 737(2008).

[9]M. Hanao, H. Hayashi, and A. Kishimoto, The mechanical and thermal properties of porous zirconia ceramics fabricated through a solid state foaming method, *J. Jpn. Soc. Powder Powder Metallurgy*, **55,** 732 737(2008).

[10]A. Kishimoto, M. Obata and H. Hayashi,Comparison of magnesia and magunesium alumina spinel on the alumina based superprastisity foam, *J. Alloys Compd.*, **471**, 32 35(2009)

[11]A. Kishimoto, M. Hanao, and H. Hayashi,Improvement in the specific strength by arranging closed pores in fully densified zirconia ceramics, *Adv. Eng. Mater.*, **11**, 96 100(2009).

[12]A. Kishimoto, M. Obata, K. Waku and H. Hayashi, Mechanical and electrical properties of superplastically foamed titania based ceramics, *Ceram. Intern.*, **35**, 1441 1445 (2009)

[13]Masatsugu Satou, Ceramic Senntannzairyou, Ohmu Sha 178(1991).

[14]Satoru Uchida: SiC kei Ceramic Shinnzairyou, Uchidaroukaku ho 90 (2001)

15Toshitake Sennno, Sennshinn Ceramics no Tukurikata to Tukaikata, Nikkann Kougyou Shinnbunn Sha 169 (2005).

[16]Toshitake Sennno, Sennshinn Ceramics no Tukurikata to Tukaikata, Nikkann Kougyou Shinnbunn Sha 119(2005).

THE RESEARCH OF MATERIALS LIFE CYCLE ASSESSMENT

ZuoRen Nie*, Feng Gao, XianZheng Gong, ZhiHong Wang, and TieYong Zuo
College of Materials Science and Engineering
Beijing University of Technology, Beijing, 100124

ABSTRACT

For the production, manufacture, application, and disposal of materials, numerous energy and resources are consumed, and environment deterioration occurs as a result. Life Cycle Assessment (LCA) is a technique for systematically analyzing a target from cradle to grave. It is an effective tool that not only gives a detailed information of environmental profiles of a material or a product, more importantly, the value of life cycle thinking lies in its ability to provide the decision making basis for sustainable development, making the products, industry and even the whole industry chain act more in line with the principles of sustainable development. However, life cycle assessment method itself has yet to be further improved. In the aspect of the actual application, it is required to combine with the specific conditions, such as the national level of industrial technology and the phases of economic development, before giving a full play of the realistic role that life cycle assessment can provide decision making basis for sustainable development. Some new progress, such as data quality, methodology localization, database and case study, related to materials life cycle assessment research and development in China are introduced in this article.

1 INTRODUCTION

For the resource extraction, manufacture, application, and disposal of materials, numerous energy and resources are consumed, and environment deterioration occurs with the inputs and outputs of various raw materials, energy, by products and wastes. In order to estimate the environmental impact on these processes, it is necessary to research the overall life cycle of materials or products. The usual analysis methods mainly include life cycle assessment (LCA) and materials flow analysis (MFA). LCA is extended to many aspects of production and consumption, including eco design of products, cleaner production, environment label, green purchase, resource management, wastes management and environment strategy, etc[1].

With the rapid economic development of China, the confliction between economic development and environment protection is more and more severe. It is important way to improve resources and energy efficiency and reduce pollutants emission of China's materials industry that using LCA as technical and decision making support to gain the target of energy saving and emission reducing. There are much progress made in China LCA research in recent decade, including the development of LCA methodology, basic database and software, and the establishment of environmental certification standard of typical materials under the support of the National High Tech. R&D (863) Program, the National Basic Research Development (973) Program, the National Key Technology Research and Development Program and the National Natural Science Foundation of China[2,3].

* Corresponding author (email: zrnie@bjut.edu.cn)

2 RESEARCHES ON LCA METHODOLOGY

LCA is a tool for estimating and assessing the potential environmental impacts attributable to the life cycle of a product, including raw material extraction, processing, manufacture, application, recycle and disposal. Data quality and the option of life cycle impact assessment (LCIA) methods are more concerned. The high quality data is an important premise to carry out life cycle assessment and the data reliability directly influences the capability of the results and its application. LCIA is the stage that has more differences and need to be further developed.

2.1 Data quality analysis

The data collection and calculation involved with the life cycle processes is time consumed stage. With the request of the reliability of LCA results, the definition and evaluation of data quality have already attracted much attention. In recent years, there are some international organizations devoted to the study of the data quality. SETAC put forward a qualitative evaluation framework of LCI data quality, which recommended uncertainty analysis and sensitivity analysis, as the important parts of impact assessment, to evaluate the environment impact resulting from the variety of data[4]. The revision of the ISO series 14040 proposed that the character of data should meet with the goal and scope of study, and date quality should be assessed by qualitative and quantitative methods, and data collection and combination methods, so as to correctly represent reliability of results[5].

The researches of LCI data quality method mainly focus on two aspects:

(1) To adopt data quality indicator from representative data, such as regional and temporal data, or data collection methods and so on.

(2) To adopt the uncertainty to represent the integrated data quality and analyze the dataset uncertainty related to the process to denote the uncertainty of LCI results

There are a number of methods to analyze the uncertainty in a calculation, including Gaussian error propagation formulas, Monte Carlo Simulation, stochastic simulation based on probability distribution, interval algorithm and fuzzy logic approaches[6 10] and so on.

Up to the present, appropriate methods and ideas for the data quality analysis of LCA is still not available although several methods for evaluating the consistency, continuity, sensitivity and uncertainty of the inventory data were proposed by LCA practitioners[11,12]. The primary source of LCI study in China is public statistical data of which uncertainty is difficult to be identified and quantified. Thus, the accumulation of field monitoring data is especially important for LCI data collectors to perform the uncertainty analysis.

In our study, the data quality indicators were calculated according to the grade matrix model and the expected value method. Base on the integrated methods, the detailed steps of the transformation from determinated LCA model to stochastic LCA model were illustrated and an LCI stochastic model was also established for an exemplification of eco cement production[13].

In view of the data deficiency of LCA research in China, the missing data were predicted and imputed logically based on the information known in life cycle inventory by using three methods: complete case analysis, linear regression analysis and Markov Chain Monte Carlo (MCMC) method. Moreover, an analysis on the advantage, disadvantage and scope of application of these three methods was performed. A data quality analysis system reducing the interference from the missing data was set up [14].

2.2 Characterization of abiotic resource depletion

The issues of development, utilization and depletion of mineral resources have always been an important component in life cycle assessment (LCA) system, and have always been given extensive attention. The current level of recognition for the mineral resource depletion issue, however, is still far lower than that for the issues, such as the greenhouse effect, acidification effects and so on, that arise in the process of developing and utilizing mineral resources[15]. There still exist some differences of opinion in the LCA study as to how to scientifically understand and assess the mineral resource depletion, primarily including:

(1) The recognition and understanding of the essence of resource depletion;

(2) The temporal and spatial assessment criteria for the mineral resource depletion;

(3) Socio economic issues related to the consumption of mineral resources;

(4) The method of determining the characteristic indicators and weights of the depletion of mineral resources in the LCA.

There also exists much controversy about the research on the characterization methods of the abiotic resource depletion impacts, and the focus of this controversy is largely centered on a number of fields, such as the determination of resources function parameters, the rationality of choosing the characteristic factors of resource depletion, as well as the impacts caused by resource extraction, substitution and recycling technology on resource depletion[16 18].

At present, there are mainly two commonly used characteristic models of resource depletion:

(1) Use the ratio of resources extraction volumes to reserves to measure the level of the abiotic resources depletion. These methods use a number of characteristic factors, such as $1/R$, U/R and U/R^2, among them R represents the reserves of a certain resource while U denotes the current volumes of use or extraction of this kind of resource. The CML method developed by the group of Leiden University in Netherlands, reflects this view[19,20].

(2) Use the expected results generated by resource exploitation as a basis for characterization. This point of view suggests that mankind's current extraction of high grade resources will cause more serious environmental and economic impacts when exploiting low grade resources in the future. Such kind of views are represented by the Eco indicator 99 method[21], which uses the energy demand required for exploiting low grade resources as the damage factor to measure resource depletion, and which believes that this kind of "additional energy" is able to interlink the functionality with technical development of the abiotic resources, rather than directly relying on estimates of hardly predictable resources reserves and annual consumption volumes in the future.

These studies provide a wide range of options for the assessment of mineral resource depletion. But it is an important issue facing the localization of LCA in China as to how to choose the assessment method and characterization factors that are appropriate to Chinese resource situation. In order to solve this problem, we chose the characterization model of resource depletion given by Eco indicator 99 and CML method, combined with China's characteristics of resources and statistical data, so as to modify important parameters involved in both models, and thus obtained through calculation China's characterization factor set of mineral resource depletion as well as normalization factor of resource depletion in 2004.

The comparison with the original method highlights the fact that geographical distribution differences of resources are unavoidable in the LCA study. Moreover, the Eco indicator 99 method uses a large number of theoretical assumptions, and the calculation of parameters requires the support of a large amount of data, in particular the need for continuous statistical data of ore extraction volumes and ore grade in a longer period of time. According to the current actual statistical situation of China's mineral resources, the statistical data of ore grade of the majority of non metallic mineral resources cannot be obtained, thereby limiting the general applicability of this model. Furthermore, the calculation process is relatively complicated, and has higher requirements for data quality, thus affecting the operation of the model. We, through case studies, comparatively illustrated the differences between the modified model and the CML model in the application and the causes for these differences, thereby providing a feasible basis for suggesting the modified model as the characterization method assessing China mineral resource depletion[22].

The characterization model of abiotic resource depletion was modified and improved in terms of localization in our study. However, the characteristic factors of the abiotic resource depletion need to be expanded in terms of both time span and resource category with the development of exploration technology, and the expansion of human demand, because some important parameters, such as resources reserves and extraction volumes, are regionally different and sensitively time bound. And the degree of correlation between characterization factors of abiotic resource depletion and economic social factors is still need to be further studied.

2.3 LCIA Method needs to improve

So far, the development of the methodology and the benchmark system of the life cycle impact assessment (LCIA) phase are still in progress, and there are several models that are used to calculate the characteristic indicators showing the relationship between inventory data and environmental impact categories. But, there is still not a widely accepted uniform standard. Internationally, a variety of methods have been proposed to implement impact assessment, and they can basically be divided into two types: midpoint methods[23] and endpoint methods[24]. The former focuses on the environmental impact categories and their function mechanism, using characteristic factors to describe the relative importance of various environmental disturbance factors. And the latter focuses more attention on the causality of the environmental impact issue.

Although significant progress has been made in the LCIA characteristic model and method, their scientific connotation still needs to continuously improved and enriched mainly in the following several areas[25]:

(1) To quantify the uncertainties of impact indicators. In the decision making process, the uncertainty analysis method needs to be established to improve the application scope and result interpretation.

(2) The differences of environmental impacts caused by spatial and temporal differentiation need to be identified.

(3) In accordance with the requirements for consistency and comparability, the depth and breadth of simulating environment mechanism need to be increased. The correlation between the characterization results and the environment needs to be further proved so as to enable the potential environmental impact assessment results to facilitate integrated decision making.

(4) Related disciplines need to be further developed so as to improve the development of the method for comparing the impact categories, such as the global warming, resource depletion, human

health, ecosystem protection, and so on, thereby providing better support for integrated decision making.

3 LCA DATABASE AND SOFTWARE

Generally, not only large numbers of environment burden data with high regional limitation but also different LCA methods and models are involved in LCA application. These data with the properties of universality, regionality and complexity, are the basis for each LCA study and supposed to be managed effectively. Therefore, owing to the advantage of database technology in data management area, the development of LCA database and evaluation software has become one of the most important directions of LCA research recently.

3.1 Research status of international LCA database and software

For promoting the communion of LCA information, a current format (SPOLD) for data exchange was established by Society for Promotion of Life cycle Assessment Development (SPOLD), which performed a detailed meta data division on each inventory record in order to assure the independence, handleability and procurability of life cycle inventory. Moreover, the SPOLD format is an open source and can be embedded in different LCA software for the data exchange between these tools.

Furthermore, an international standard (ISO14048) [26] for LCA data exchange is formulated by International Organization for Standardization (ISO), which put forward a normative information format including process information, model information and management information. Whereas, more detailed criterions for data selection and technology requirement are demanded for actual LCA study.

The efficiency of LCA implement can be improved and the cost of time and manpower is reduced by the application of LCA software which is often divided into three groups: general software for LCA experts and consultants, professional software for the decision of engineering design, sale or environment and waste management, application software for specific users (mainly the enterprise users). At present, the amount of LCA software related to material and production is more than twenty worldwide, the environment database exceeds one thousand, and over three thousand commercial softwares with embedded default database are sold, in which some famous tools (such as Simapro[27], Gabi[28], Team[29], etc) have been widely applied in LCI, LCIA, Eco design and cost analysis.

3.2 Research status of Chinese LCA database

In recent year, the research of LCA in China developed rapidly due to the high attention from the public and government, although the study started relatively late. In the support of National 863 Program, which initiated by Beijing university of technology (BJUT) and co operated with other colleges, research institutes and material corporations, the environment burden data of main material production (steel, cement, aluminum, engineering plastics, architectural coatings, ceramic, etc.) was collected and processed, and based on these data a basic MLCA database and related software with independent intellectual property were also developed[30]. As a result of the exploration and development within recent ten years, a research and consultation platform of LCA with the biggest data quantity and covered the widest range of materials in China was established in BJUT, involving six servers, firewalls and routers, eleven workstations and some professional evaluation software (Gabi4.0, Simapro7.0, UmberTo4.0[31], Team3.0, etc.). Furthermore, the website of Center for National Materials Life Cycle Assessment (CNMLCA, www.cnmlca.com.cn) has been opened to society and public in order to propagandize the origin, development and application of life cycle assessment, introduce the latest research trend and result at home and abroad, promote the formation and widely development of ECO material think and evaluation, and most important, support the LCA and ECO design performance in China. By far, the practice of life cycle assessment in China has attracted attention with several international LCA institutions. The research center has widely cooperated with ISO and PRé Consultants, and participated in the establishment of global LCA union.

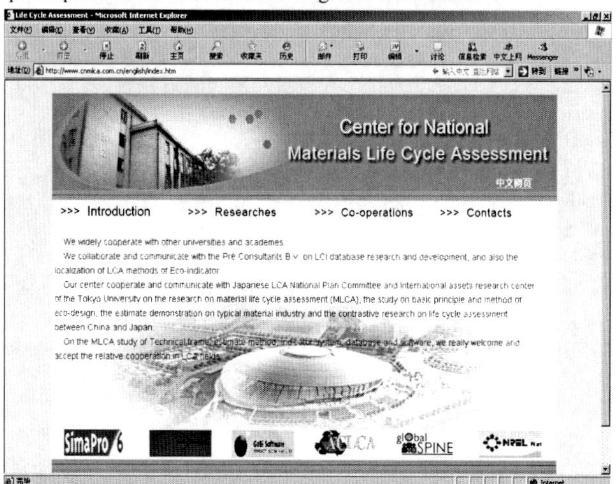

Fig.1 The website of Center for National Materials Life Cycle Assessment

SinoCenter Database established by Windows Advance Server and MSQL Server Enterprise version is an internet oriented platform that focused on research and development, which consisted of several connected sub database as standard database, system frame database, basic substance database, method database, unit database, literature database, project database, material property database, regional material flow database[32], etc. Presently, more than one hundred thousand records are involved and the concrete classification is shown as followed:

- Power supply: thermal power, hydropower and nuclear power;
- Primary energy: raw coal, crude oil and natural gas;
- Secondary energy: fuel oil, gasoline, diesel, coal gas;
- Transportation: pipeline transportation, road transportation, shipping and railway transportation;
- Freshwater resources: rivers and lakes water system;
- mineral resources: ferrous metal, nonferrous metals and inorganic nonmetals;
- Materials: ferrous metal materials (steel, aluminum, magnesium, etc), building materials (cement, glass, ceramic, concrete, plastic steel door and window, admixture, coating, carpet, floor coiled material, wallpaper, wooden furniture, adhesives, wood based panel, etc), chemical materials (Ethylene, HDPE, PVC, PP, ABS, etc) , connecting material (solder, etc);
- LCA Methods: Eco indicator 99, CML 2001, EDIP 2003, etc;

- Standards: steel, cement, etc.

4 REPRESENTATIVE CASE STUDY

The development of China economy will spur a significant growth in energy and raw materials industries. Using LCA methods to adjust industrial layout and to choose, optimize and design technique processes, such as energy supply as well as the production and manufacturing of materials, will be able to provide scientific decision making and technical guidance for China materials industry to achieve cleaner production and to carry out energy saving and emission reducing targets.

4.1 Primary energy

The life cycle inventory (LCI) of the production of primary energy and the major secondary energy is the fundamental data to carry out LCA for the materials industry and even all industrial products. In order to further develop China materials LCA database, we have worked out the inventories of primary energy, including the energy consumption and environmental emissions involved in the extraction process of coals, crude oils and natural gas, and have compiled a full data inventory from "cradle to gate" of several major downstream products derived from coals and crude oils, such as cleaned coal, coke, gas, petrol, diesel oil and fuel oil[33]. We also studied the energy consumption of power generation, as well as the emissions of gaseous pollutants, liquid pollutants and solid wastes[34]. These fundamental energy inventories have already been applied in China's environmental impact assessment of materials and products as well as in international comparative studies.

4.2 Design of Eco cement and structural adjustment of Beijing cement industry

According to the current development situation and future goals of the Chinese cement industry, we used the newer dry preheater/precalciner kiln system as the subject of our study to analyze the environmental impacts caused by this system in different situations, like not burning wastes, burning wastes, as well as using wastes to replace cement raw materials and fuels, among which the environmental impact acidification potential and human toxicity potential generated by the method of using wastes to replace 20% of cement raw materials is 90% down compared to the method that didn't burn wastes, thereby providing data support for designing eco cement products[35].

Combining the process LCA analysis and the regional material flow analysis (MFA), under the premise that guarantees both the cement demands of Olympic construction and environmental protection requirements, we put forward the 2008 Beijing's cement industry layout adjustment program. Under the circumstance that keeps the cement output basically unchanged, the overall consumption volume of materials and energy in Beijing's cement plants, through adjustment, combination and technological upgrading, was basically the same as in 2001, but the emissions of atmospheric pollutants, including soot, fumes and sulfur dioxide, were decreased 50%, 11% and 2%, respectively, compared to 2001. This program provides an extremely important reference for significant improvement of the atmospheric environment quality in Beijing and for the phased objective achievement of reducing and controlling air pollution[36].

4.3 LCA analysis of civilian buildings

The green building system currently advocated is to consider, from the perspective of sustainable development, the impacts on resources, energy and environment during a whole life cycle of buildings. China now has buildings with a total floor area of more than 40 billion square meters, and the direct energy consumption during the construction and use of buildings accounts for 30% of the total amount consumed by the whole society. In order to meet the urgent demand for the development of energy saving buildings and green materials, we have cooperated with Canada Wood Group to jointly analyze the environmental impacts produced by three different types of construction structures of multi story and multi residential civilian buildings in Beijing, including concrete structure, light steel structure and wood structure, during the stages of building materials production, construction and use. In this analysis, the life cycle inventory of a wide range of materials, including metal materials, gypsum materials, cement and concrete materials, materials for doors and windows and vinyl materials, as well as of fossil energy, electricity and transportation are all derived from the SinoCenter database. Through the calculation of characteristic indicators of 11 types of environmental impacts as well as the uncertainty and sensitivity analysis of the results, we determined that, among these three different kinds of construction structures, wood structure shows very obvious advantages because it ranks lowest in 8 kinds of environmental impact category, especially in the climate change, radiation effects, ozone depletion and land resources damages, all of 4 are closely related to human survival and life.

4.4 Iron and steel

Environmental load data of iron and steel materials derives from research into the production situation of more than 70 major Chinese iron and steel manufacturing plants as well as from industry statistical reports. The scope of the data covers the life cycle stages ranging from "cradle to gate", representing the environmental load of enterprises in different regions and with different levels of technology. Through the assessment of energy saving and wastes recycling and reuse technology during the iron and steel production process, a program for large scale integrated iron and steel enterprise to carry out the practice of recycling economy was put forward[37]. For a large scale integrated iron and steel enterprise with an annual production output of 10 million tons, the use of the new recycled iron and steel production process is able to, annually, absorb from the market 1.2 million tons of scrap steels and 200,000 tons of waste plastics, generate 9 billion kWh of electrical power, and produces 3 million tons of high grade cement through digesting wastes produced by itself, thereby yielding huge economic and social benefits.

4.5 Aluminum

It is a complex system for primary aluminum production. The development of this industry has been restrained to a large extent by the issues related to resources, energy and environment. With the initiation and promotion by the International Aluminum Institute (IAI), life cycle assessment (LCA) has been introduced into the environmental assessment system for aluminum and aluminum products[38 45]. At present, China has become the world's largest aluminum producer. Because of the characteristics of nationwide bauxite resources and the energy consumption, especially the structure of the electricity industry, the specific overall energy consumption of China aluminum production is 50% higher than that of the world average level, thereinto the aluminum smelting 45% higher, and alumina production 56% higher.

The life cycle analysis result showed that the GWP of China primary aluminum production in 2003 is nearly 1.7 times higher than that of the world average level in 2000, and the contribution of the process is also different. The efforts of raising the control level of electrolytic pots and reducing the coefficient of anode effects from 0.5 to 0.2 enabled the GWP caused by PFCs to decrease by 75% in 2006 compared to that in 2003. When the overall energy consumption for alumina production is reduced to 700 kg coal eq per ton, GHG emissions will be able to basically reach the world average level in 2000. With the decline of overall electricity consumption for electrolytic aluminum, the GHG emissions from the aluminum industry in 2010 and 2020, in comparison with that in 2006, will decrease by 6.2% and 12.3%, respectively[46].

4.6 Magnesium

Since 1990s, China magnesium industry has gained a rapid development, and China is the largest primary magnesium producer and supplier in the world. Magnesium production with the Pidgeon process was resources and energy intensive and leads to relatively severe environment pollution, and this situation has already attracted much attention of the local government and enterprises. So far, the international LCA research on both the production of primary magnesium and the magnesium products is still underway[47 49], and it also needs to further study the environmental impacts on the extensive use of the magnesium products.

According to the actual situation of China magnesium production, we analyzed the environmental impacts on the directly coal burning technics. The results showed that the reduction process accounts for 50% for the global warming potential, it is followed by the calcination process, being at 45%. For the acidification potential, the contribution of the refinement process and the reduction process accounts for 56% and 35%, respectively. The human toxicity potential mainly occurs in the reduction process which accounts for 95%. Based on different fuel use strategies, the environmental impacts of three scenarios were analyzed and compared. The direct coal burning showed the poorest environmental performance. The overall environmental load of producer gas as the main fuel decreased 1.9%, but the GWP were 14% higher compared with the direct coal burning. The environmental load using coke oven gas as major fuel for magnesium production decreased 17.5% than that of the direct coal burning. This means that a positive and environment friendly improvement can be achieved by integrated a sort of production networks based on the materials flow and supply, by product exchange and the waste heat utilization[50,51].

4.7 Other materials

LCA of the materials industry is a cross/multi disciplinary research area, involving several aspects of science, including materials science, environmental science and management. With respect to the case study of specific materials or products, there will still be a large number of useful research and exploration, such as environmental impacts produced by pyrometallurgical and hydrometallurgical process of copper production[52], as well as environmental impacts during the smelting process in typical lead and zinc plants[53,54]. Furthermore, the accumulated fundamental data and case studies of LCA for a wide range of materials, including several kinds of macromolecule materials[55], glass, ceramics, lead free solders and biodegradable plastics, play a positive role in the improvement of the LCA database and its application in the materials industry.

5 CONCLUSION

As a quantitive tool, life cycle assessment plays an important role in materials and products life cycle analysis, eco design, cleaner production, decision making and industry structure layout. However, in general, the applications of LCA were still limited in several demonstration fields. And there is a wider gap between evaluation results and the criterion people expected. Therefore, the development of LCA study should not only extend the application range of industry and agriculture fields, but also improve LCA methodology in economic and social aspects, for example, introducing life cycle costing[56] and social life cycle assessment[57,58] to establish quantitive indicators of sustainable development based on three elements of environment, economy and society.

Although the research of LCA methods and its application still need to make a deep investigation, it has been accepted and made important progress in the definition of goal and scope, framework, and the challenge of LCA. It is an effort for China materials industry to perform the work of energy saving and emission reducing, but we think that LCA method be a great potential tool to provide the decision making and technical support for the achievement of the target.

ACKNOWLEDGEMENT

This work was financially supported by the National Natural Science Foundation of China (NSFC, Project No. 50525413) and National Basic Research Program of China (973 Program, Project No.2007CB613706), and Beijing Natural Science Foundation (Project No.2081001).

REFERENCES

[1]G. Rebitzera, T. Ekvallb, R. Frischknecht, et al, Life cycle assessment part 1: Framework, goal and scope definition, inventory analysis, and applications, *Environment International*, 30, 701 720 (2004)

[2]ZR Nie, TY Zuo, Ecomaterials research and development activities in China, *Current Opinion in Solid State and Materials Science*, 7, 217 223 (2003)

[3]ZR Nie, F Gao, ZH Wang, et al, Development and Application of Methodology of MLCA in China, *Proceedings of The Seventh International Conference on EcoBalance*, Tsukuba ,Japan ,503 506 (2006)

[4]R. Bretz, SETAC LCA workgroup, Data availability and data quality, *Int J Life Cycle Ass*, 3(3), 121 123 (1998)

[5]ISO, ISO 14044, Environmental management life cycle assessment requirements and guidelines, *International Standard Organization*, Geneva (2006)

[6]A. Ciroth, G. Fleischer, and J. Steinbach, Uncertainty calculation in life cycle assessments A combined model of simulation and approximation, *Int J Life Cycle Ass*, 9(4), 216 226 (2004)

[7]M. Hung, and H. Ma, Quantifying system uncertainty of life cycle assessment based on Monte Carlo simulation, *Int J Life Cycle Ass*, 14(1), 19 27 (2009)

[8]S. Lo, H. Ma, and S. Lo, Quantifying and reducing uncertainty in life cycle assessment using the Bayesian Monte Carlo method, *Science of the Total Environment*, 340, 23 33 (2005)

[9]R. Heijungs, and R. Frischknecht, Representing statistical distributions for uncertainty parameters in LCA, *Int J Life Cycle Ass*, 10(4), 248 254 (2005)

[10]G. Geisler, S. Hellweg, and K. Hungerbuhler, Uncertainty analysis in Life Cycle Assessment Case study on plant protection products and implications for decision making, *Int J Life Cycle Ass*, 10(3), 184 192 (2005)

[11]A. Ciroth, Uncertainty in life cycle assessment, *Int J Life Cycle Ass*, 9(3), 141 142 (2004)

[12]S. Ross, D. Evans and M. Webber, How LCA studies deal with uncertainty, *Int J Life Cycle Ass*, 7(1), 47 52 (2004)
[13]XZ Gong, Basic database researches for the life cycle assessment of materials, *Dissertation of Doctor's Degree,* Beijing University of Technology (2006) (in Chinese)
[14]Y Liu, XZ Gong, ZH Wang, et al, Multiple imputation for missing data in life cycle inventory, *Materials Science Forum*, 610 613, 21 27 (2009)
[15]G. Finnveden, The resource debate needs to continue, *Int J Life Cycle Ass*, 10 (5), 372 (2005)
[16]B. Steen, Abiotic Resource depletion different perceptions of the problem with mineral deposits, *Int J Life Cycle Ass*, Spec. issue 1, 49 54 (2006)
[17]J. Potting, and M. Hauschild, Spatial differentiation in life cycle impact assessment, *Int J Life Cycle Ass*, Spec. issue 1, 11 13 (2006)
[18]M. Stewart, and B. Weidema, A consistent framework for assessing the impacts from resource use, *Int J Life Cycle Ass*, 10(4), 240 247 (2005)
[19]L. Oers, A. Koning, G. Jeroen, et al, Abiotic resource depletion in LCA, *Leiden, Road and Hydraulic Engineering Institute* (2002)
[20]J. Guinée, M. Gorrée, R. Heijungs, et al, Life cycle assessment An operational guide to the ISO standards, *Kluwer Academic Publishers*, Netherlands (2001)
[21]M. Goedkoop, and R. Spriensma, The Eco indicator 99 A damage oriented method for life cycle impact assessment methodology report Third Edition, *PRé Consultants*, the Netherlands (2001)
[22]F Gao, ZR Nie, ZH Wang, et al, Characterization and Normalization Factors of Abiotic Resource Depletion for Life Cycle Impact Assessment in China, *Science in China (Series E, Technological Sciences)*, 52(1), 215 222 (2009)
[23]O. Jolliet, R. Mueller Wenk, J. Bare, et al, The LCIA midpoint damage framework of the UNEP SETAC Life Cycle Initiative, *Int J Life Cycle Ass*, 9(6), 394 404 (2004)
[24]J. Bare, P. Hofsterter, D. Pennington, et al, Life Cycle Impact Assessment Midpoints vs. Endpoints the Sacrifices and the Benefits, *Int J Life Cycle Ass*, 5(6), 319 326 (2000)
[25]D. Penningtona, J. Pottingb, G. Finnveden, et al, Life cycle assessment part 2: Current impact assessment practice, *Environment International*, 30, 721 739 (2004)
[26]ISO/CD 14048.2, Environmental management Life Cycle Assessment LCA data documentation format, *International Standard Organization*, (2001)
[27]Pré consultants, Simapro 6.0 ® professional Version (2003)
[28]IKP University of Stuttgart and PE Europe GmbH Leinfelden Echterdingen, Gabi 4.0® (2003)
[29]The Ecoblian Group, Team 3.0® (2003)
[30]XZ Gong, ZR Nie, ZH Wang, et al, Research and Development of Chinese LCA Database and LCA Software, *Rare Metal*, 25(Spec. issue), 101 104 (2006)
[31]ifu Hamburg GmbH, UmberTo 4.0® (2003)
[32]XZ Gong, ZR Nie, ZH Wang, et al, Research and Development of Chinese LCA Database, *Proceedings of The Seventh International Conference on EcoBalance*, Tsukuba ,Japan ,435 438(2006)

[33]BR Yuan, Measurement method for sustainable development of chemical industry and its application, *Dissertation of Doctor's Degree*, Beijing University of Technology (2006) (in Chinese)
[34]XH Di, ZR Nie, BR Yuan, et al, Life Cycle Inventory for Electricity Generation in China, *Int J Life Cycle Ass*, 12(4), 217 224 (2007)
[35]XZ Gong, ZR Nie, ZH Wang, Environmental Burdens of Beijing Cement Production, *J WuHan Univ. Technol.*, 28(3), 121 14 (2006) (in Chinese)

[36]XQ Chen, YQ Guo, SP Cui, et al, Material energy metabolism and environmental implications of cement industry in Beijing, *Resource science*, 27(5), 40 46 (2005) (in Chinese)
[37]HM Zhou, TM Wang, WC Hao, et al, Analyses of environmental impact assessment on the development model of circulation economy in Chinese steel industry, *International Materials Research Conference*, Beijing, 24 30 (2006)
[38]IPAI Life Cycle Working Committee, Life cycle inventory of the worldwide aluminum industry with regard to energy consumption and emissions of greenhouse gases, http,//www.world aluminum.org/iai/publications/documents/full_report.pdf (2000)
[39]K. Hayashi, M. Okazaki, N. Itsubo, et al, Development of damage function of acidification for terrestrial ecosystems based on the effect of aluminum toxicity on net primary production, *Int J Life Cycle Ass*, 9(1), 13 22 (2004)
[40]H. Reginald, H. Hsien, An LCA study of a primary aluminum supply chain, *J Cleaner Prod*, 13, 607 618 (2005)
[41]R. Gerald, B. Kurt, The role and implementation of LCA within life cycle management at Alcan, *J Cleaner Prod*, 13, 1327 35 (2005)
[42]K. Martchek, Modelling more sustainable aluminium, *Int J Life Cycle Ass*, 11(1), 34 37 (2006)
[43]T. Norgate, S. Jahanshahi, and W. Rankin, Assessing the environmental impact of metal production processes, *J Cleaner Prod*, 15, 838 848 (2007)
[44]N. Frees, Crediting Aluminium recycling in LCA by demand or by disposal, *Int J Life Cycle Ass*, 13 (3), 212 218 (2008)
[45]J. Gatti, G. Quieroz, and E. Garcia, Recycling of aluminum can in terms of life cycle inventory (LCI), *Int J Life Cycle Ass*, 13(3), 219 225 (2008)
[46]F Gao, ZR Nie, ZH Wang, et al, Greenhouse Gas Emissions and Reduction Potential of Primary Aluminum Production in China, *Science in China (Series E, Technological Sciences)*, 52 (2009) (in press)
[47]S. Ramakrishnan, P. Koltun, Global warming impact of the magnesium produced in China using the Pidgeon process, *Resources, Conservation and Recycling*, 24, 49 64 (2004)
[48]M. Hakamada, T. Furuta, and Y. Chino, Life cycle inventory study on magnesium alloy substitution in vehicles, *Energy*, 32, 1352 60 (2007)
[49]F. Cherubini, M. Raugei, and S. Ulgiati, LCA of magnesium production Technological overview and worldwide estimation of environmental burdens, *Resources, Conservation and Recycling*, 52, 1093 1100 (2008)
[50]F Gao, ZR Nie, ZH Wang, et al, Resource Depletion and Environmental Impact Analysis of the Magnesium Produced Using the Pidgeon Process in China, *The Chinese Journal of Nonferrous Metals*, 16(8), 1456 61 (2006)
[51]F Gao, ZR Nie, ZH Wang, et al, Assessing the Environmental impact of Magnesium Production using

the Pidgeon Process in China, *Transactions of Nonferrous Metals Society of China*, 18(3), 749 754 (2008)
[52]JL Jiang, JF Dai, WJ Feng et al, Life cycle assessment of metallic copper produced by the pyrometallurgical and hydrometallurgical processes, *J Lanzhou Univ. Technol.*, 32(1), 19 21 (2006) (in Chinese)
[53]X Xiao, SW Xiao, XY Guo, et al, LCA Case Study of Zinc Hydro and Pyro Metallurgical Process in China, *Int J Life Cycle Ass*, 8(3), 151 155 (2003)
[54]QH Li, XY Guo, SW Xiao, et al, Life cycle inventory analysis of CO_2 and SO_2 emission of imperial smelting process for Pb Zn smelter, *J Cent. South Univ. Technol.*, 10(2), 108 112 (2003)
[55]H Chen, WC Hao, F Shi, et al, Life cycle assessment of several typical macromolecular materials, *Acta Scientiae Circumstantiae*, 24(3), 545 549 (2004) (in Chinese)
[56]G. Rebitzer, and D. Hunkeler, Life Cycle Costing in LCM, Ambitions, Opportunities, and Limitations, Discussing a Framework, *Int J Life Cycle Ass*, 8(5), 253 256 (2003)
[57] L. Dreyer, M. Hauschild and J. Schierbeck, A Framework for Social Life Cycle Impact Assessment, *Int J Life Cycle Ass*, 11(2), 88 97 (2006)
[58]B. Weidema, The integration of economic and social aspects in life cycle impact assessment, *Int J Life Cycle Ass*, Special Issue 1, 89 96 (2006)

MODELING DUAL AND MgO SATURATED EAF SLAG CHEMISTRY

Kyei Sing Kwong, James Bennett, Rick Krabbe, Art Petty, Hugh Thomas
National Energy Technology Laboratory, US DOE,
Albany Oregon, USA 97321

ABSTRACT

Foamy slag practice has been widely adopted by the EAF industry because it shields the electrical arcs, increases yield, lowers noise levels, prevents radiant energy loss, eliminates arc flares, saves overall energy and extends refractory service life. Foamy slag requires the control of slag chemistry and viscosity to sustain gas bubbles during processing. This is accomplished through the formation of magnesium wűstite particles in the slag at the operating temperature. A thermodynamic program, Factsage™, was utilized to compute the dual (saturated with CaO and MgO containing phases) and MgO saturated EAF slag chemistry for a 5 component system of MgO CaO FeO SiO_2 Al_2O_3, at steelmaking oxygen partial pressures, temperatures, and slag basicity. The computational results will be analyzed to explore oxygen pressures, temperatures, slag basicity, and Al_2O_3 effects for creating a model to predict the dual and MgO saturated slag chemistry.

INTRODUCTION

Foamy slag practice has been widely adopted by EAF industry because it shields the electrical arcs, increases yield, lowers noise levels, prevents radiant energy loss, eliminates arc flares, saves overall energy, and extends refractory service life. Slag chemistry plays a critical role on the quality of foaming; which includes the size, quantity, and duration of gas bubbles. Ideal foaming EAF slag requires MgO saturation, because only MgO saturated slag can yield optimum viscosity with the presence of a suspended second phase particle [MgO · FeO magnesium wustite (MW)] at the operating temperature to allow long lasting gas bubbles. Thin slag can not sustain gas bubbles, while thick slag discourages their formation. The saturated MgO slag will be not only foam better, but will also decrease refractory wear.

EAF slags typically contain five major oxides: CaO, MgO, SiO_2, FeO and Al_2O_3. Pretorius(1) studied EAF foaming slag and proposed isothermal stability diagrams (ISD) expressed at constant basicity and temperature, shown in Figure 1, which was separated into four regions; molten slag, magnesium oxide saturated, calcium oxide saturated, and dual saturated. Point (D) is bounded by these four regions in Figure 1, and is called the dual saturated point in this paper. A straight line separating the molten slag and magnesium oxide saturated region is called the magnesia saturated line. Slag in the magnesium oxide saturated region has precipitates of magnesium oxide containing phases such as magnesium wűstite [(Mg,Fe)O], while slag in the calcium oxide saturated region has precipitates of calcium oxide containing phases such as Ca_2SiO_4. Therefore, slag at the dual saturated point (D) contains the initial precipitates of magnesium and calcium oxide containing phases. Pretorius built a computer model based on a mass balance approach to design target slag compositions, however, detailed information on his algorithm is unknown.

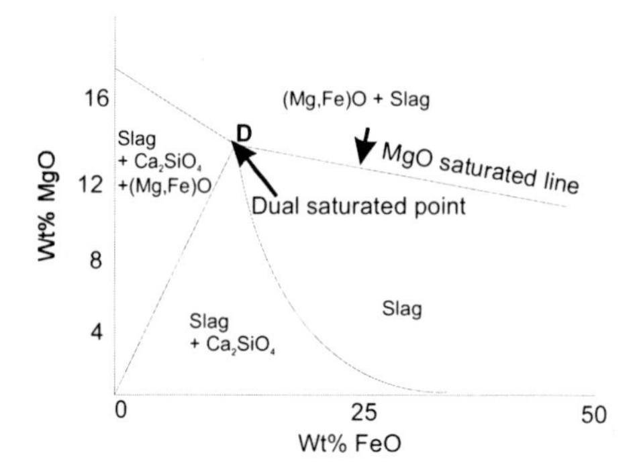

Figure 1 Isothermal stability diagram proposed by Pretorius (1).

Figure 2a shows the saturation lines of MgO based solid solution phases (Fe,Mg)O and $(Fe,Mg)_2SiO_4$ in the CaO FeO MgO SiO_2 system in contact with metallic iron at 1600°C (2). Figure 2b shows the saturation lines of the CaO based phase (Ca_2SiO_4) for the same phase system, environment, and temperature (2). Both diagrams show contour lines of MgO above 6 wt%. Dual saturation [Ca_2SiO_4 and (Fe,Mg)O] points can be determined by overlapping both figures. Dual saturation points at a line 6 11 12 13 represent the slag chemistry with different C/S ratios at 1600°C. Several linear (straight line) relationships were found between basicity, acidity, and constituents of dual saturated slags from diagrams as shown in Table I. Based on these linear relationships, a computer model was built to calculate dual saturated and MgO saturated EAF slag chemistry for specific basicity indexes (3).

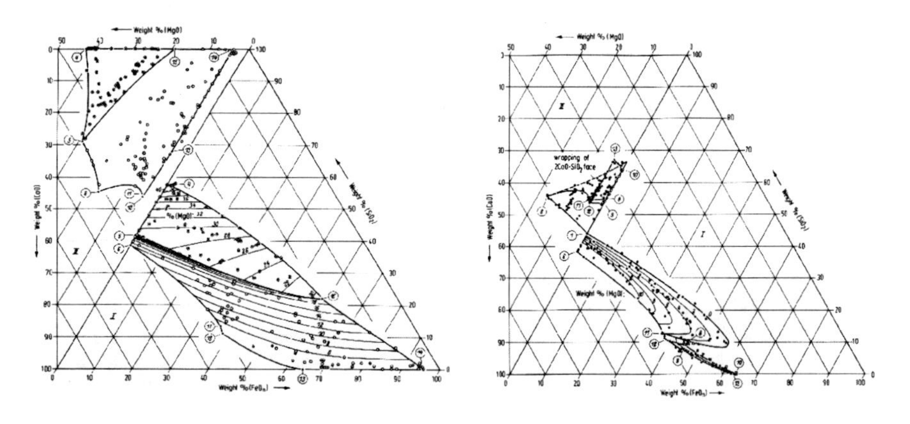

Figure 2 a) The saturation lines of MgO based solid solution phases (Fe,Mg)O and $(Fe,Mg)_2SiO_4$
b) The saturation lines of CaO based phase (Ca_2SiO_4) in the CaO FeO MgO SiO_2 system in contact with metallic Iron at 1600EC (2).

Table I The regression fitting quality (R square) between atomic percentages of oxides and acidicity (S/C) or basicity (C/S)

Relationships	R Square	Relationships	R Square
S / C SiO_2	0.98	C / S CaO	1.0
S / C MgO	0.99	MgO FeO	0.99
S / C FeO	0.98		

THERMODYNAMIC CALCULATION CONDITIONS

A thermodynamic program, FactSage™, was utilized to compute the dual (saturated with CaO and MgO containing phases) and MgO saturated EAF slag chemistry for a 5 component system of MgO CaO FeO SiO_2 Al_2O_3, at different oxygen partial pressures, temperatures, and slag constituents. An isothermal stability diagram can be plotted by inputting $Ca_xSi_yAl_zO_w$, MgO, and FeO in the reactant windows of FactSage's phase diagram module. The subscripts x, y, and z are specific designed values for Ca, Si, and Al separately. The subscript w depends on the value of x, y and z to balance the valence (zero charge) of the calcium aluminum silicates. By this way, the basicity (CaO/SiO_2) can be kept constant, which is required when using isothermal stability diagram to express the MgO saturation line or dual saturation point.

Oxygen partial pressure has a large influence on the formation of FeO or Fe_2O_3, and consequently, on the liquidus temperature of slag, and the saturation level of MgO. A review of literature (4,5) and field samples from EAFs indicated that oxygen partial pressure in foaming slag varies depending on the operational conditions, location in the EAF, and feedstock. Six temperature values, from 1600 to 1700°C, and five values of oxygen partial pressure per temperature were selected for thermodynamic studies. The lowest PO_2 value was set to the interface of Fe FeO formation (figure 3). The highest PO_2 value was set to the interface of FeO Fe_3O_4 formation. The medium value was the average of the high and low values. Similarly, the medium high/medium low values were the average of the medium and high/low values. Temperature also affects the formation of Fe, FeO and Fe_2O_3; therefore, these PO_2 values were varied with temperatures. Because of these variations, the X axis showing FeO wt% from Figures 4 to 11 were the converted values for Fe, FeO and Fe_2O_3. However, Figure 13 will distinguish between Fe_2O_3 and FeO in the slag. 7500 cases were computed varying oxygen pressure, temperatures, and slag chemistry in order to study their effects on dual and MgO saturated slags (table II).

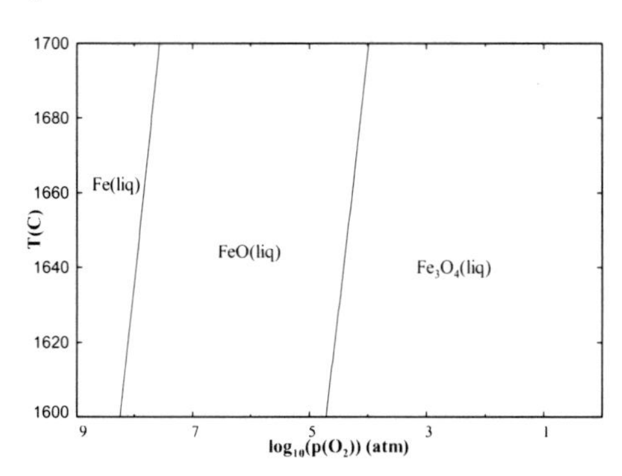

Figure 3 The stability diagram of Fe O compound at the oxgyen partial pressure from 10^{1} to 10^{9}.

Table II A list of variables and their values for thermodynamic calculation used in this study

Variables	Values Studied				
X for Ca	1.3 to 2.5 in 0.2 increments				
Y for Si	0.847, 0.806, 0.769, 0.735, 0.704.				
Z for Al	0.152, 0.193, 0.23, 0.264, 0.295				
Temperature °C	1600 to 1700 in 20 °C increments				
$Log(PO_2)$	High	Medium high	Medium	Medium low	Low
1600 °C	4.7215	5.6061	6.4907	7.3752	8.2598
1620 °C	4.5742	5.4613	6.3484	7.2355	8.1226
1640 °C	4.4269	5.3165	6.2062	7.0958	7.9854
1660 °C	4.2796	5.1717	6.0639	6.9561	7.8482
1680 °C	4.1323	5.0270	5.9217	6.8163	7.7110
1700 °C	3.9850	4.8822	5.7794	6.6766	7.5738

RESULTS AND DISCUSSION

The Dual Saturated EAF Slag Chemistry Derived by Thermodynamic Calculations

A dual saturated EAF slag chemistry was not able to be computationally obtained, if the specified slag chemistry was too acidic (higher SiO_2/CaO ratio) or basic (higher CaO/SiO_2). About 47% slag chemistries evaluated were found to have a dual saturated point. For simplification and clarification, only these cases will be discussed in this paper.

Figure 4 shows the effect of basicity (CaO/SiO_2) on the dual saturated EAF slag chemistry for cases of $Ca_xSi_{0.847}Al_{0.23}O_w$, where x=1.3 to 2.3 for every 0.2 increment, at a constant temperature (1640°C), and high oxygen partial pressure ($\log(PO_2)$= 4.4269). This figure shows that slag with increasing basicity is dual saturated with linearly decreasing MgO and increasing FeO concentration when MgO content is more than 6 wt.%. Table III indicates that the constituents of dual saturated slags, from cases studied in figure 4, have linear relationships with basicity or acidity similar to the results obtained using the FeO MgO CaO SiO_2 phase diagram from the Slag Atlas. Figure 4 and table III also show that some relationships are non linear when MgO concentration in a dual saturated slag is below 6%; lowering regression reliability for predicting the dual saturated slag chemistry.

Figures 5 and 6 show the effect of CaO, SiO_2 and Al_2O_3 on the dual saturated EAF slag chemistry. Figure 5 is for ($Ca_{1.5}Si_yAl_zO_w$) where y and z are variables at a constant temperature (1620°C) and high oxygen partial pressure ($\log(PO_2)$= 4.5742). Figure 5 shows dual saturated slags with increasing Al_2O_3 or SiO_2 require more MgO and less FeO to maintain dual saturation, in a linear relation. Figure 6 is for cases where Y_{Si} is a constant (0.769), while X_{Ca} and Z_{Al} content are varied at 1600°C and an oxygen partial pressure of $\log(PO_2)$= 4.7215. Figure 6 shows dual saturated slags with increasing CaO require less MgO and more FeO to maintain dual saturation, other relationships between Al_2O_3 with FeO, Al_2O_3 with MgO, SiO_2 with FeO, and SiO_2 with MgO for dual saturated slag chemistry were examined, and in all cases relationships were linear.

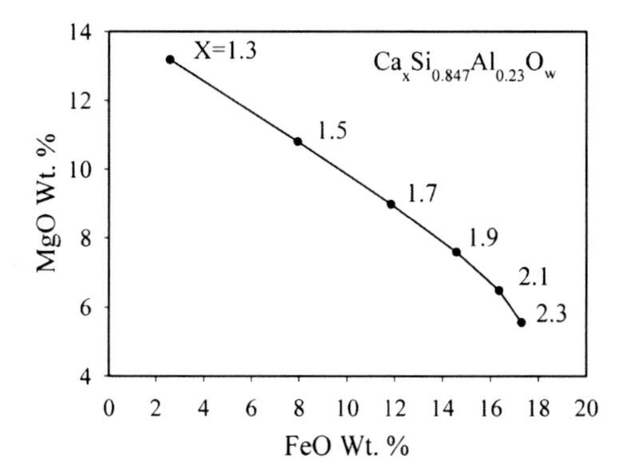

Figure 4 The effect of basicity on the dual saturated EAF slag chemistry, for cases which $Ca_xSi_{0.847}Al_{0.23}O_w$ where x=1.3 to 2.3 for every 0.2, at a constant temperature (1640°C), and high oxygen partial pressure ($\log(PO_2)$= 4.4269)

Table III The regression reliability of linear relationships with the consideration of all cases and when the MgO content in a dual saturated slag is more than 6 wt%.

Relationships	R square (all)	R square (MgO > 6 wt. %)
FeO MgO	0.9911	0.9971
SiO_2 S/C	0.9999	1.0000
FeO S/C	0.9894	0.9961
CaO C/S	0.9979	0.9996
MgO S/C	0.9999	0.9999
Al_2O_3 MgO	0.9999	1.0000
Al_2O_3 FeO	0.9915	0.9965

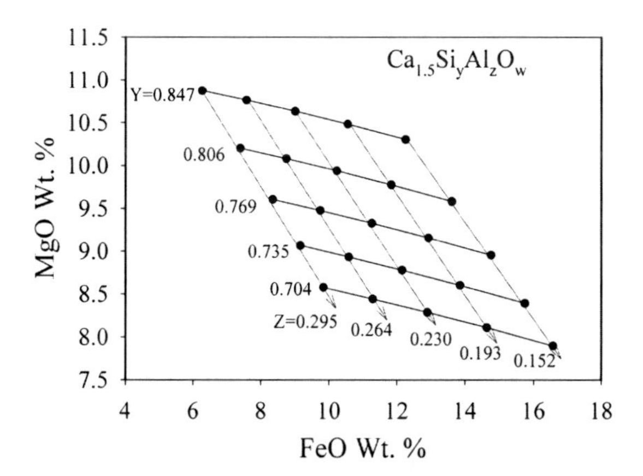

Figure 5 The MgO FeO diagram for dual saturated EAF slag cases $Ca_{1.5}Si_yAl_zO_w$ where y and z are variables at 1620°C and high oxygen partial pressure (PO_2= 4.5742).

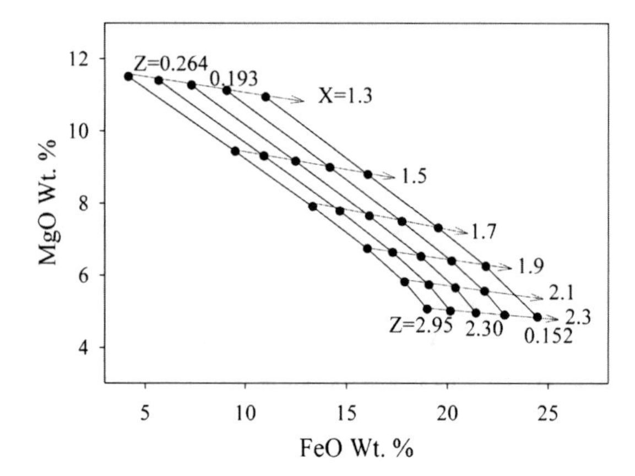

Figure 6 The MgO FeO diagram of dual saturated EAF slag cases $Ca_xSi_yAl_zO_w$ where y is a constant (y=0.769) and x and z are variables at 1600°C and high oxygen partial pressure ($\log(PO_2)$= 4.7215).

Oxygen partial pressure also affects the dual saturated slag chemistry because of the varied amount of Fe, FeO and Fe_3O_4 and their basicity. Figure 7 shows the values of FeO and MgO content of dual saturated slags with different oxygen partial pressure for two slags. This figure indicates that higher MgO and less FeO content is needed for dual saturated EAF slag in a higher oxygen partial pressure environment.

Dual saturated EAF slag chemistry is also critically affected by temperature. Figure 8 shows the temperature effects on dual saturated EAF slag chemistry for cases of $Ca_{1.3}Si_{0.704}Al_zO_w$ in a medium oxygen partial pressure. Figure 8 indicates that higher concentration of MgO and less of FeO are needed to keep slags saturated with CaO and MgO at a constant basicity as temperature increases. Regression analysis in Table IV show a good fit exists between constituents of dual saturated slags and temperature for cases of dual slag chemistry $Ca_{1.3}Si_{0.806}Al_{0.152}O_{3.1409}$ at the highest oxygen partial pressures. Therefore, interpolation is an acceptable way to predict dual saturated slag chemistry at any temperature between 1600 and 1700°C.

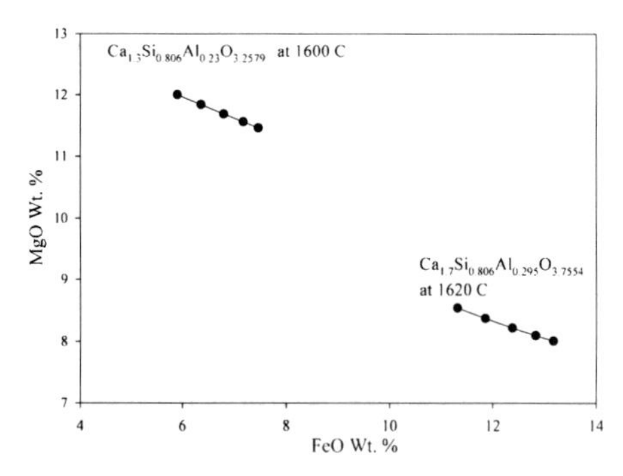

Figure 7 The MgO FeO diagram demonstrating the effect of oxygen partial pressure on the two sets of dual saturated slag chemistry.

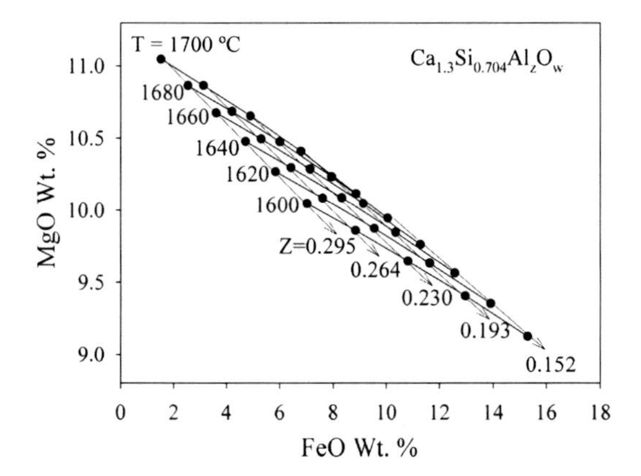

Figure 8 Temperature effect on dual saturated slag cases $Ca_{1.3}Si_{0.704}Al_zO_w$ where z is a variable under medium oxygen partial pressure. Z's values have been indicated in the figure.

Table IV Regression reliability between oxide concentration in dual saturated slag and temperature for cases of dual slag chemistry $Ca_{1.3}Si_{0.806}Al_{0.152}O_{3.1409}$ in the highest oxygen partial pressure.

Relationships	R square	Relationships	R square
CaO Temperature	1	FeO Temperature	1
SiO_2 Temperature	1	MgO Temperature	0.997
Al_2O_3 Temperature	1		

The MgO Saturated Slag Chemistry Derived by Thermodynamic Calculations

Previous discussion demonstrated that the dual saturated EAF slag chemistry can be predicted using linear relationships among oxide constituents, their basicity (or acidity), oxygen partial pressure and temperature. The effect of slag constituents, temperature, and oxygen partial pressure on a MgO saturated slag chemistry are studied and discussed below. FeO content in the EAF slags is rarely above 50wt%. The reason figures 9 to 11 shows FeO ranging from 0 to 100 wt.% on the x axis is for the scientific predictions of MgO saturated slag chemistry.

The Effect of CaO, SiO_2, and Al_2O_3

Figure 9 shows MgO saturated lines for cases with different concentrations of CaO at 1640°C in a medium low oxygen partial pressure ($10^{7.0958}$). This figure indicates that slags with high CaO content are saturated with less MgO content. However, as the iron oxide content in a slag increases, high CaO containing slags require additional MgO to stay saturated. Conversely, slags with low CaO content can be saturated with less additional MgO when the FeO content increases.

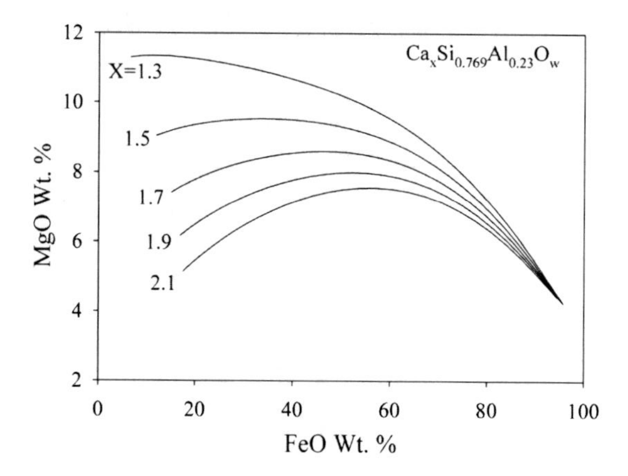

Figure 9 The effect of CaO on MgO saturated EAF slag chemistry for cases $Ca_xSi_{0.769}Al_{0.23}O_w$ at 1640ºC in a medium low oxygen partial pressure. X is a variable with its values indicated in the figure.

Figure 10 illustrates the effect of SiO_2 on MgO saturated lines for cases $Ca_{1.5}Si_yAl_{0.193}O_w$ at 1620 ºC in a medium oxygen partial pressure ($10^{6.3484}$). This figure indicates that slags with increasing SiO_2 require more MgO to be saturated at a constant amount of FeO.

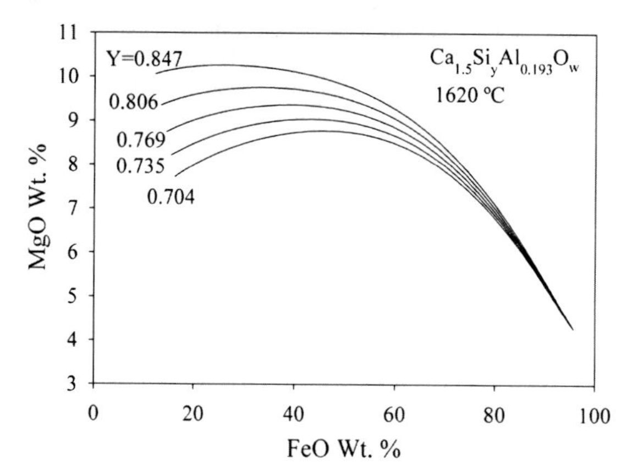

Figure 10 The effect of SiO_2 on MgO saturated EAF slag chemistry for cases $Ca_{1.5}Si_yAl_{0.193}O_w$ at 1620 ºC in a medium oxygen partial pressure ($10^{6.3484}$). Y is a variable and its value has been indicated in the figure.

Figure 11 illustrates the effect of Al_2O_3 on MgO saturated lines for cases $Ca_{1.7}Si_{0.806}Al_zO_w$ at 1680 ºC in a medium high oxygen partial pressure ($10^{5.027}$). This figure indicates that slags with increased quantities of Al_2O_3 require more MgO to be saturated at a constant amount of FeO.

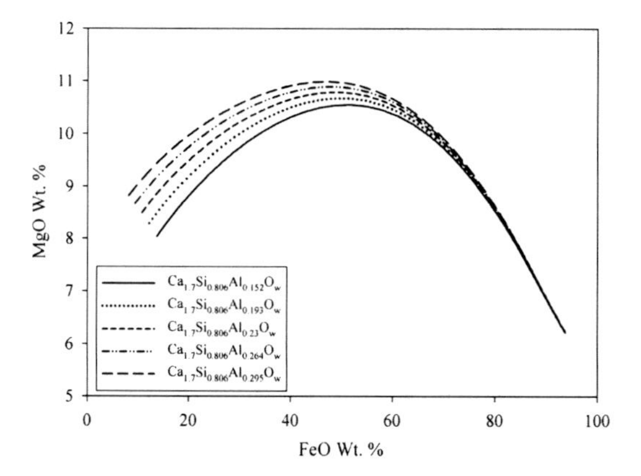

Figure 11 Al_2O_3 effects on MgO saturated EAF slag chemistry for cases $Ca_{1.7}Si_{0.806}Al_zO_w$ at 1680 °C in a medium high oxygen partial pressure ($10^{5.027}$). Z is a variable for the content of Al_2O_3 in slags and its values have been indicated in the figure.

The Effect of Temperature and oxygen on MgO Saturated Lines

The effect of temperature on the chemistry of MgO saturated slags was studied for $Ca_{1.5}Si_{0.769}Al_{0.152}O_{3.2644}$ at medium oxygen partial pressure. The slag constituents were plotted against temperature with the assumption of a constant FeO content (30 Wt. %). The results are plotted in Figure 12, and indicate a linear relationship between oxide concentrations and temperature. Therefore, the interpolation method of predicting the slag chemistry along the MgO saturated line between temperatures from 1600 to 1700°C is possible.

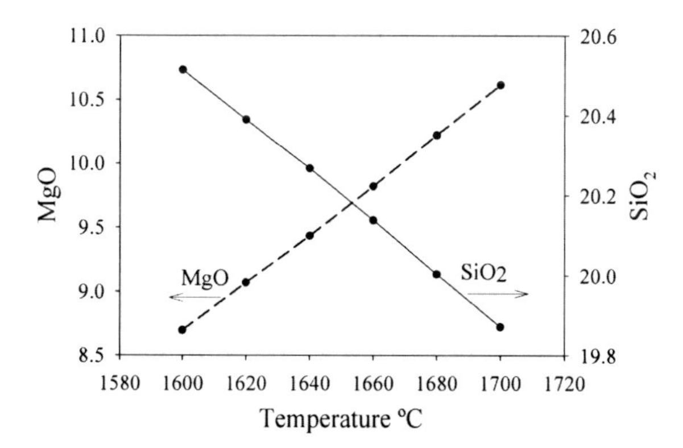

Figure 12 The effect of temperature on slag chemistry for the case $Ca_{1.5}Si_{0.769}Al_{0.152}O_{3.2644}$ at medium oxygen partial pressure and constant FeO content.

As previously discussed, the oxygen partial pressure in an EAF is unknown, and depends on the operational conditions and the location within the slag. Slag near the surface or close to the oxygen lance has a higher oxygen partial pressure, while slag near the molten metal has a lower oxygen partial pressure. In addition, the turbulence of slag and molten metal in the furnace may also dynamically change the oxygen partial pressure in the EAF. Analysis of oxygen partial pressure effects on the MgO saturated EAF slags indicated that higher MgO levels are needed in a higher oxygen partial pressure environment to maintain saturation. Therefore, slag with a higher oxygen partial pressure may dissolve more of the EAF refractory (MgO/C refractory) than low oxygen partial pressure.

It has been observed that the constituents Al_2O_3, SiO_2, and CaO in MgO saturated slags have a good linear relationship with FeO, but not with MgO. If a linear relationship among Al_2O_3, SiO_2, CaO, and FeO was recognized, they should have a linear relationship with MgO. Further FactSage™ calculations were conducted for slag chemistry along the MgO saturated line. It was found that Fe2O3 increased with FeO content in the slag until FeO content reaches a threshold level of 30 wt. %; not including Fe_2O_3 and at that point Fe_2O_3 concentration remained nearly constant. Other constituent relationships in the slag also change from a linear to non linear with FeO content (figure 13). If the relationship of FeO and Fe_2O_3 in different slags and oxygen partial pressure can be found, then accurate predictions of the MgO saturation line would be possible.

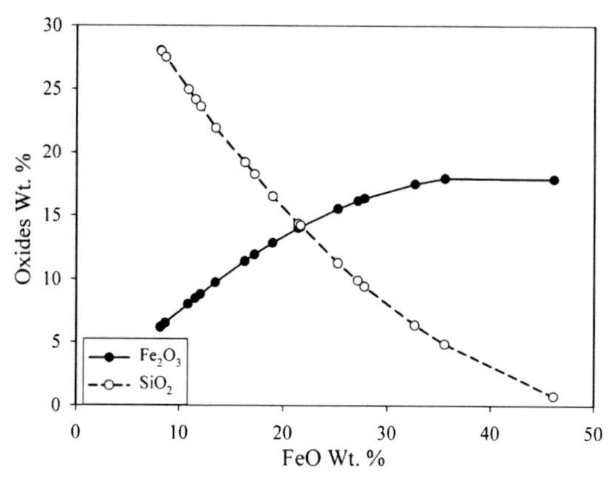

Figure 13 The formation of Fe_2O_3 with increasing FeO for a slag chemistry of $Ca_{1.7}Si_{0.806}Al_{0.193}O_{3.6024}$ at 1680°C in a medium high oxygen partial pressure.

CONCLUSIONS

A five oxide compoent system (MgO CaO FeO SiO_2 Al_2O_3) has been studied by a thermodynamic program, Factsage™ to compute the MgO content in the dual (saturated with CaO and MgO) and MgO saturated EAF slags with variations of oxygen partial pressure, operational temperature, and constituents. Linear relationships among oxide components, and slag constituents are found, making the prediction of the dual saturation slag chemistry possible. However, as FeO increased to a threhold level along the MgO saturation line, the formation of Fe_2O_3 did not follow FeO increase, causing the prediction of MgO saturated slag chemistry to be more difficult. If the relationship of FeO Fe_2O_3 in different slag chemistry and oxygen partial pressure can be predicted, then the prediction of MgO saturated slag chemistry should be possible.

REFERENCES

1) E. Pretorius "Slag Fundamentals"; in An Introductions to the Theory and Practice of EAF Steelmaking, Iron & Steel Society short course book, New Orleans, LA Nov. 1998
2) M. Kowalski, P. J. Spencer, and D. Neuschutz, "Phase Diagrams"; in Slag Atlas, 2nd ed. Edited by Verein Deutscher Eisenhüttenleute, Verlag Stahleisen Gmbh, 21 215 (1995).
3) K. S. Kwong, and J. P. Bennett, "Balancing MgO for Foamy Slag and Refractory Protection" D.L. Schroeder & Associates, Twenty Second Annual Symposium, Process Systems for Electric Furnace Steelmaking, Orlando, Florida, November 2000.
4) R. J. Fruehan, "Optimization of Post Combustion in Steelmaking" DOE report for work performed under Cooperative Agreement No. DE FC36 97ID13554 March 2004.
5) M. Kirschen, V. Veiikorodov, and et al. "Off gas Measurements at the EAF Primary Dedusting System" Stahl und eisen 124 (11), 73 89 (2004)

Advanced Powder Processing

AQUEOUS PROCESSING OF TiC PREFORMS FOR ADVANCED CERMET PREPARATION

R. Bradley Collier and Kevin P. Plucknett*
Dalhousie University, Materials Engineering Program, Department of Process Engineering and Applied Sciences, 1366 Barrington Street, Halifax, Nova Scotia, B3J 2X4, Canada.

ABSTRACT

Advanced cermets based on titanium carbide (TiC), with a ductile nickel aluminide (Ni_3Al) binder, have shown significant promise for use in a variety of demanding wear environments, due to a combination of high strength and good corrosion behavior. A unique feature of TiC/Ni_3Al cermets is that they show increasing strength from room temperature up to ~1,000°C, while current materials such as tungsten carbide/cobalt (WC/Co) show significant strength degradation above ~500°C. In the present study, aqueous colloidal forming methods have been applied to process TiC preforms. Without a dispersant, the TiC powder used shows an isoelectric point at ~ pH 2.6. The use of an ammonium salt of polymethacrylate (Darvan C) does not significantly alter the isoelectric point. TiC preforms were prepared by slip casting suspensions of up to 50 vol. % solids content. After drying, the TiC based cermets were processed by melt infiltration with the Ni_3Al alloy (IC 50) at 1475°C. Ni_3Al content was varied between 20 and 50 vol. % using this approach, resulting in final densities that exceeded 98 % of theoretical.

INTRODUCTION

The aqueous processing of titanium carbide (TiC) has been studied as far back as the mid 1960's.[1] TiC has many very attractive properties which make it ideal for high temperature high wear applications. With its high melting point of 3065°C,[2] low density of 4.93 g/cm^3,[3] and high Vickers hardness of >1450, it is not hard to see why TiC is an attractive alternative to the heavier, more expensive WC based materials. Traditionally these materials have been processed via dry pressing techniques. Dry pressing has several drawbacks; agglomeration and reduced homogeneity, residual porosity, geometry limitations etc. Colloidal processing is very attractive, as complex geometries can be cast. The cast product also has higher green densities, therefore lower porosity, which results in increased mechanical properties.[4]

TiC on its own is brittle and has low fracture toughness. To overcome these limitations a ductile binder phase is introduced to improve the facture toughness. Ni, Co, and Fe based binders have been successfully used to produce a composite with adequate mechanical properties. In the work presented here nickel aluminide (Ni_3Al), has been used as the ductile binder phase. This composite material has been shown to have excellent high temperature properties, with not only retention of its strength up to ~950°C but a significant increase in its fracture strength with increasing temperature.[5] This is in contrast to WC Co which loses most of its strength above ~500°C.[6] This unique behavior, coupled with its excellent corrosion and oxidation resistance, make TiC Ni_3Al composites extremely well suited for high temperature applications. In this paper the colloidal processing conditions and characterization of the slips will be examined, along with the resulting composites produced by pressureless melt infiltration performed on the slip cast preforms.

EXPERIMENTAL PROCEDURES

Raw Materials Characterization

The TiC powder used in the experiments performed was sourced from Pacific Particulate Materials Ltd. (TiC 2012 Batch# 20125339) with a quoted average particle size of 1.30 μm, and 0.11% retained free carbon. This powder from the above mentioned batch was used throughout the

* Corresponding Author

experimental trials presented in this paper. The TiC powder was first analyzed as received to verify the particle size distribution using acoustic attenuation spectroscopy (Zeta APS, Matec Applied Sciences). The particle morphology and size were also examined using scanning electron microscopy (Hitachi S 4700, Hitachi Co.) to verify the particle size data. The surface area of the TiC powder was measured using surface nitrogen adsorption (Horiba SA 6201). The Ni_3Al powder used in the melt infiltration processing was obtained from Ametek (alloy IC 50 Batch# 0412399), and had a manufacturer quoted particle size of 325 mesh. This powder was also examined for morphology and particle size, but was not subjected to further analysis as it plays no part in the colloidal processing of the preforms, only in the subsequent melt infiltration procedure.

The deflocculent used was an ammonium salt of polymethacrylate (PMA NH_4) produced in 25 wt% aqueous suspension (Darvan C; R.T. Vanderbilt Co.). All water used in processing of the powders was double distilled.

Characterization of Aqueous Slips

To understand the stability regions and to determine the optimal processing conditions for the slips, several methods were employed. The first method used to identify the broad stability region for the suspensions was sedimentation trials. For these tests a 10 vol. % TiC suspension was sonicated for ~2 min. at 300 W (Model 500 Sonic Dismembrator, Fisher Scientific) to break up any agglomerates and divided up into 15 graduated test tubes. The polyelectrolyte concentration was varied from 0 0.25 wt. %. These were left to settle for 10 days after which the sediment volume was measured. The more stable a suspension the better the particle packing, resulting in a minimal sediment volume. This allowed for a smaller Darvan C concentration range to be investigated in subsequent analysis.

Acoustophoresis (Zeta APS, Matec Applied Sciences) was then used to track changes in the zeta potential with respect to pH and Darvan C concentration. 2 wt. % suspensions of TiC were prepared and sonicated for 2 min. at 300 W to break up any agglomerates. Zeta potential titrations were run for pH values between 11.5 2.0 with a step size of 0.2 pH units. In this instance the polyelectrolyte concentration was varied from 0 to 0.05 wt. %. The suspensions were auto titrated using 1M NaOH and HNO_3.

Finally the slips were rheologically analyzed using a controlled stress rheometer (TA Instruments, AR 2000) using a standard 28 mm concentric cylinder geometry. The suspensions were tested for steady state flow characteristics on a range of shear rates from 0.1 s^{1} to 1500 s^{1}. The affect of polyelectrolyte concentration and solids loading on the apparent viscosity of the slips was examined.

Preform Production and Melt Infiltration/Sintering

Production of the preforms was accomplished by directional slip casting of the TiC suspensions using PTFE rings on plaster of Paris plates. The suspensions were mixed with double distilled water and the appropriate polyelectrolyte concentration, then hand mixed and subsequently sonicated for 2 minutes. The slip was then degassed under vacuum for 5 min. to remove any entrained air in the slip. To ensure that no bubbles were introduced into the cast pieces, an anti foaming agent was used (Tri n butyl phosphate, 98 % Alfa Aesar) in 0.025wt% concentration. The suspension was then poured into the PTFE moulds. The casting pieces were then allowed to dry at room temperature; after approximately an hour the cast disc would shrink away from the mould and could be removed and dried overnight in a drying oven (Precision 27, Thelco) at 40°C.

The melt infiltration process has previously been successfully used for the production of TiC Ni_3Al composites.[5, 7] The process followed in these current works is similar to past approaches. The TiC preforms are accurately weighed and the appropriate amount of Ni_3Al powder to add is calculated based on desired final volume fraction in the composite. The preform is placed in an alumina crucible on a thin bed (2 3 mm) of bubble alumina. The Ni_3Al powder is then placed on top of the preform. The crucible is covered and placed in the furnace (Vacuum Furnace/Hot Press, MRF Research Furnaces).

The infiltration is pressureless and occurs due to the excellent wetting of the Ni_3Al TiC system. The infiltration process works so effectively that "upward" melt infiltration, whereby the preform is placed on top of the infiltrating powder and it is drawn up into the porous preform, is also possible with this system and has been successfully demonstrated.[8] The furnace is evacuated to a vacuum of <10 millitorr, then ramped up to the hold temperature of 1475°C at a rate of 10°C/min. The samples are held at this temperature for 1 hour, and then cooled at a rate of 25°C/min. to room temperature. The infiltrated composites densities were then determined by the Archimedes immersion method in water. The hardness values were obtained on the polished samples using a Vickers diamond pyramid indenter; 50kg load (V 100A, Leco). The samples were also examined microstructurally using optical microscopy and SEM.

RESULTS AND DISUSSION

TiC Powder Characterization

The results from the acoustic particle sizing and the SEM observation of the as received TiC powder are presented in Figure 1. The particle size distribution shows a bimodal powder, one narrow peak at about 0.1 μm and a broad main peak at 1.0 μm. The average calculated particle size was ~1.2 μm. The SEM observation agrees with the particle sizing data; the large majority of the particles seen are ~1.0 μm, with the fines also clearly visible. The morphology of the particles is also revealed by the SEM images. The particles are very jagged and have somewhat of a plate like habit. The observed morphology can be explained by the powder production route; the TiC is ball milled by the manufacturer to produce the desired particle distribution. This would explain the fractured irregular particles and the abundance of fine fragments. The large number of fines, although a smaller fraction of the cumulative total, represents a large surface area. This could have an affect on the dispersant concentrations required to adequately coat the particles. The measured average surface area was 5.3 ± 0.3 m^2/g for the as received TiC powder.

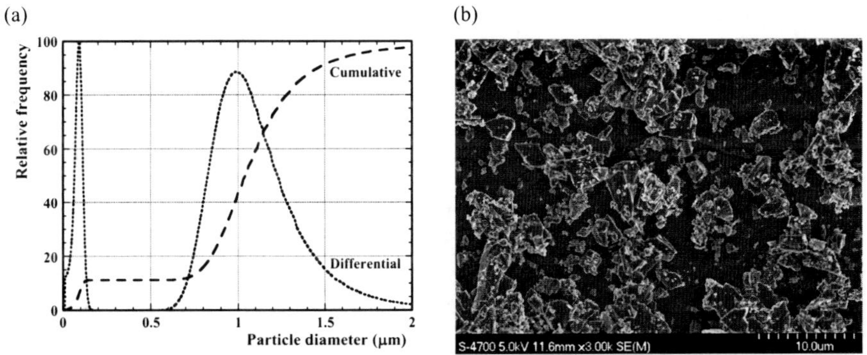

Figure 1. (a) Acoustic Particle Size Distribution for TiC powder as received. (b) SEM micrograph of the TiC powder as received.

Slip Characterization and Stabilization

The results of the sedimentation trials are shown in Figure 2. The data shown are for 10 days settling duration, at a pH of 7.0. The concentration required to achieve optimal stabilization was reached very early. With no polyelectrolyte the packing is poor, resulting in higher sediment volumes. With the addition of a small amount (0.01 0.02 wt. %) of PMA NH_4 polyelectrolyte the sediment volume drops off sharply and then begins to quickly climb again until a plateau is reached at approximately 0.06 wt. %.

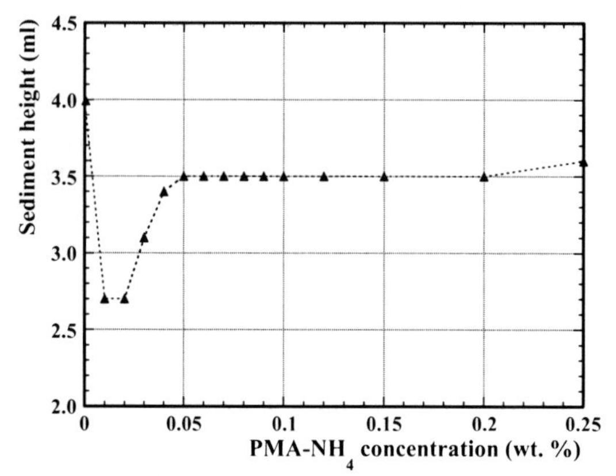

Figure 2. The sedimentation volume for TiC powder in water pH 7.0 after settling for 10 days.

The results of several zeta potential trials on the TiC powder are displayed in Figure 3. The powder with no polyelectrolyte added shows a distinct isoelectric point of pH ~2.6. It can also be seen the addition of the polyelectrolyte has no distinguishable effect on the isoelectric point as it remains significantly unchanged. This is in contrast to some other studies which have shown the same isoelectric point with out any polyelectrolyte, however with the addition of a polyelectrolyte, the suspension showed no isoelectric point at all.[9]

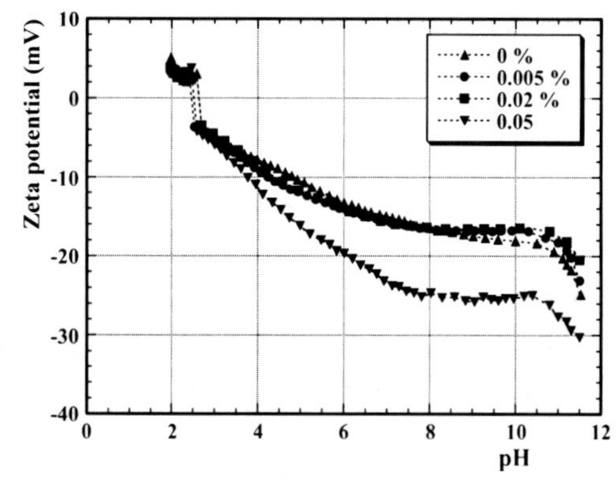

Figure 3. Zeta potential titration curves for TiC in water with varying wt. % PMA NH_4 polyelectrolyte concentrations.

The zeta potential information clearly shows a threshold below which a critical concentration the polyelectrolyte does not appear to coat the particles sufficiently to have a significant effect on the surface charge. Then, in the same region indicated by the sedimentation trials, the zeta potential begins to increase in magnitude with increasing pH; the surface charge transitions from 18 mV at pH of 8.5 to 25 mV with the addition of 0.0125 wt. % polyelectrolyte. However, the increase displays the same plateau from about pH 7 onward. This would seem to indicate that the stabilization mechanism is primarily eletrosteric and not electrostatic. In this case a dispersant which was primarily ionic in nature may provide increased stabilization at higher pH values, as preparing slips with a pH greater than ~8.5 shows no benefit to stability. There is some increase in the magnitude of the zeta potential above pH ~10.5. However, operating in this region is impractical as the amount of reagent is excessive and such basic conditions may have a detrimental effect on subsequent processing and analysis equipment. It should also be noted that further increasing the polyelectrolyte concentration above 0.0125 wt. % does not show a marked increase in zeta potential. It actually starts to degrade the properties of the slip shown by the sedimentation tests and in the following section describing the rheological observations.

Once the surface charge characteristics of the powder had been determined, together with the optimal range of operational pH values and polyelectrolyte concentrations, the slip rheology was analyzed for a variety of solids loadings and polyelectrolyte concentrations (all tests were performed at 20°C). Under all the examined conditions the suspensions show shear thinning behavior, with a transition to shear thickening at higher shear rates (~600 s^{1}). The shear thinning to thickening transition appears to shift to lower shear values with decreasing solids loading. The transition remains largely unchanged with changing polyelectrolyte concentration. Figure 4 shows the viscosity trends observed at a shear rate of ~200 s^{1}.

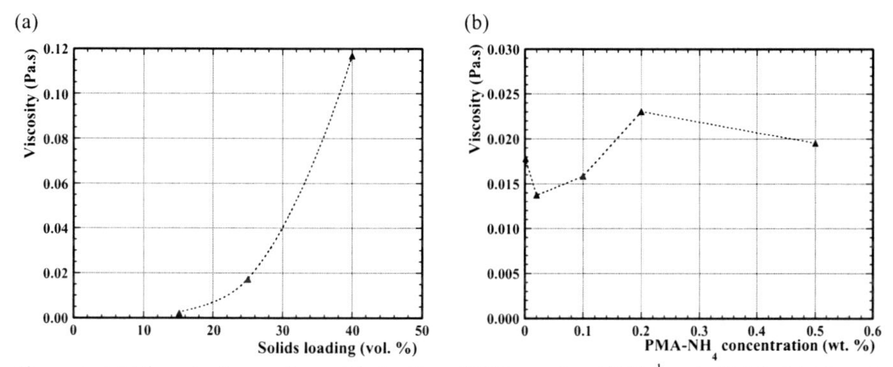

Figure 4. (a) Viscosity for varying solids loading of TiC powder at 200s^{1} and pH 8.5. (b) Viscosity versus polyelectrolyte concentration at 200 s^{1} with 30 vol. % solids loading (pH 8.5).

As expected the viscosity increases with increasing solids loading. The minimum slip viscosity which is also an indication of good stability is again shown in the 0.01 0.02 wt. % PMA NH_4 region. The viscosity then begins to increase as the dispersant concentration is increased. The subsequent reduction of the viscosity at high concentrations is most likely do to the fact significant liquid is being added to the suspension which is effectively reducing the solids loading and therefore the viscosity. Figure 5 shows the viscosity with respect to shear rate for various solid loadings and polyelectrolyte concentrations. As mentioned earlier, the shear thinning response is clearly visible for all of the suspensions that were examined. This can be generally explained by the particle morphology and, in the present instance, the somewhat plate like habit of the TiC particles (see Figure 1). As the shear rates increase the particles will begin to align in the shear field, thereby facilitating the particles to slip past one another resulting in a reduction of the overall viscosity. Eventually a critical point is reached which the suspensions transition to shear thickening. A detailed explanation of shear thinning and thickening in ceramic systems can be found in other works.[10] The viscosity trend for the PMA NH_4 polyelectrolyte corroborates the previously presented data, displaying the same optimized dispersant concentration region for suspension stability.

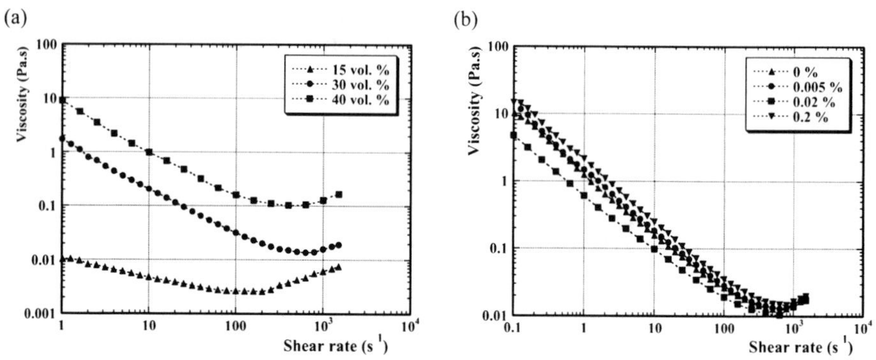

Figure 5. (a) Viscosity versus shear rate for varying TiC solids loadings (shear rates from 1 1500 s^{-1} at pH 8.5). (b) Viscosity versus shear rate for varying polyelectrolyte concentrations (at pH 8.5).

Sintering/Melt Infiltration

The TiC preform discs were created using the optimal slip conditions, determined by the previous experimentation, via directional slip casting onto plaster of Paris plates. Once dried, the samples were sintered using the procedure outlined previously in the Experimental Procedures section, then subsequently polished and examined. Figure 6 shows typical SEM micrographs of the final composite produced from the melt infiltration process. The TiC grains are clearly visible embedded in the Ni_3Al matrix. There is only limited porosity present in the lower Ni_3Al content samples and very good dispersion of the TiC particles within the matrix, with little to no agglomeration of particles above 20 vol. %. Higher Ni_3Al content samples show essentially no residual porosity. In the lower volume fractions there is insufficient liquid to completely fill all the interparticle voids. Melt infiltrations were successfully performed with volume fractions of Ni_3Al ranging from 20 to 50 vol. %.

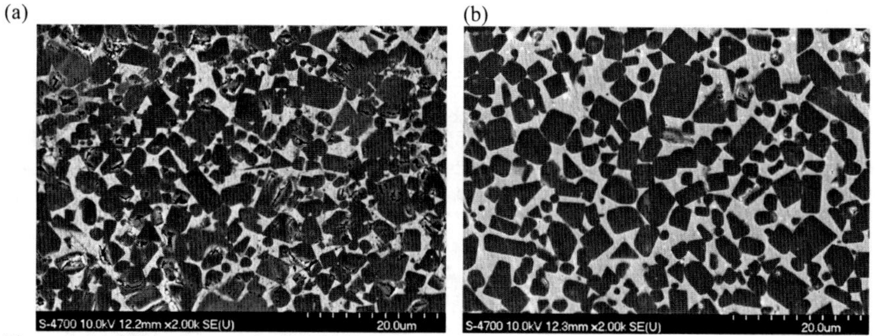

Figure 6. (a) SEM micrograph of TiC Ni_3Al 20 vol. % composite. (b) SEM micrograph of TiC Ni_3Al 45 vol. % composite.

For direct comparison of basic mechanical properties, in this case hardness, a series of dry pressed samples were also processed using exactly the same sintering conditions as for the slip cast samples. This allows for a direct comparison and quantification of the benefits of aqueous processing over conventional dry pressing methods. The resulting reduction in voids and increased particle packing should result in superior final properties in the infiltrated composites when prepared using slip cast preforms. Figure 7 shows the results from the Vickers indentation tests for both the dry pressed and slip cast samples with respect to the volume fraction of Ni_3Al. The trend of decreasing hardness with increasing binder content is essentially linear in both instances. This reduction in hardness with increasing binder is to be expected, based on the significantly lower elastic modulus of the intermetallic binder. Although not examined in the current work, it is has been shown that the fracture toughness will follow the opposite trend and increase with increasing binder content.[11] The hardness on average is increased by ~4 % in the slip cast specimens relative to those that were dry pressed.

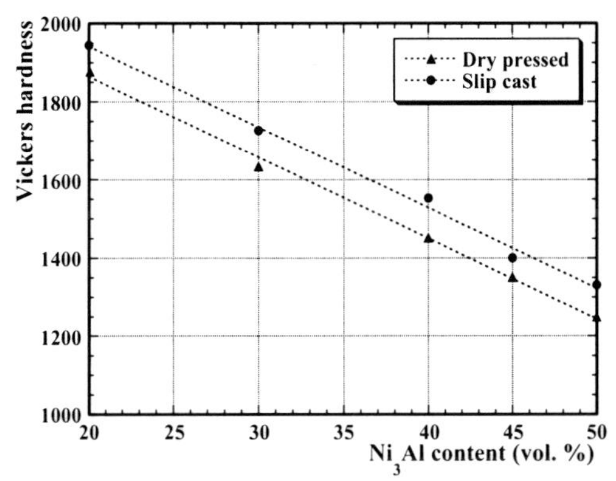

Figure 7. Vickers hardness versus Ni_3Al volume fraction for both slip cast and dry pressed samples.

It can be seen that the aqueous processing approach produced measurable improvements in the final composite material produced. The average dry pressed density was 98.8 % of theoretical in comparison to 99.7 % of theoretical in the slip cast samples. As seen in Figure 7 this amounts to significant gains in the materials hardness.

CONCLUSIONS

TiC slips were produced and the optimal deflocculent concentration was determined, along with the ideal pH region for slip casting with this system. The effect of over use of the polyelectrolyte was also observed with a marked loss of slip stability and the resulting higher viscosity. The slips exhibited significant shear thinning throughout the majority of shear rates tested only transitioning to shear thickening at high shear rates which would most likely not be found in typical processing conditions. The slip cast preforms that were produced could be successfully infiltrated with Ni_3Al using the pressureless melt infiltration method. The Vickers hardness of the resulting aqueous processed composites were compared to conventional dry pressed samples, sintered under the same conditions, and were demonstrated to be ~ 4 % higher for all Ni_3Al contents that were examined.

ACKNOWLEDGEMENTS

The National Science and Engineering Research Council of Canada (NSERC) are gratefully acknowledged for funding of this work through the Discovery Grants program. We would also like to thank the supporters of the Institute of Materials Research (IRM) who provided access to the SEM and the vacuum hot press, these include; the Canadian Foundation for Innovation, the Atlantic Innovation Fund, and several other supporting organizations.

REFERENCES

[1]A.G. Dobrovol'skii, "Slip Casting of Titanium Carbide Components," *Powder Metall. Metal Ceram.*, **4** 731 (1965).

[2]E.K. Storms, *The Refractory Carbides*; pp. 285. Academic Press, New York (1967).

[3]D.B. Miracle and H.A. Lippsitt, "Mechanical Properties of Fine Grained Substoichiometric Titanium Carbide," *J. Am. Ceram. Soc.*, **66** [8] 592 (1983).

[4]J.A. Lewis, "Colloidal Processing of Ceramics," *J. Am. Ceram. Soc.*, **83** [10] 2341 (2000).

[5]P. Becher and K.P. Plucknett, "Properties of Ni_3Al bonded Titanium Carbide Ceramics," *J. Eur. Ceram. Soc.*, **18** [4] 395 (1998).

[6]W. Acchar, U.U. Gomes, W.A. Kaysser and J. Goring, "Strength Degradation of a Tungsten Carbide Cobalt Composite at Elevated Temperatures," *Mater. Charact.*, **43** [1] 27 (1999).

[7]K. Plucknett and P.F. Becher, "Processing and Microstructure Development of Titanium Carbide Nickel Aluminide Composites," *J. Am. Ceram. Soc.*, **84** [1] 55 (2001).

[8]Y. Pan and K.W. Sun, "Preparation of TiC/Ni_3Al Composites by Upward Melt Infiltration," *J. Mater. Sci. Tech.*, **16** [4] 387 (2000).

[9]C. Yeh and M.H. Hon, "Dispersion and Stabilization of Aqueous TiC Suspension," *Ceram. Int.*, **21** [2] 65 (1995).

[10]L. Bergström, "Shear thinning and shear thickening of concentrated ceramic suspensions," *Colloids Surf. A*, **133** [1 2] 151 155 (1998).

[11]K. Plucknett, P.F. Becher and S.B. Waters, "Flexure strength of melt infiltration processed titanium carbide/nickel aluminide composites," *J. Am. Ceram. Soc.*, **81** [7] 1839 (1998).

THE EFFECT OF PRECIPITATOR TYPES ON THE SYNTHESIS OF $La_2Zr_2O_7$ POWDERS BY CHEMICAL COPRECIPITATION METHOD

Jing Wang, Shuxin Bai, Hong Zhang, Changrui Zhang
College of Aerospace and Materials Engineering, National University of Defense Technology, Changsha, 410073, China

ABSTRACT

Gel like precursors were synthesized by co precipitation method using ammonia and oxalate ammonium as precipitators, respectively. The process and crystallography characters of these precursors were studied by means of DSC TGA and XRD. The particle morphology and size distribution were researched by SEM and Laser Particle Size Analyzer. The results show that the hydrate precursor decomposes into amorphous composites of oxides at 400 750°C followed by crystallization of these oxides. The single phase $La_2Zr_2O_7$ is obtained just at 800 °C for 2.5h. And the oxalate precursor decomposes into mixture of amorphous oxides at 400 650°C followed by the solid state reaction between La_2O_3 and ZrO_2. The pure $La_2Zr_2O_7$ from oxalate precursor is not achieved until be calcined at 1450°C for 2.5h. The resultant powders from hydrate and oxalate precursor exhibit flaky and cuboid like morphology and the average particle sizes are 14.10μm and 9.58μm, respectively. Oxalate ammonium is prone to obtain dense and homogenous $La_2Zr_2O_7$ powders compared with ammonia.

INTRODUCTION

$La_2Zr_2O_7$ is becoming a very promising candidate of yttria stabilized zirconia (YSZ) for new TBCs because of its excellent thermal stability, low thermal conductivity, chemical resistance and low sintering rate [1 4].

Different methods can be used to prepare $La_2Zr_2O_7$ and these include solid state reaction, nitric acid dissolution route, sol gel technique, self propagating high temperature synthesis (SHS), hydrothermal routes and coprecipitation [4 9]. Co precipitation is widely used to prepare powders with uniform composition on a nanostructure level because of its simple equipments, low energy consumption, inexpensive material and short production period etc.

The precipitant agents used in co precipitation usually include ammonia, oxalic acid, salvolatile and hydrazine, etc [10 13]. However, the oxalate ammonium used as precipitator for preparing LZ has not been investigated up to now. In this research, $La_2Zr_2O_7$ ceramic were synthesized by the coprecipitation method, using different precipitant agents, ammonia and oxalate ammonium. The resultant precursors have been characterized by DSC TGA and XRD to examine the effect of precipitator on the formation processes and conditions. The resultant powders have been characterized by various techniques to examine the effect of the precipitators on the morphology and particle size distribution.

EXPERIMENTAL

Preparation of $La_2Zr_2O_7$ powders

In the present study, lanthanum oxide powder and zirconium oxychloride were chosen as the

reactants. Ammonia and oxalate ammonium were used as precipitators respectively. Firstly, the lanthanum oxide was dissolved in nitric acid and zirconium oxychloride was dissolved in distilled water. The concentration of all cations was 1mol/L. And then these solutions were mixed in appropriate proportions and stirred for 30 min, which were named as solution A.
Then two portions of solution A were added dropwise to ammonia (1:1) and oxalate ammonium solution (30% excessive), separately, with stirring, to obtain two different kinds of gel like precipitates. During the whole process, the pH was kept at 10 by adjusting the flow of ammonia. The mixture was stirred vigorously for 30 min. Then the precipitates were filtered and washed with distilled water and ethanol. The washed precipitates were dried at 120°C for 12 h and the densified cake was milled. The hydrate precursors were calcined at different temperatures from 700 to 1100 °C for 2.5h. The oxalate precursors were calcined at different temperatures from 850 to 1450 °C for 2.5h.

Characterization techniques

Thermal decomposition characteristics of the different samples were studied in air with a differential scanning calorimeter (NETZSCH STA 449C, Germany) operating in a temperature range between 25°C and 1300°C at a heating rate of 10 K/min in air.
The crystalline phase structure was determined by a Japan RigaKu D/MAX 2550 VB/PC diffractometer with Cu K_α radiation (λ=0.154056 nm). Continuous scans in the range 10 85° were used.
The density and particle size of the powder was measured by the Archimedes method with an immersion medium of distilled water and by laser particle size analyzer (MS 2000, U.K.).
The particle morphology was recorded on ultra high resolution scanning electron microscope (S 4800, Japan).

RESULTS AND DISCUSSION

Thermal analysis

Fig.1 shows the results of thermal analysis (DTA and TGA) of the hydrate precursor (a) and oxalate precursor (b) prepared through coprecipitation.

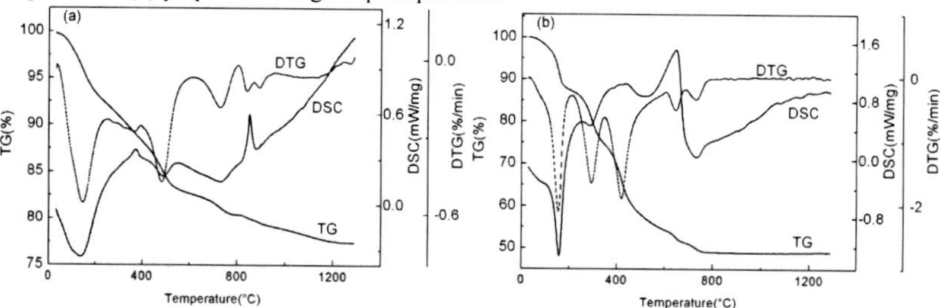

Figure 1. DSC TGA curves of La Zr precursor: (a) Hydrate, (b) Oxalate

There are three endothermic peaks and two exothermic peaks in the DSC curve of Fig.1a. The endothermic peak at 140°C is due to the expulsion of physically bonded water. The exothermic peak at 372°C is almost neglectable, accompanied by a small weight loss. The temperature of sublimation for NH_4Cl is 340°C, so this exothermic reaction is probably due to the decomposition of NH_4Cl residual in hydrate precursor. Increasing the temperature, the DSC trace shows two endothermic peaks, one centred around 488°C, and the other centred around 734°C, corresponding with some weight loss(7.87%). The removal of the hydroxyl groups of metal hydrate complexes has been described as being endothermic [14]. So the reactions which were taken place between 400 and 800°C are that, the hydroxides decompose into amorphous oxides and H_2O. The abrupt exothermic peak at around 852°C, approaching that of J. NAIR et al., given as 812°C [14], is probably attributed to the crystallization of above mentioned amorphous oxides and is also accompanied by a small weight loss.

There are three endothermic peaks and two exothermic peaks in the DSC curve of Fig.1b. When lanthanum oxalate and zirconium oxalate are heated, they undergoes dehydration followed by decomposition into theirs oxide. So the endothermic peaks stand at 160°C and 288°C, corresponding to weight losses as seen from the TG and DTG, are due to the loss of physically adsorbed water in the gel. Increasing the temperature, the DSC trace shows two exothermic peaks, one centred around 420°C, and the other centred around 646°C, corresponding to the decomposition of the oxalate complex. The significant weight loss probably comes of the evolution of CO and CO_2. Around 733°C, an endothermic peak appears, corresponding with small weight loss (2.24%). This phenomenon agrees well with that in Fig.1a. The reaction has been described as dehydration of hydroxide. And it indicates that hydroxide also exists in oxalate precursor. Further analysis will be given out infra. After 800 °C, no marked change in mass or energy can be observed. Due to limitations of the measurement apparatus, the DSC TGA is determined only from ambient temperature to 1300°C.

XRD studies

The hydrate precursor (a) and oxalate precursor (b) prepared through coprecipitation have been calcined at different temperature, and the crystallography characters of ceramic powder are studied by XRD (Fig.2).

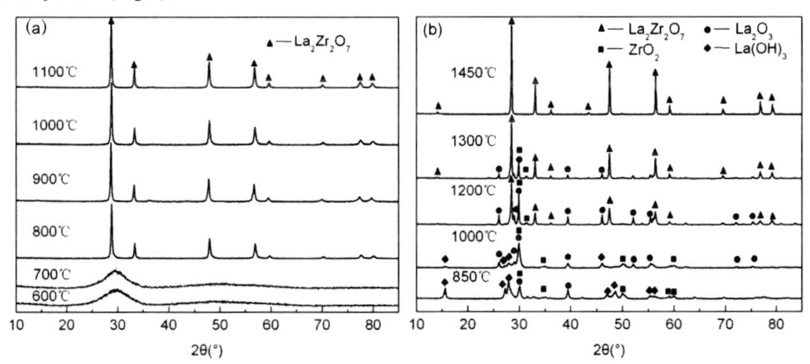

Figure 2. XRD Patterns of lanthanum zirconate ceramic powders treated at different temperatures: (a) Hydrate, (b) Oxalate

The results show that, when heated, the hydrates undergo decomposition into theirs amorphous composite oxides, which exist even at 700°C. The XRD lines of $La_2Zr_2O_7$ begin to appear under 800°C and their intensities increased up with increasing the heating temperature. It is remarkable that the presence of ZrO_2 or La_2O_3 is not observed throughout the heating process.

The reaction processes of oxalate precursors leading to the formation of $La_2Zr_2O_7$ are more intricate than that of hydrates. As shown in Fig.2b, the oxalates decompose into mixture of amorphous oxides under 650°C. The crystallization of amorphous oxides was observed with increasing temperature. At 850°C, the XRD lines of ZrO_2 and La_2O_3 are inapparent, and then the latter's intensity increased up to 1200°C, followed by decreasing gradually at 1200 to 1300°C. The specimen heated at 1200°C also gives the characteristic XRD pattern of $La_2Zr_2O_7$, and the formation is observed with the temperature increasing. The results suggest that the $La_2Zr_2O_7$ is formed by the diffusion of Zr^{4+} towards La_2O_3.The single phase $La_2Zr_2O_7$ is obtained in the specimen heated at 1450°C for 2.5h. The XRD patterns show that, after 700°C, three reactions, such as the decomposition of $La(OH)_3$, the crystallization of mixture of amorphous oxides and the solid state reaction between La_2O_3 and ZrO_2, take place synchronously. The reactions above mentioned are endothermic, exothermic and endothermic, respectively. Here, the unmarked change of energy in DSC curve in Fig.1a can be explained soundly.

Fig.2b also confirms the existence of little $La(OH)_3$ crystal in oxalate precursors which are calcined under 1200°C. The essence of the existing of $La(OH)_3$ is that, the La^{3+} will not be deposited completely by oxalate ammonium because of its strong alkalescence, and residual La^{3+} will react with the ammonia which was used for adjusting the PH of precursor solution. From the result of XRD, we can conclude that the decomposition temperature of $Zr(OH)_4$ is lower than that of $La(OH)_3$, given as about 488°C and 734°C.

When all the results are consisted, the formation process and mechanism are different by the precipitant agents. The heat temperature for single phase LZ from hydrate precursors is much lower. La^{3+} and ZrO^{2+} mixed with NH_4OH to form La Zr composite hydrate, in which the La^{3+} and Zr^{4+} achieve homogeneous at the level of atom. So the single $La_2Zr_2O_7$ can be obtained by heating at low temperature (800°C). However, it is difficult for the lanthanum oxalate and zirconium oxalate to mix at the level of atom but to react at the interface of solid state because of their great volumes. Undergoing heat, oxalates decompose to amorphous oxides, which then transform into ZrO_2 and La_2O_3. The mechanism of reaction between ZrO_2 and La_2O_3 to form LZ is similar to that of solid state reaction. Therefore the formation of pure $La_2Zr_2O_7$ can not achieved until heating at 1450°C.

Powder characteristics of $La_2Zr_2O_7$

The LZ powders from hydrate and oxalate precursors which were calcined at 1100°C and 1450°C, respectively, were named PI and PП. The bulk density of PI and PП were 5.37g/cm^3 and 6.00g/cm^3, respectively, corresponding to 88.82% and 99.24% of theoretical density (6.046 g/cm^3) [13]. Fig.3 and Fig.4 show the SEM micrographs for $La_2Zr_2O_7$ powders. As can be seen, precipitators have significant effect on particle morphology in the co precipitation method, and the as prepared PI particles are lacunaris and had flaky morphology (Fig.3a), while PП exhibit dense and cuboid like morphology (Fig.4a). High synthesize temperature allow more time and energy for the conversion from oxalate precursors to single phase LZ, and result in the increase of crystal and the sintering of powders, which would increase the density of particles (Fig.3b and Fig.4b). The SEM observations agree well with trend of density by Archimedes method.

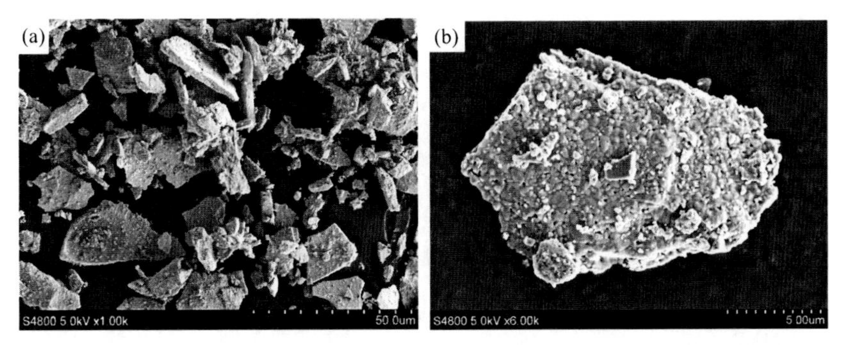

Figure 3. SEM images of lanthanum zirconate ceramic powders obtained from Hydrate at (a) ×1,000 and (b) ×6,000.

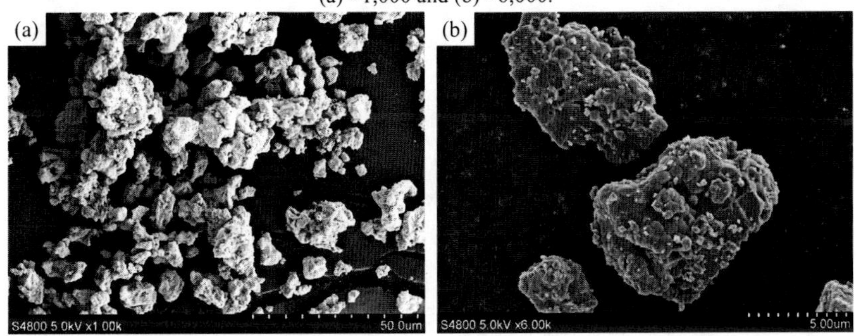

Figure 4. SEM images of lanthanum zirconate ceramic powders obtained from Oxalate at (a) ×1,000 and (b) ×6,000.

In order to support this observation, further investigation was carried out using laser particle size analyzer. And the particle size distributions are shown in Fig.5. It reveals that the mean particle size of PΠ is more homogeneous and smaller than that of PI, which are 9.58μm and 14.10μm, respectively. Apparently, the particle size of PI from SEM observations corresponding to that from Laser Particle Size Analyzer demonstrates that it is easier for oxalic acid to produces finer particles than ammonia.

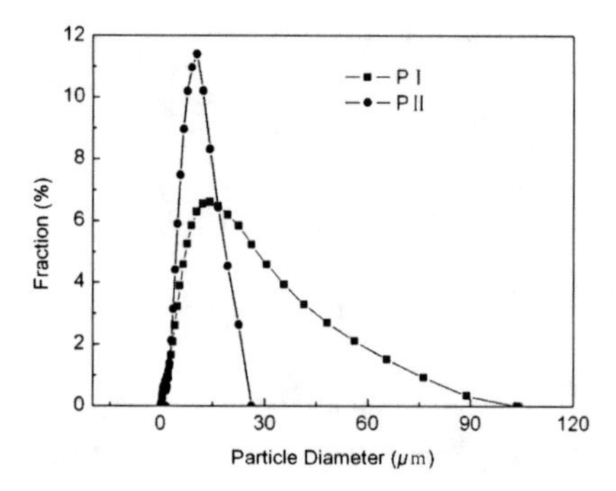

Figure 5. Particle size distribution of lanthanum zirconate ceramic powders obtained from (■) Hydrate and (●) Oxalate

CONCLUSIONS

Single $La_2Zr_2O_7$ (LZ) with pyrochlore structure can be synthesized by co precipitation method using ammonia and oxalate ammonium as precipitators, respectively. And this work demonstrates that the type of precursor has significant effects on synthesize process, particle size and morphology. Pure $La_2Zr_2O_7$ can be obtained at lower temperature while using ammonia as precipitator. The resultant powders from hydrate and oxalate precursors exhibit flaky and cuboid like morphology and the average particle sizes are 14.10μm and 9.58μm, respectively. Besides, the LZ powders from oxalate precursor behave more dense and homogenous.

REFERENCE

[1] D.R .Clarke, S.R. Phillpot, Thermal Barrier Coating Materials, *Mater. Today* **6**, 22 29 (2005).

[2] X.Q. Cao, R. Vassen, D. Stoever, Ceramic Materials for Thermal BarrierCoatings, *J. Eur. Ceram. Soc.*, **24**, 1 10 (2004).

[3] D. Stöver, G. Pracht, H. Lehmann, M. Dietrich, J.E. Döring and R. Vaβen, New Material Concepts for the Next Generation of Plasma sprayed Thermal Barrier Coatings, *J. Therm. Spray Technol.*, **13**, 76 83 (2004).

[4] H.M. Zhou, D.Q. Yi, Z.M. Yu and L.R. Xiao, Preparation and Thermophysical Properties of CeO_2 Doped $La_2Zr_2O_7$ Ceramic for Thermal Barrier Coatings, *J. Alloys Compd.*, **438,** 217 21 (2007).

[5] J.Y. Li, H. Dai, X.H. Zhong, Effect of the Addition of YAG ($Y_3Al_5O_{12}$) Nanopowder on the Mechanical Properties of Lanthanum Zirconate, *Mater. Sci. Eng. A*, **460 461**, 504 8 (2007).

[6] F.W. Poulsen, N. Puil, Phase Reactions and Conductivity of Sr and La Zirconates, *Solid State Ionics*, **53 56**, 777 83 (1992).

[7] H. Kido, S. Komarneni, R. Roy, Preparation of $La_2Zr_2O_7$ by Sol Gel Route, *J. Am. Ceram. Soc.*, **74**, 422 24 (1991).

[8] Y.P. Tong, Y.P. Wang, Z.X. Yu, Preparation and Characterization of Pyrochlore $La_2Zr_2O_7$ Nanocrystals by Stearic Acid Method, *Mater. Lett.*, **62**, 889 91 (2008).

[9] S. Komarneni, Hydrothermal Preparation of Low expansion NZP Family of Materials, *Int. J. High. Tech. Ceram.*, **4**, 31 9 (1988).

[10]S. Lutique, P. Javorsk, R.J.M. Konings, Low Temperature Heat Capacity of $Nd_2Zr_2O_7$ Pyrochlore, *J. Chem. Thermodyn.*, 35, 955 65 (2003).
[11]C. Bo, J.B. Yi, Y.S. Han and J.H. Dai, Effect of Precipitant on Preparation of Ni Co Sipinel Oxide by Coprecipitation Method, *Mater. Lett.*, **58**, 1415 1418 (2004).
[12]C.H. Li, K.K. Yao, J. Liang, Study on the Features of Multiwaled Carbon Nanotube Supported Nickel Aluminum Mixed Oxides, *Appl. Catal. A Gen.*, **261**, 221 24 (2004).
[13]Y. Matsumura, M. Yoshinaka, K. Hirota and O. Yamaguchi, Formation and Sintering of $La_2Zr_2O_7$ by the Hydrazine Method, *Solid State Commun.*, **104**, 341 45 (1997).
[14]J. Nair, P. Nair, E.B.M. Doesburg, J.G. Van ommen, J.R.H. Ross and A.J. Burggraaf, Preparation and Characterization of Lanthanum Zirconate, *J. Mater. Sci.*, **33**, 4517 23 (1998).

THE STUDY OF PREPRATION OF BLUE V ZIRCON PIGMENT BY USING ZIRCON AND SULPHURIC ACID

M. Riahi, M.A. Faghihi Sani

Department of Materials Science and Engineering, Sharif University of Technology, Tehran, Iran.

ABSTRACT

Depending on the type of dopant metal, many kinds of zircon based pigments are produced and blue vanadium pigment is the most important one of them. In this study, blue zircon ceramic pigments were synthesized from intermediate product, resulting from decomposition of zircon sand with NaOH. In this regard, various amounts of sulfuric acid, water, NH_4VO_3 as colorant, NaF as mineralizer and extra quartz were added to the prepared Na_2ZrSiO_5. Role of quartz was to omit the repercussions of presence of free zirconia in the composition. The main objective of this work is to assess various reactions at different temperatures during blue pigment synthesis. Phase analysis was done by X ray diffraction and infrared spectroscopy. Color quality was determined by CIE Lab. The best composition was studied by DTA to determine temperatures of the reactions. Powders heated at various temperatures were studied by X ray diffraction and infrared spectroscopy to investigate crystallization behavior. It was concluded that V^{+5} doped tetragonal zirconia has first been formed at temperatures lower than 600°C. Then, from 820 to 890°C, an endothermic reaction of formation of monoclinic zirconia from tetragonal zirconia and an exothermic reaction of formation of zircon from monoclinic zirconia and quartz occur, respectively.

Keywords: Zircon Sodium silicozirconate Vanadium Sulfuric acid Tetrahedral

INTRODUCTION

Pigments based on zircon ($ZrSiO_4$) are widely used in ceramic industries because of their superior stability under the adverse conditions of high temperatures and corrosive environments encountered in molten glazes. Furthermore, doped and inclusion zircon based pigments yield a variety of colors and shades such as blue, yellow, coral and red[1,2].

The principal structural unit cell of zircon is a chain of alternating edge sharing SiO_4 tetrahedral and ZrO_8 triangular dodecahedral, characterized by the presence of empty octahedric cavities. The metal dopants can be accommodated into the zircon network in tetrahedral interstitial positions as well as Zr^{4+} or Si^{4+} substitutional lattice positions. In several cases controversial conclusions are reached in the literatures about the actual location of the metal ion and its valence in the zircon lattice[2,3]. The colourants can also form an encapsulated phase to prepare zircon inclusion pigments.

Recently, it has been reported that these pigments can also be prepared at lower temperatures by non conventional methods such as aerosol hydrolysis[4], sol gel[5,6] and coprecipitation routes[7] than solid state methods. Since ZrO_2 is obtained by complex and expensive purification process from zircon sands, synthesis of a pigment by using zircon or intermediate compounds in the process of making ZrO_2, might have a significant economical advantage. Several numbers of patents have claimed that intermediate products in the process of making ZrO_2 from zircon can be used to produce zircon pigments with more regular and intense hue[1,8,9]. In this regard, as the first step, zircon is heated in the presence of NaOH to produce sodium silicozirconate[1,7]. During the second step, the silicozirconate get decomposed to silica and zirconia by using different kinds of acids such as sulfuric acid. Then, during heating, silica is attacked by the mineralizer such as sodium fluoride (NaF) and transported to the ZrO_2 grains and reacts with them to form $ZrSiO_4$. The dopant ion is also trapped in the lattice during the zircon formation process[10 12].

Two objectives of the present work are to determine the best condition to prepare sodium silicozirconate and to assess various reactions occurring during blue V zircon pigment synthesis.

EXPERIMENTAL PROCEDURE

Decomposition of zircon into the intermediate products

It is known that zircon decomposes at high temperature before direct melting. The decomposition temperature decreases in presence of fluxing agents such as NaOH, Na_2CO_3 or $CaCO_3$. In this study, to decompose zircon sand, NaOH was chosen as a fluxing agent due to the fact that the decomposition reaction occurs at a lower temperature in comparison with the others[13]. In this regard, various molar ratios of NaOH (Merck) to zircon (cookson with 98.5% purity and less than 5 μm powder size) were chosen (as 2, 2.1 and 2.2). In order to determine approximate temperature of zircon decomposition in the presence of sodium hydroxide, simultaneous thermal analysis (STA; Model 409PC, NETZSCH) was conducted on the mixture. According to the result, the mixtures were heated at 900°C for two hours. To identify the crystalline phases present in the yield materials, XRD patterns were obtained using conventional powder diffraction technique in a Bruker diffractometer with Ni filtered, Cu Kα radiation. After choosing the best molar ratio as 2.2, thermal treatment was carried out at temperatures between 750 and 950°C for 1, 1.5, 2 and 3 h by applying a heating rate of 600°C/h. In order to identify the best heating condition, X ray diffraction patterns of the heated mixtures were obtained.

Preparation and characterisation of pigments

To synthesize zircon pigment, sodium silicozirconate was mixed with excess SiO_2 (Merck), used to omit repercussions of presence of extra zirconia in composition, together with NH_4VO_3 (Aldrich) and NaF (Merck), as colorant and mineralizer, respectively, according to Table 1.

Table I. Pigment compositions prepared with sodium silicozirconate (weight part)

code	Na_2ZrSiO_5	water	H_2SO_4	Extra SiO_2	NaF	NH_4VO_3	t_c (h)*	T_c (°C)**
B1	50	70	25.6	1.25	2.12	2.31	2	1000
B2	50	50	25.6	1.25	2.12	2.31	2	1000
B3	50	50	19.7	1.25	2.12	2.31	2	1000
B4	50	50	30.73	1.25	2.12	2.31	2	1000
B5	50	50	25.6	1.25	2.12	2.65	2	1000
B6	50	50	30.73	1.25	2.56	2.31	2	1000
B7	50	50	30.73	1.25	3.42	2.31	2	1000
B8	50	70	25.6	2.5	2.12	2.31	2	1000
B9	50	50	25.6	2.5	2.12	2.31	2	1000
B10	50	50	19.7	2.5	2.12	2.31	2	1000
B11	50	50	30.73	2.5	2.12	2.31	2	1000

The mixtures with various amounts of water were ground for 30 minutes to obtain homogeneous slurries. Then, sulphuric acid (Merck with 96% concentration) was added to decompose sodium silicozirconate into SiO_2 and ZrO_2. The formed gel was then dried and calcined in an electric furnace at 1000°C for 2 h by applying a heating rate of 600°C/h.

The prepared pigments were wet ground and washed with water and sulphuric acid to remove undesirable soluble salts.

L*, a* and b*, as colour parameters of the pigment powders, were measured using a spectrophotometer (Model Colour Eye 7000A, GretagMacbeth) in the wavelength range from 400 to 750 nm with an illuminant D65.

To study mechanism of the blue pigment formation, one of the dried gels, resulted in the best quality blue pigment, was first analyzed by differential thermal analysis (Model 409PC, NETZSCH). Then, XRD and IR spectroscopy were conducted on the heated mixtures at critical temperatures, derived from the DTA result.

The IR spectra were measured in the wave number range from 380 to 2000 cm^{-1}, using a Fourier transformation infrared (FTIR) spectrometer (Model Mahson 1000, Unicam). The potassium bromide (KBr) pellet method was employed and the pigment powders were diluted in the KBr.

RESULTS AND DISCUSSION

Sodium silicozirconate synthesis

Zircon decomposition reaction in the presence of alkaline hydroxide occurs on the basis of the following equation[9]:

$$ZrSiO_4+2NaOH \rightarrow Na_2ZrSiO_5+H_2O \quad (1)$$

In order to verify a suitable temperature range for the zircon decomposition at the presence of sodium hydroxide, STA was conducted on the raw mixture. Figure 1 shows that zircon decomposition starts at 750°C and finishes at 950°C.

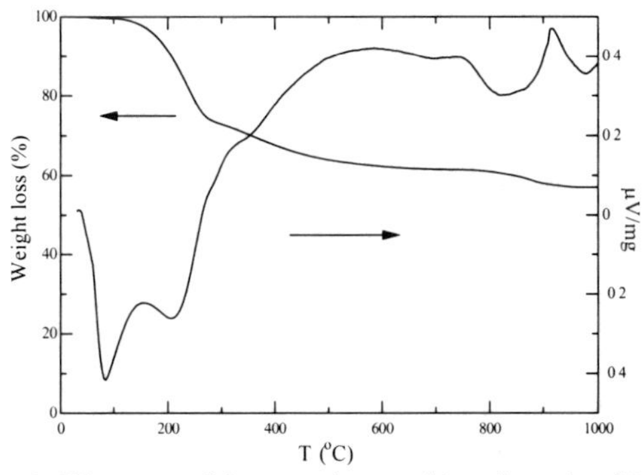

Figure 1. STA curve of the raw mixture with molar ratio of 2

According to the reaction 1 and STA results (Figure 1), it can be expected that raw mixture with molar ratio of 2 might be transformed to Na_2ZrSiO_5 around 900°C, because the exothermic peak (related to formation of Na_2ZrSiO_5) occurs around this temperature. But XRD results in Figure 2 show presence of unreacted zircon in the yield materials, probably resulting from evaporating and escaping some parts of NaOH. Figure 2 indicates that the peaks of unreacted zircon disappear with increase of n up to 2.2.

XRD patterns of the heated samples at 750, 800, 850, 900 and 950°C for an hour are presented in Figure 3. When temperature increases, intensity of the unreacted zircon peaks decreases.

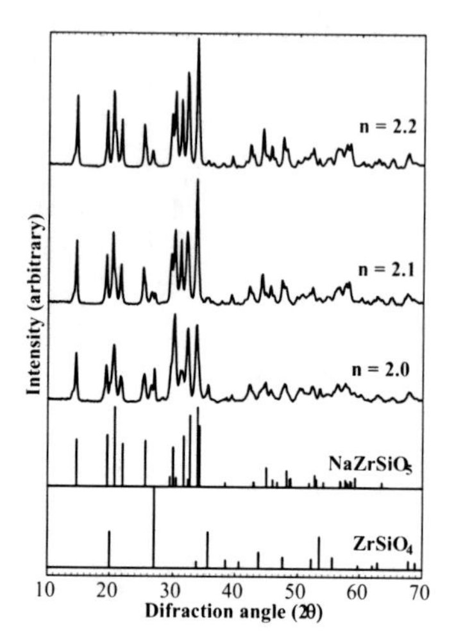

Figure 2. XRD patterns of Na_2ZrSiO_5 synthesized by various molar ratios of NaOH to zircon after heating at 900°C for 2 hours

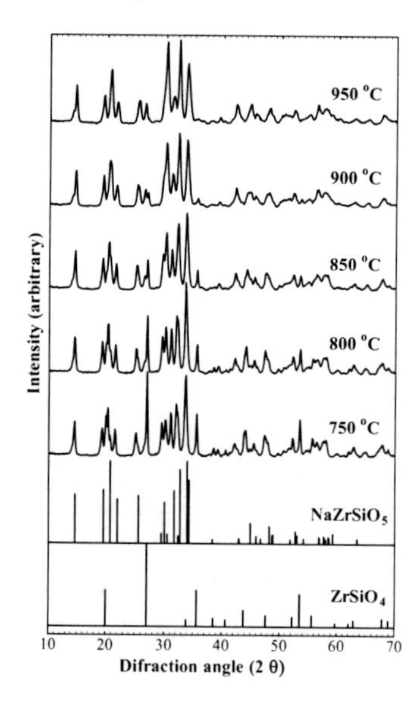

Figure 3. XRD patterns of Na_2ZrSiO_5 heated at various temperatures for 1 hour (n=2.2)

To find out effect of the heating time on the reaction, samples were heated at 900°C for various heating times. XRD patterns of these samples are illustrated in Figure 4. When heating time increases, intensity of zircon peaks decreases and their peaks completely disappears after heating 2 hours. Therefore, the best heating condition was assigned as 900°C for 2 hours.

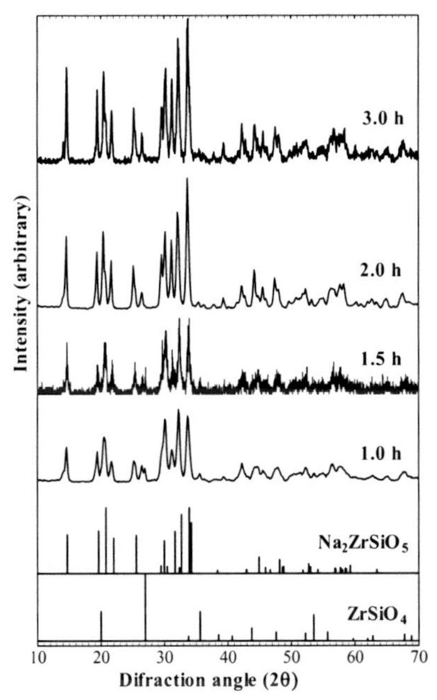

Figure 4. XRD patterns of the heated mixtures at 900°C for various heating times (n=2.2)

Pigment preparation

Table II shows L* a* b* values of the prepared pigments. According to Table II sample B4, as the best pigment, was studied to identify mechanism of the blue pigment formation.

Table II. The L* a* b* values of the prepared pigments

code	L*	a*	b*	code	L*	a*	b*
B1	76.7	9.4	11	B7	62.9	19.2	21.3
B2	64	16.5	22	B8	78	8.5	10
B3	64	16.6	20.3	B9	65.8	15.4	21.3
B4	57.7	16	22.2	B10	65.9	15.4	19.8
B5	62.7	16.4	22.1	B11	64.3	15.8	20.1
B6	63	17.1	22	RCP***	70.1	16.7	19.6

Infrared spectra of the mixture, including Na_2ZrSiO_5, water, NH_4VO_3, NaF, after adding sulphuric acid, are shown in Figure 5.

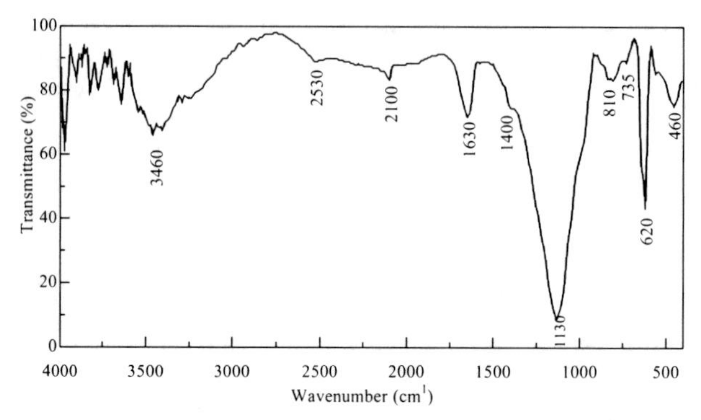

Figure 5. Infrared spectra of the raw mixture used to prepare B4

In this curve, IR absorption peak around 3460 cm $^{-1}$ is O H stretching mode and the 2530 cm $^{-1}$ peak is related to the O=S=O band in Na_2SO_4 compound. The peak 2100 cm $^{-1}$ belongs to the H O H, the peak 1630 cm $^{-1}$ is associated with O H stretching vibration mode, the peak 1400 cm $^{-1}$ is related to the Zro , the peak 1130 cm $^{-1}$ is relevant to O=S=O in Na_2SO_4, the peak 810 cm $^{-1}$ is relevant to the V_2O_5 compound and the peak 460 cm $^{-1}$ is related to Si O Si bending mode[4, 14, 15,16].

As it is indicated in the above figure, there is no peak relevant to the sodium silicozirconate compound. This means that the sodium silicozirconate has been totally decomposed with sulphuric acid. Presence of peak relevant to the O H band shows hydration of the compound.

Figure 6 indicates DTA curve of raw mixture of sample B4. The endothermic peak at 113.5°C is related to removal of water. There is another weak endothermic peak at 256.1°C, which might be due to the dehydration (as indicated in the IR spectra in Figure 5) and also the decomposition of NH_4VO_3 to V_2O_5 [9].

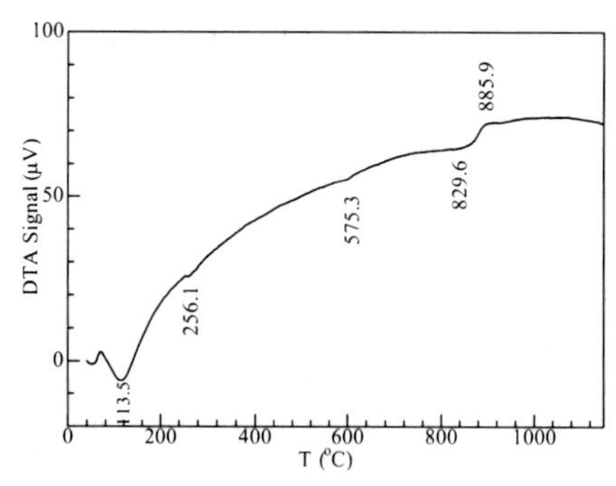

Figure 6. DTA curve of sample B4 (heating rate: 600°C/h)

The endothermic peak around 575.3°C must be related to the quarts transformation. According to the previous study[14], tetragonal zirconia can also be formed in presence of NH_4VO_3 around this temperature. To confirm formation of zirconia, the raw sample after heating at 600°C for 30 minutes was analyzed by XRD and IR spectroscopy. The results are presented in Figure 7 and 8. As Figure 7 shows, both tetragonal and monoclinic zirconia peaks are presented in the XRD pattern, which confirms the previous works. Vanadium can stabilize zirconia in the form of tetragonal at lower temperature[14].

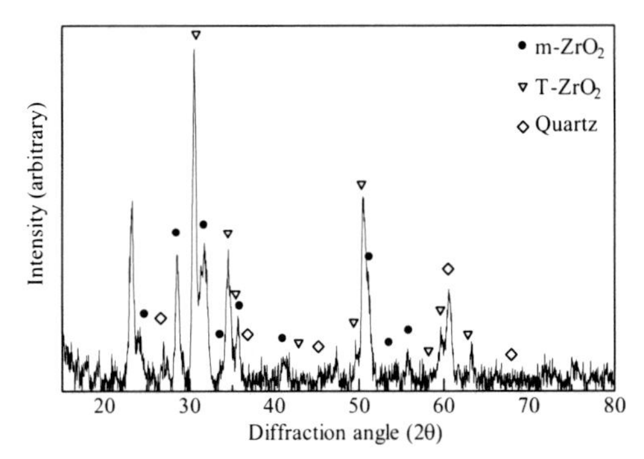

Figure 7. XRD pattern of the sample B4 after heating at 600°C for 30 minutes

The peak 3450 cm^{-1} in Figure 8 is O H stretching mode, the peak 1130 cm^{-1} relevant to the O=S=O band in Na_2SO_4, the peak 975 cm^{-1} is related to the SiO , the peak 750 cm^{-1} belongs to the Zr O stretching mode, the peak 620 cm^{-1} is relevant to V_2O_5 compound and the peak 470 belongs to the Si O Si bending mode [4, 14, 15 16].

According to these results, one can conclude that at 600 °C there are very few hydrated compounds remained in the sample, also V^{5+} in the form of amorphous V_2O_5 exists in the system, causing the sample to seem yellow.

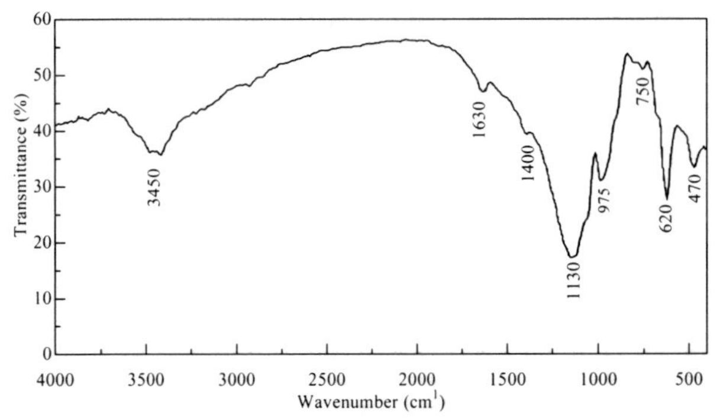

Figure 8. Infrared spectra of mixture of the sample B4 after heating at 600°C for 30 minutes

The endothermic and exothermic strong peaks around 829.6°C and 885.9°C in the DTA curve (Figure 6) are resulted from the transformation of monoclinic to tetragonal zirconia and reaction of silica and monoclinic zirconia to produce zircon, respectively. These two reactions are usually carried out simultaneously[14].

Figure 9 shows the XRD pattern of the synthesized blue pigment. Zircon is the only crystalline phase in this pattern.

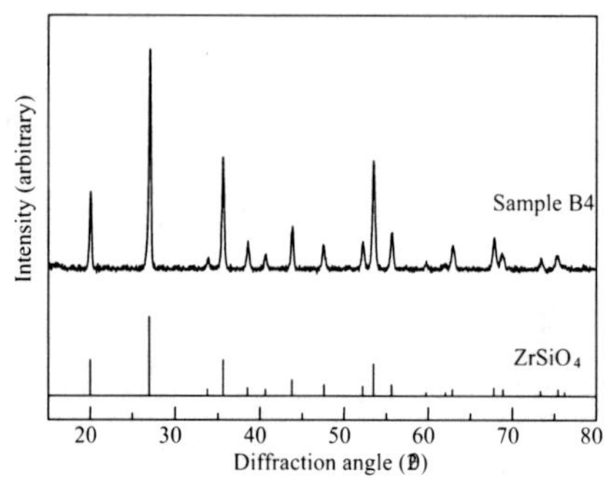

Figure 9. XRD pattern of sample B4 after heating at 1000°C for 2 hours and washing

Figure 10 illustrates IR spectra of synthesized blue pigment. The peak 1000 cm^{-1} is Si O Si stretching vibration mode, the peak 900 cm^{-1} is relevant to the Si O band in SiO (nonbridging oxygen), the peak 615 cm^{-1} is related to the Zr O in ZrO_8 octahedral position, the peak 535 and 440 cm^{-1} belong to the Zr O stretching mode, which all relevant to the zircon compound [4, 14, 15 16]. Also, there is another peak at 800cm^{-1} related to the V^{4+} ions, located in tetrahedral positions. In several cases controversial conclusions are reached in the literature about the actual location of the V^{4+} ion in the zircon lattice. However, S. Ardizzone et. al. claimed that V^{4+} ion is likely located in tetrahedrally coordinated interstitial site[3, 17].

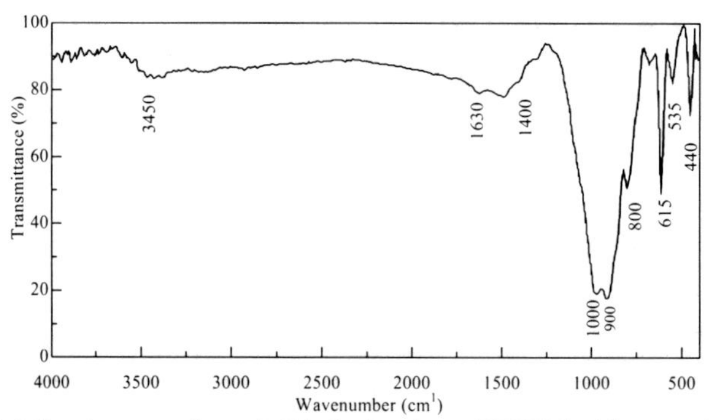

Figure 10. Infrared spectra of sample B4 after heating at 1000°C for 2 hours and washing

CONCLUSION

In this work, a blue V zircon pigment, comparable with the commercial one, was synthesized by a simple and economic method. In the first step, zircon sand was thermally decomposed into the sodium silicozirconate by reacting with NaOH. Effects of molar ratio of NaOH/zircon as well as calcination temperature and time on the decomposition reaction were investigated and the best conditions were found as 2.2, 900°C and 2 hours, respectively.

To synthesis the blue pigment, sodium silicozirconate was mixed with excess SiO_2 (used to omit repercussions of presence of extra zirconia in composition) together with NH_4VO_3 as colorant and NaF as mineralizer. Then, sulfuric acid was added to the slurry to decompose sodium silicozirconate. The prepared gel was finally dried and heated at 1000°C for 2 hours to obtain the blue pigment. Investigating mechanism of pigment formation and reactions showed that tetragonal zirconia was formed in presence of NH_4VO_3 around 600°C. The transformation of monoclinic to tetragonal zirconia and reaction of silica and monoclinic zirconia to produce zircon occur around 829.6 °C and 885.9 °C, respectively. IR spectroscopy results confirmed the previous studies concerning localization of V^{4+} ions inside tetrahedral positions.

FOOTNOTES

* Calcination time

** Calcination temperature

*** Reference commercial pigment

REFERENCES

[1]E. Ozel , S. Turan, Production of coloured zircon pigments from zircon, Journal of the European Ceramic Society, 27, 1751 1757 (2007).

[2]Giulia Del Nero, Giuseppe Cappelletti, Silvia Ardizzone, Paola Fermob and Stefania Gilardoni, Journal of the European Ceramic Society, 24, 3603 3611 (2004).

[3]Silvia Ardizzone, Giuseppe Cappelletti, Paola Fermo, Cesare Oliva, Marco Scavini and Fabio Scime, Structural and Spectroscopic Investigations of Blue, Vanadium Doped ZrSiO4 Pigments Prepared by a Sol Gel Route, J. Phys. Chem. B, 109, 22112 22119 (2005).

[4]Tartaj, P., Serna, C. and Ocana, M., Preparation of blue vanadium zircon pigments by aerosols hydrolysis. J. Am. Ceram. Soc., 78, 1147 1152 (1995).

[5]Shoyama, M., Hashimoto, K. and Kamiya, K., Iron zircon pigments prepared by the sol gel method. J. Ceram. Soc. Japan, 107, 534 540 (1999).

[6]Gair, I., Jones, R. and Airey, A., The nature of the chromospheres centre in an iron zircon pigment prepared by conventional and sol gel routes. Qualicer, 63 65 (2000).

[7]Llusar, M., Calbo, J., Badenes, J., Tena, M. and Monros, G., Synthesis of iron zircon coral by coprecipitation routes. J. Mater. Sci., 36, 153 163 (2001).

[8] M. Trojan, Synthesis of a Green Blue Zirconium Silicate Pigment, Dyes and Pigments, 14, 9 22 (1990).

[9]M. Trojan, Synthesis of a Blue Zircon Pigment, Dyes and Pigments, 9, 231 232 (1988).

[10]Eppler, R. A., Mechanism of formation of zircon stains. J. Am. Ceram. Soc., 53, 457 462 (1970).

[11]Eppler, R. A., Kinetics of formation of an iron zircon pink color. J. Am. Ceram. Soc., 62, 47 49 (1979).

[12]Berry, F. J., Eadon, D., Hollaway, J. and Smart, L. E., Iron doped zircon: the mechanism of formation. J. Mater. Sci., 34, 3631 3638 (1999).

[13]Ayala, J. M., Verdeja, L. F., Garcia, M. P., Llavona, M. A. and Sancho, J. P., Production of zirconia powders from the basic disintegration of zircon, and their characterization. J. Mater. Sci., 27, 458 463 (1992).

[14]Carla Valentin, Mari Carmen and Javier Alarcon, Synthesis and Characterization of Vanadium Containing ZrSiO4 Solid Solutions from Gels, Journal of Sol Gel Science and Technology, 15, 221 230 (1999).

[15]J.B. Vicent, J. Badenes, M. Llusar, M.A. Tena and G. Monr ′os, Differentiation between the Green and Turkish Blue Solid Solutions of Vanadium in a Zircon Lattice Obtained by the Sol Gel Process, Journal of Sol Gel Science and Technology, 13, 347 352 (1998).
[16]Manuel OcanÄ, Alfonso Caballero, AgustõÂ n R. GonzaÂ lez Elipe, Pedro Tartaj, Carlos J. Sernab and Rosa I. Merinoc, The Effects of the NaF Flux on the Oxidation State and Localisation of Praseodymium in Pr doped Zircon Pigments, Journal of the European Ceramic Society, 19, 641 648 (1999).
[17]A. Niesert, M. Hanrath, A. Siggel, M. Jansen and K. Langer Theoretical study of the polarized electronic absorption spectra of vanadium doped zircon, Journal of Solid State Chemistry, 169, 6 12 (2002).

PREPARATION OF BLUE CERAMIC PIGMENTS BY REACTION BONDING

Enrique Rocha Rangel, Imelda Villanueva Baltazar and Lucia Téllez Jurado
Departamento de Ingeniería Metalúrgica, ESIQIE IPN
UPALM, Av. IPN s/n, San Pedro Zacatenco, México, D. F. 07738

Elizabeth Refugio García
Departamento de Materiales, Universidad Autónoma Metropolitana
Av. San Pablo # 180, Reynosa Tamaulipas, México, D. F., 02200,

ABSTRACT

Synthesis of $CoAl_2O_4$ based blue ceramic pigments following the reaction bonding process was studied. An intense mechanical powder mixture of metallic cobalt and Al_2O_3 was prepared. The resulting powders from the milling period were heat treated in air in order to oxide; first the metallic cobalt with the oxygen present in the furnace atmosphere, and then, attain $CoAl_2O_4$ blue powders by a solid reaction between fresh CoO and initial Al_2O_3. The evolution of solid reactions was follows by TG, XRD, FT IR and UV Vis analysis. The obtained results from TG analysis indicate that Co oxidation starts at 200 °C and finishes at 700 °C. XRD patterns show that the formation of $CoAl_2O_4$ is completed at 850 °C. On the other hand, FT IR and UV Vis studies identify those different interactions between CoO and Al_2O_3 are taking place at temperatures in the region of 800 to 850 °C forming the $CoAl_2O_4$ spinel. Finally, the feature of obtained samples after solid reactions is of an intense blue color.

INTRODUCTION

The cobalt aluminate powders ($CoAl_2O_4$) are agents with ceramic character and they are widely employed as pigments of blue color, these ceramics also are used in the heterogeneous catalytic field[1, 2]. The $CoAl_2O_4$ presents the structure of a normal spinel, in where Co^{2+} ions are located in tetrahedral positions, whereas, Al^{3+} ions are located in octahedral positions. These kinds of components can be prepared by different routes as are; Reactions between cobalt and aluminum oxides at temperatures among 800 and 1300 °C[3]; by co precipitation of cobalt and aluminum sales and subsequent calcination between 400 and 800 °C[4]; by calcination of alumina powders immerse in a cobalt nitrate solution at 1200 °C[2]; this compound also has been prepared by the calcination of alumina powders the same that were covered by a thing film of metallic cobalt at 1000 °C during 20 h[5]. However, most of these processes are expensive, difficult in their operation and present low productivity. For these reasons, in this study it is desired take as a reference point for the ceramic pigment production, the process named reaction bonding aluminum oxide (RBAO)[6]. In this process it starts from an alumina and aluminum mixture powders the same that is heated at very slow rates in order to oxidize all the aluminum for obtaining new alumina, which will bond with the old alumina in the mixture. Authors of this process have documented that alumina obtained by this method presents better physical and mechanical characteristics than alumina produced conventionally[6 8]. At the same time new alumina presents very small grain size and a good size grain distribution in the crystalline material, independently that this new process is cheaper and easy in its operation. In this study a homogeneos cobalt and alumina mixture is used as precursor materials in order to obtain $CoAl_2O_4$ blue ceramic pigments through heat treatment in where first all cobalt is oxidized, and then new CoO formed will reacts with original alumina, looking for the advantages that offer the production of oxide ceramics through RBAO process.

EXPERIMENTAL PROCEDURE

The starting materials were Al_2O_3 powder (99.9%, 1 μm, Sigma, USA) and Co powder (99.9%, 1 μm, Aldrich, USA). The amount of each material employed was in agreement with the stoichiometry of reaction 1. Powder blends of 30 g were prepared in ball mill with ZrO_2 media, the rotation speed of the mill was of 250 rpm during 12 h. The milled powder mixture was warm up in air at different temperatures between 100 and 900 °C, the heating speed was fixed at 1 °C/min. Characterization was performed as follows; Thermogravimetric analysis were carried out on a Simatzu SS345 (TG DTA). X ray diffraction measurements were carried out on a Siemens D5000 diffractometer. UV VIS spectra were recorded on a HP8543 spectrophotometer. FTIR spectra were recorded on a Bruker IFS60v spectrometer.

$$Co + \frac{1}{2} O_2 + Al_2O_3 \rightarrow CoAl_2O_4 \qquad (1)$$

RESULTS AND DISCUSSION

Thermogravimetric analysis

Figure 1 presents the thermogravimetric curve resulted after warm up powders since 100 until 900 °C. In the curve it can be seen that between 100 and 200 °C there are not changes in the samples' weight. However, over 200 °C the powders start to obtain an important weight gain that reaches a maximum in weight gain at 700 °C. This weight gain is due to cobalt's oxidation present in the powders mixture. The CoO recently formed in agreements with the reaction 2 and than can be verified in the FT IR analysis is the precursor for the obtaining of the cobalt aluminate ($CoAl_2O_4$) that corresponds with the desired blue ceramic pigment.

$$Co + \frac{1}{2} O_2 \rightarrow CoO \qquad (\Delta G = \ 280 \text{ KJ/mol}^{9}) \qquad (2)$$

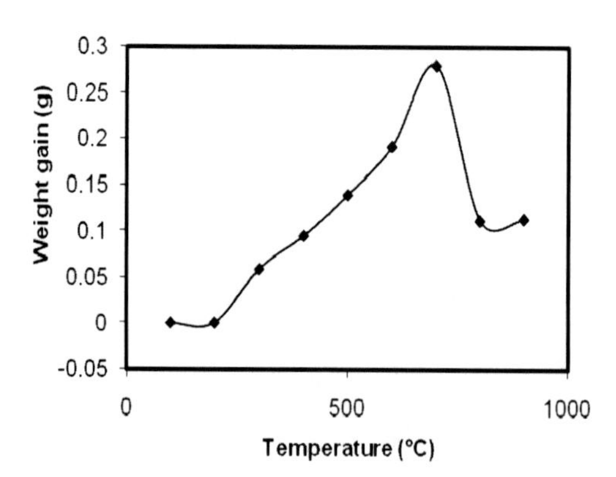

Figure 1. Graphic of the thermogravimetric analysis.

X ray diffraction

Figure 2 shows the X ray diffraction patterns of the Co and Al_2O_3 powders mixture warm up at different temperatures. In this figure it is observed that in original sample there are present the precursor materials Co and Al_2O_3. Nevertheless, when the powder sample is warmed up is observed that the peaks of both elements begin to vanish, giving rise to the formation of new peaks in the sample, which in all the cases corresponds to the compound $CoAl_2O_4$. As the treatment's temperature increases the advance of the formation of the cobalt aluminate is more complete, because of the intensity of the characteristics peaks of this compound becomes more intense in the diffraction patterns. Apparently at 800 °C the reaction of formation of the desired product has happened, since this temperature the existence of the peaks corresponding to the cobalt or alumina are not observed. The peak that appears in all the patterns at 81.3° corresponds to used platinum as supported of the powder sample.

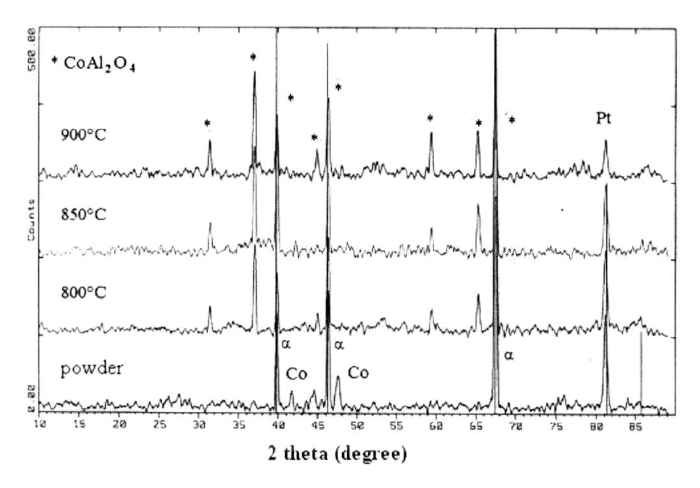

Figure 2. X ray diffraction patterns of the mixture of powders ($Co + Al_2O_3$) warmed up at different temperatures.

Aspect of the heated samples

In Figure 3 it is presented two photographies taken with a digital camera of two of the prepared samples. The picture (a) corresponds to a sample before being put under the cycle of reaction and picture (b) corresponds to the same sample after to be warmed up at 900°C. As it can be seen in these photographies the samples conserve their cylindrical shape after being heated. Nevertheless, the chemical reaction that happened during the heating brought about in the first place a shrinkage of 20% of the sample. But most important it is the change of color that is had in the same, since before to be heated the sample it was of a white color, whereas after to have been heated the same is of an intense blue color. This change of color in the sample is indicative of which the reaction of formation of the $CoAl_2O_4$ took place *in situ* during the heating of the original sample.

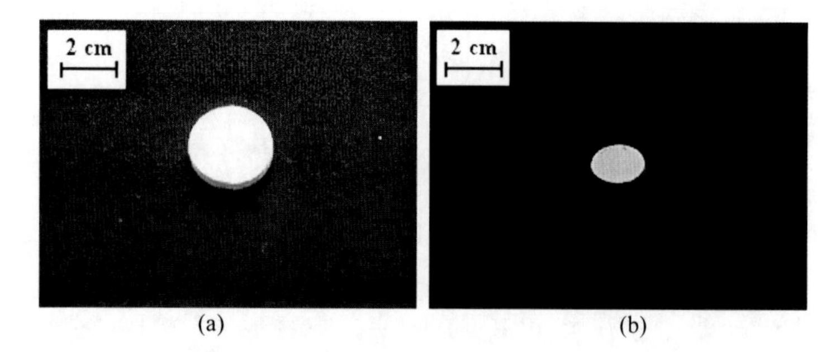

Figure 3. Samples taken with a digital camera.
Photo (a) corresponds to sample before being warmed up.
Photo (b) corresponds to the same sample after to be warmed up.

FT IR spectroscopy

FTIR measurements were used to identify and characterized the resulting spinel powders. In the Figure 4 they are possible to be observed, the spectra obtained at different temperatures. The IR spectra of the warmed up sample at 25 °C, which is the alumina with cobalt mixture, shows five located main bands of absorption in 3450 cm^{-1}, 1635 cm^{-1}, 1150 cm^{-1}, 664 cm^{-1} and 568 cm^{-1}. The first band of little intensity and very amplitude (3450 cm^{-1}) corresponds to the vibration of group OH in H_2O (H OH). Also another band of small intensity is observed and that is associated to groups OH of the water which is located in 1635 cm^{-1}. This water corresponds to water absorbed by the KBr, used for the dilution of the sample in the IR analysis. The most important bands of oxides can be observed between 1000 and 400 cm^{-1}, where they are located the fourth and fifth band previously commented (664 cm^{-1} and 568 cm^{-1}), which correspond to metal oxygen in Al_2O_3 and metal oxygen in CoO, respectively. In the IR spectra of the samples treated at 800 and 850 °C observed also in this figure, exactly the same described bands can be observed as in the IR spectra of the treated sample at 25 °C. The main difference in these spectra is the appearance of a shoulder located in 795 cm^{-1} and another shoulder of smaller intensity in 448 cm^{-1}, that correspond to the vibrations of M O bond (M = Al and Co) in the spinel, which are increased with the temperature of treatment. The appearance of these shoulders is attributed to that these temperatures the alumina and cobalt oxide react to form the cobalt spinel ($CoAl_2O_4$), in agreement with the reaction (1). On the other hand, in the treated sample at 900 °C, is observed major amplitude and intensity for the shoulder in 795 cm^{-1} and major intensity for the band that appears at 448 cm^{-1}. Also it is possible to be observed in the spectra of the sample, the displacement of the band located in 568 cm^{-1} (IR spectra of samples treated at 25, 800 and 850 °C) to majors' frequencies 591 cm^{-1}. Finally, equal the diminution (until almost disappearing) of the band located in 664 cm^{-1} is observed and the appearance of a band with median intensity in 637 cm^{-1}, the displacement and appearance of these bands (vibrations of the bond metal oxygen) correspond to the formation of the cobalt spinel at this temperature, from corresponding oxides (alumina and cobalt oxide), which is in agreement with the X ray diffraction analysis.

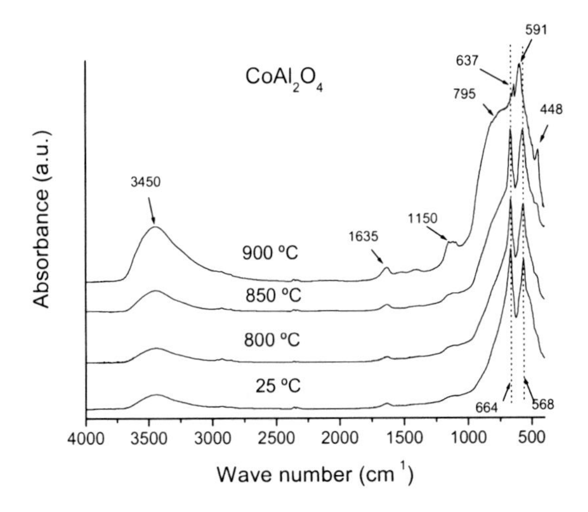

Figure 4. FT IR spectra obtained at different temperatures.

UV Vis Spectra of the samples

Figure 5 shows the UV Vis spectra of the obtained samples treated at different temperatures. As it is observed in the figure, the UV Vis spectra show changes as the temperature is increased. In the spectra that corresponds to the treated sample at 25 °C as well as in the spectra of the warmed up sample at 800 °C, two absorption bands are only observed (252 and 379 cm^{-1}) which correspond to the precursor materials (alumina and cobalt), indicating that the spinel phase not yet is observable at these temperatures by this technique. In the spectra that correspond to treated sample at 850 and 900 °C, it can be observed the appearance of three absorption bands with maximums in 543, 580 and 630 cm^{-1}, which are attributed to the presence of the cobalt spinel ($CoAl_2O_4$). These observation are also in aggrement with the previous observation in the TG, XRD and FTIR analysis and with the change of color of the sample before and after it has been warmed up, confirming in this way the formation of the $CoAl_2O_4$ by the proposed method.

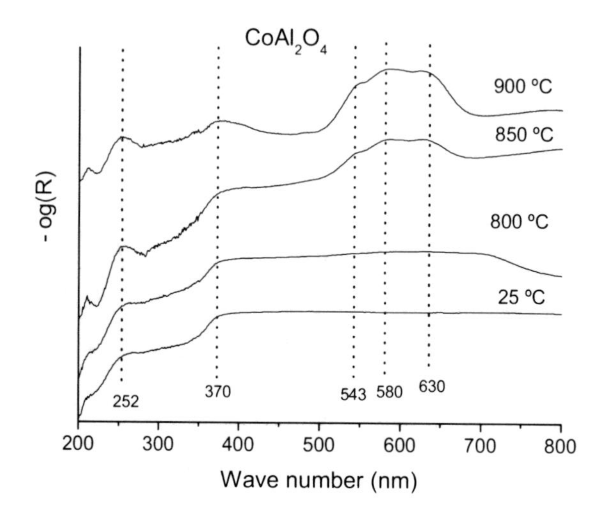

Figure 5. UV Vis spectra of the obtained samples treated at different temperatures.

CONCLUSIONS

- Through the propose methodology blue ceramic pigments were obtained by means of reactions in solid state of a cobalt and alumina powder mixture.
- The thermogravimetric analysis indicates that the oxidation of cobalt starts at 200 °C and finishes at 700 °C. The diffraction analysis of X rays indicates that the total reaction between CoO and alumina to form the spinel it happens at 850 °C
- The analysis by infrared spectroscopy indicates the formation of certain absorption bands between 800 and 850 °C they indicate that the chemical species CoO and alumina are interacting for the formation of the $CoAl_2O_4$ spinel. Whereas, the analysis by ultraviolet spectroscopy indicates that the formation of the $CoAl_2O_4$ spinel happens between the temperatures of 850 and 900 °C.

ACKNOWLEDGMENT

Authors would thank Universidad Autónoma Metropolitana for technical and financial support from 2260235 project and ESIQIE IPN for technical support.

REFERENCES

1. Zhou, Y., Liang, X.,Yu, F., Wang, F.,Luo, H., Pan, M, Preparation of Nanometer $CoAl_2O_4$ Powders by Sol Gel Hydrothermal Method, Journal of the Chinese Ceramic Society 34(10), pp. 1259 1262, 2006.

2. Shin, D. Y., Kim, K. N., Han, S. M, "Synthesis and Characterization of Nano Size $CoAl_2O_4$ Spinel Powder Sing Sol Gel Process", Materials Science Forum 544 545, pp.869 872, 2007.
3. Azurdia, J., Marchal, J., Laine, R.M., "Characterization of Mixed Metal Oxide Nanopowders Along the CoOx Al2O3 Tie Line Using Liquid Feed Flame Spray Pyrolysis", Journal of the American Ceramic Society 89 (9), pp. 2749 2756, 2006.
4. Zayat Marcos and Levy David, "Blue $CoAl_2O_4$ Particles prepared by the Sol Gel and Citrate", Instituto de Ciencia de los Materiales de Madrid, Cantobalnco Madrid Spain., pp. 2763 2769, 2000.
5. Melth Casarin, El Habra N., Natali M., Rossetto G., Sada C. and Tondello E. " MOCVD Deposition of $CoAl_2O_4$ Films Carta G", Italia, pp 4592 4599, 2000.
6. S. Wu, D. Holz and N. Claussen, "Mechanisms and Kinetics of RBAO Ceramics", J. Am. Ceram. Soc., 76, pg. 970 980, 1993.
7. N. Claussen, S. Wu and D. Holz, "Reaction Bonding of Aluminum Oxide (RBAO) Composites: Processing, Reaction Mechanisms and Properties", J. Eur. Ceram. Soc., 14, pg. 97 109, 1994.
8. D. Holz, S. Wu, S. Scheppokat and N. Claussen, "Effect of Processing Parameters on Phase and Microstructure Evolution in RBAO Ceramics", J. Am. Ceram. Soc., 77, pg. 2509 2517, 1994.
9. J. F. Shackelford and W. Alexander, Materials Science and Engineering Handbook, CRS Press, Boca Raton Florida, pg. 1 95, 2001.

COLLOIDAL CHARACTERIZATION AND AQUEOUS GEL CASTING OF BARIUM TITANATE CERAMICS

Cameron D. Munro and Kevin P. Plucknett*
Dalhousie University, Materials Engineering Program, Department of Process Engineering and Applied Science, 1360 Barrington Street, Halifax, Nova Scotia, B3J 2X4, CANADA

ABSTRACT

Ferroelectric ceramics such as barium titanate ($BaTiO_3$) are typically prepared by dry pressing of powders. However, a number of studies have assessed the influence of colloidal processing methods. In the present work, aqueous colloidal forming of $BaTiO_3$ powders has been investigated using an environmentally benign gelling agent, namely agar. $BaTiO_3$ suspensions have been characterized using zeta potential, sedimentation and rheological measurements for powders with nominal particle sizes of 100, 200 and 500 nm. Blending of 500 and 100 nm mixtures resulted in a minimum viscosity at 55 vol. % total solids loading. High concentration agar suspensions, with up to 8 wt. % biopolymer in water, have then been mixed with the $BaTiO_3$ suspensions, giving total agar concentration varied between 0.5 to 1 wt. % in the mixed suspension. Gel casting was conducted using a variety of shaped moulds. After gel casting, it was also demonstrated that automated green machining is possible for the dried performs. The resulting sintering and densification behavior are also described.

INTRODUCTION

The ferroelectric/dielectric properties of barium titanate ($BaTiO_3$) have been studied since the early 1940's. As a consequence this material represents a well known substitute for common lead based systems.[1] Like most advanced ceramics, ferroelectrics stand to gain much from the enhanced homogeneity and reduced porosity afforded by aqueous colloidal processing.[2] Colloidal consolidation techniques are also intriguing as they are capable of producing a variety of geometries that would otherwise be difficult to form using the traditional dry pressing route. The aqueous colloidal processing of barium titanate powders has been investigated to some degree, with attention focused primarily on slip casting,[3] tape casting,[4] gel casting,[5] and electrophoretic deposition.[6] Of these forming methods, gel casting is especially attractive as it is a near net shape forming technique and results in bodies that are usually strong enough for subsequent machining in the green state. However, while colloidal forming offers significant potential benefits over more conventional dry forming methods, it is important to ensure high powder packing densities to minimize drying shrinkage and therefore retain dimensional stability. In particular, benefits can be achieved through blending powders in distinct particle size ranges, such that one powder can pack interstitially, with a consequent increase in packing density.[7,8]

In the present work, issues associated with the aqueous colloidal processing of commercially available $BaTiO_3$ powders are investigated. Three powders of different particle size were studied, in order to determine both the conditions necessary for stability and to quantify any benefits associated with the mixing of two differently sized powders. The consolidation method of aqueous gel casting, using a benign biopolymer agent (i.e. agar), was also qualitatively and quantitatively assessed with respect to rheological behavior, drying shrinkage, green density, and green machining capabilities.

EXPERIMENTAL PROCEDURE

Three $BaTiO_3$ powders, nominally displaying particle sizes of 500 nm (Lot# IAM6068BTO5), 200 nm (Lot# IAM12084BTO2) and 100 nm (Lot# IAM3286BTO1), were investigated (Inframat Advanced Materials, Farmington, CT). These powders were characterized in terms of particle size using acoustic attenuation spectroscopy (Model Zeta APS, Matec Applied Sciences, Northborough, MA) and specific surface area via nitrogen adsorption (Model SA 6201, Horiba, Irvine, CA). Particle morphology and crystalline phase were determined by scanning electron microscopy (SEM; Model S 4700, Hitachi Co., Tokyo, Japan), and X ray diffraction (XRD; Model D8 Advance, Bruker, Madison, WI), respectively. An ammonium salt of poly(methacylate) (PMA NH_4; Darvan C, R. T. Vanderbilt Co., Norwalk, CT) supplied as a 25wt% aqueous solution was used as a dispersant, while the agent used for gel casting was a general purpose grade agar powder (Lot# 17543A; Stock #A10752, Alfa Aesar, Ward Hill, MA). Double distilled water was used in preparing all suspensions.

Three techniques were used to determine the conditions required for suspension stability. First, acoustophoresis (Model Zeta APS, Matec Applied Sciences, Northborough, MA) quantified the zeta potential of individual suspensions as a function of pH and deflocculant concentration. Suspensions were deagglomerated using an ultrasonic horn operating at 240 W for 30 seconds (Model 500 Sonic Dismembrator, Fisher Scientific, Ottawa, ON), and were then titrated with 1M solutions of NaOH and HNO_3. Zeta potential measurements were taken approximately every 0.2 of a pH unit. Sedimentation tests were likewise done on individual suspensions. Well dispersed, stable suspensions are known to settle to a higher density sediment than unstable, and so suspension stability could be inferred by measuring the sediment volume of various suspensions left standing for extended periods (~weeks). The flow properties of concentrated suspensions, both of individual powders and powder mixtures, were investigated using a controlled stress rheometer (Model AR 2000 Advanced Rheometer, TA Instruments, New Castle, DE) fitted with concentric cylinder testing geometry (a 28 mm diameter rotor and a 1 mm annulus). Suspensions were subjected to shear rates that varied from 1500 s^{1} down to 1 s^{1}, and the apparent viscosity was calculated at each shear rate.

In order to understand the evolution of viscoelastic properties during gel casting, combined agar/ceramic suspensions were subjected to oscillatory, rather than rotational, rheological tests. Prepared agar solutions were mixed with ceramic suspensions at 60 ± 2°C by hand and ultrasonically (80 W for ~1 minute). A 40 mm stainless steel plate and a temperature controlled Peltier plate constituted the oscillatory geometry. Both surfaces contacting the sample were covered with adhesive backed, 600 grit silicon carbide paper, in order to reduce slippage at the interface and a solvent trap and geometry cover were used to limit evaporation. The suspensions were cooled from 60°C to 5°C at a rate of 10°C/min, sampled at a frequency of 1 Hz and a strain of 0.5 %.

Two robust consolidation techniques, slip casting and gel casting, were performed on concentrated, stabilized suspensions. Unidirectional slip casting of discs was performed in plastic rings on a porous plaster of Paris slab. Drying of the castings occurred at room temperature, during which time the discs shrank away from the plastic rings. Gel castings were made in simple circular moulds to yield discs, and also into more complex Teflon/PVC moulds to produce industrially practical geometries. A syringe loaded with the heated combined suspension was used for mould filling, and the filled moulds were placed into a refrigerator for 1 hour to accelerate and ensure complete gelation. After demoulding, the castings were left to dry slowly by loosely covering with glass or plastic vessels. Generally drying was completed after 72 hours at room temperature. Dried gel cast green bodies were machined in the green state using a desktop computer numerical controlled (CNC) milling machine (Model MDX 15, Roland DG, Irvine, CA, USA). The unit utilizes a 10 W DC motor running at 6500 rpm for cutting. Tungsten carbide bits of either 1/8" or 1/16" diameter, with ball or mill ends, were used with a variety of milling head speeds. Both slip and gel cast bodies were sintered in air for 2 hours at 1300°C in an electric furnace (Model H16, MHI Inc., Cincinnati, OH) in covered alumina crucibles. The bulk densities of sintered bodies were determined via immersion in distilled water and compared to a theoretical density of 6.02 g/cm^3.

RESULTS AND DISCUSSION

Powder Characterization

Table I includes the results of particle sizing and surface area analysis, as well as XRD phase determination, while detailed particle size distributions for each of the $BaTiO_3$ powders are presented in Figure 1(a). The PSD of each powder exhibits one dominant peak approximating the nominal particle size, although all powders are multimodal. SEM investigation agreed with the particle sizing data and showed the morphology to be nominally spherical with low porosity. An example micrograph of the 500 nm $BaTiO_3$ powder is provided as Figure 1(b).

Table I. $BaTiO_3$ powder characteriztics.

	500 nm	200 nm	100 nm
Dominant Peak Centre (nm)	540	162	107
Surface Area (m^2/g)	2.6 ± 0.4	4.9 ± 0.1	12.2 ± 0.3
Crystalline Phase	Tetragonal	Tetragonal	Cubic

Suspension Stabilization

Acidic titrations of aqueous suspensions of both the 500 nm and 100 nm powders exhibited no isoelectric points (IEPs) in the absence of polyelectrolyte, as shown in Figure 2.[9] Rather, the zeta potentials were positive at all pH levels studied. This result has been noted before during acoustophoretic studies on $BaTiO_3$,[3] although zeta potential tests utilizing electrophoresis have yielded a variety of IEPs.[10] The purely positive zeta potential has been attributed to dissolution of Ba^{2+} ions, either from the surface of $BaTiO_3$ particles or from discrete $BaCO_3$ particles, which then become incorporated in the electric double layer surrounding the particles. Acoustophoretic studies require concentrated suspensions and so the statistical probability of incorporation of Ba^{2+} ions is higher than in the very dilute suspensions necessary for electrophoresis. Therefore the effects of barium dissolution, though theoretically present during all studies, are only noted during testing of concentrated suspensions. It is known that this dissolution of barium ions is enhanced under acidic conditions,[11] and that free barium ions can lead to inhomogeneity and exaggerated grain growth upon consolidation and sintering.[12] In addition, the excess Ba^{2+} ions in solution at lower pH values can also increase the suspension pH during aging.[13] Thus, low pH levels are to be avoided during the aqueous processing of barium titanate powders.

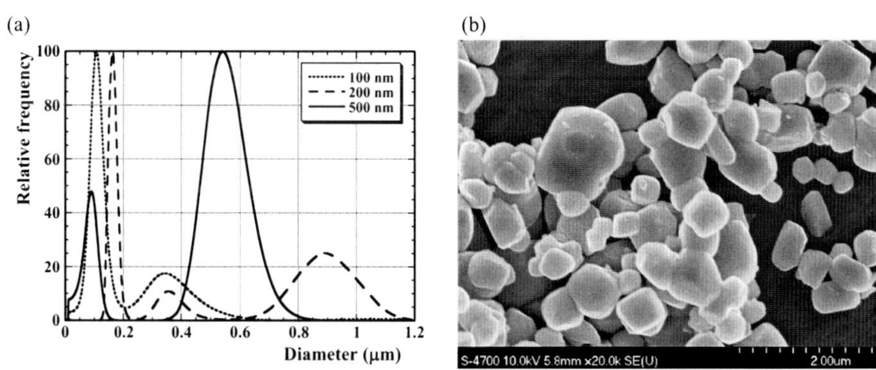

Figure 1. (a) Measured particle size distributions for 500 nm, 200 nm and 100 nm $BaTiO_3$ powders (b) SEM micrograph of the 500 nm $BaTiO_3$ powder used in the present study.

The addition of deflocculant, PMA NH_4, imparts a negative zeta potential at the basic end of the pH scale, and results in an IEP that shifts to lower pH levels with increasing polyelectrolyte. Since the adsorbed species is the negatively charged PMA molecule, both of these effects are as expected. The zeta potential curves give an indication of the polyelectrolyte concentration at which surface coverage begins to be complete. Beyond 0.1 wt. % and 0.3 wt. % for the 500 nm and 100 nm powders, respectively, additional polyelectrolyte does not largely increase zeta potential. The final aspect of interest in Figure 2 is the fact that zeta potential values, even under very basic conditions, stay relatively small, not exceeding 40 mV.

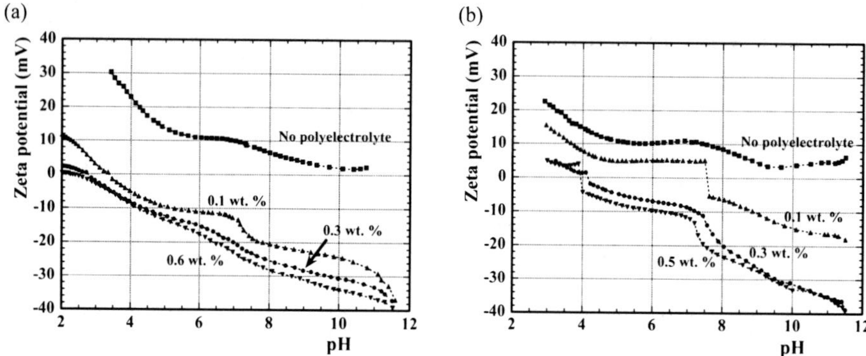

Figure 2. Zeta potential curves of aqueous suspensions of (a) 500 nm and (b) 100 nm $BaTiO_3$ powders (prepared at 2.5 and 1 vol. % respectively). In each case, varying amounts of PMA NH_4 deflocculant are used.

Based on the dissolution of barium evident previously reported in the literature and zeta potential curves,[3,9] subsequent sedimentation and rheological tests were limited to pH levels in the basic regime. Figure 3 compares the sediment densities to polyelectrolyte concentration at three basic pH levels for both the 500 and 100 nm powders, namely pH 7, pH 9.3, and pH 12. As was evident in zeta potential testing, the sediment densities showed that there is insufficient electrostatic repulsion for stability at useful pH levels without the use of deflocculant.

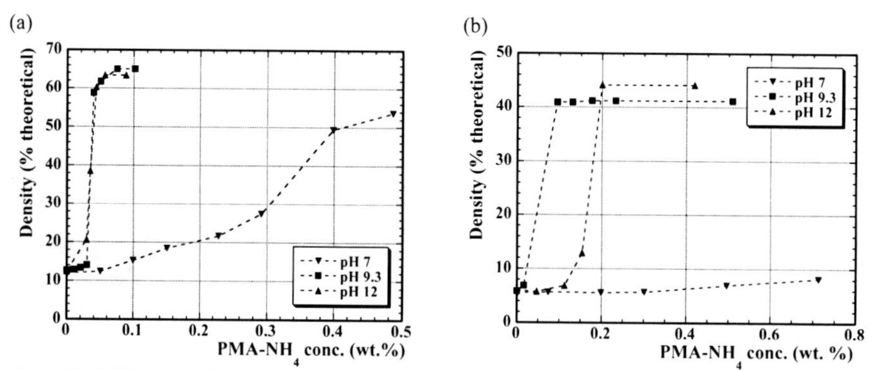

Figure 3. Sediment densities of 10 vol. % suspensions of (a) 500 nm $BaTiO_3$ after 1 week and (b) 100 nm $BaTiO_3$ after 4 weeks.

As shown in Figure 3(a), significantly lower polyelectrolyte concentrations are required to successfully stabilize the basic suspensions than the suspension at pH 7. It can also be seen that the pH 9.3 and pH 12 suspensions exhibit roughly the same behavior for the 500 nm sized powder, both being stabilized with less than 0.1 wt. % deflocculant. This data supports the conclusion of zeta potential testing that 0.1 wt. % corresponds to nearly complete surface coverage. At this polyelectrolyte concentration, zeta potential values are relatively low (approximately 23 mV at pH 9.3) and so it can be said that stabilization under these conditions occurs by an electrosteric, rather than purely electrostatic, mechanism. This conclusion has also been reached in the work of others.[13] The response of the 100 nm powder is generally similar, although there is a more discernable difference between the behavior at pH 9.3 and pH 12 (Figure 3(b)), which likely arises due to the higher surface area combined with the instability of $BaTiO_3$ in aqueous suspensions.[9] Based on these general zeta potential and sedimentation observations, for the remaining discussion of suspension behavior, the focus will be predominantly based on suspensions prepared at pH 9.3.

Rheological Testing

Concentrated $BaTiO_3$ slips displayed shear thickening and/or shear thinning behavior at each of the pH levels examined.[9] Figure 4(a) gives a representative plot of concentrated $BaTiO_3$ suspension rheology at pH 9.3. Unstabilized suspensions were shear thinning for all shear rates investigated, while well dispersed suspensions exhibited shear thinning at low shear rates and shear thickening regimes at higher shear rates. For a detailed discussion of shear thickening, the reader is directed to more formidable texts.[14] The critical shear rate for shear thickening increased as particle size decreased, and the shear thickening behavior was not detrimental to the preparation or deagglomeration of concentrated suspensions. Taking a shear rate that corresponds the same behavior for all suspensions, it is possible to further determine the conditions necessary for suspension stability.

A comparison of the viscosity vs. polyelectrolyte concentration behavior is provided for the 500 nm powder in Figure 4(b). It is clear that suspension preparation at pH 9.3 leads to the lowest viscosity, and that a polyelectrolyte concentration of 0.1 wt. % yields superior stability at this value of pH. Based on rheological behavior, similar plots were also developed for the 200 nm and 100 nm powders (not shown) so subsequent suspension preparation could be optimised when blending two different sized powders in a nominally bimodal suspension.

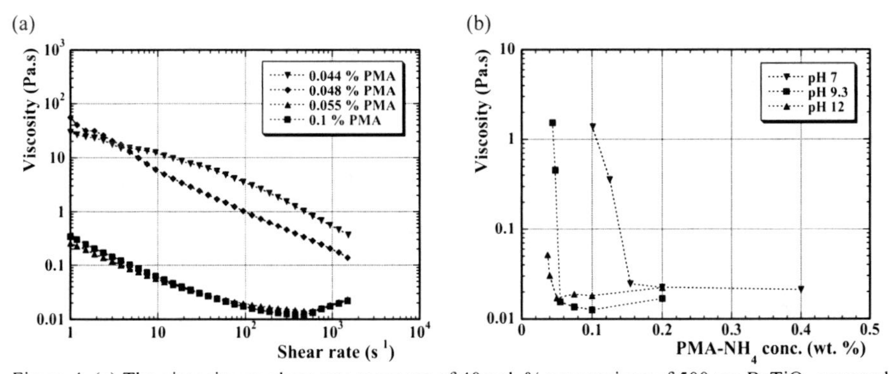

Figure 4. (a) The viscosity vs. shear rate response of 40 vol. % suspensions of 500 nm $BaTiO_3$ prepared at pH 9.3 and (b) the viscosity (at 299.3 s^{1}) of 40 vol. % suspensions of 500 nm $BaTiO_3$ as a function of dispersant concentration.

The mixing of particles of distinct sizes has been shown to be beneficial to suspension rheology and particle packing in ceramic systems.[7,8] In the present work, combinations of 500 nm and 200 nm (i.e. a nominal size ratio of 5:2), as well as 500 nm and 100 nm (i.e. a nominal size ratio of 5:1) underwent rheological testing to investigate these potential benefits for colloidal forming of $BaTiO_3$ based ceramics. The viscosities of suspensions of the 500:200 nm and 500:100 nm particle mixtures are provided in Figure 5(a) and (b), respectively. In the case of mixtures of 500 nm and 200 nm powders, the viscosity essentially follows a rule of mixtures behavior, and the endpoints signify the maximum and minimum viscosities possible via mixing. This is either due to an insufficient difference in particle size, such that true interstitial packing is affected, or to some complex effect resulting from the multimodal nature of both raw powders. However, in the case of 500 nm and 100 nm powders a distinct advantage of powder mixing is evident. This is particularly apparent when preparing suspensions at moderately high total solids loading (i.e. 50 vol. %). Suspensions exhibit a clear minimum viscosity when 85% coarse powder is mixed with 15% fine. This feature is important, as it allows the preparation of higher solids content suspensions. In this instance the suspensions can still be processed in the colloidal state, so they readily flow, but there will ultimately be less drying shrinkage and warping upon consolidation due to the high solids content.

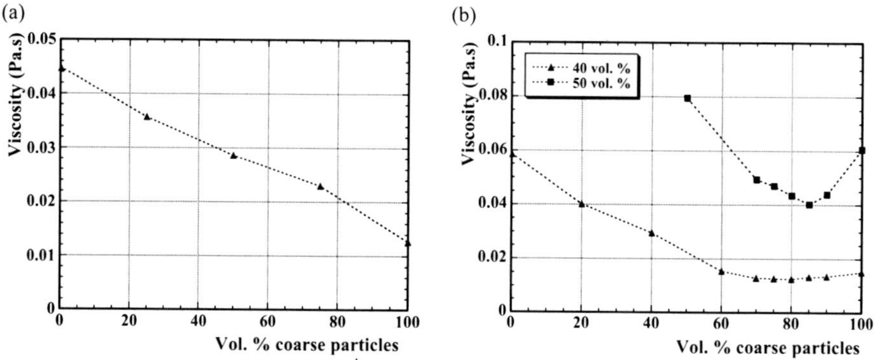

Figure 5. (a) The viscosity (at 299.3 s^{-1}) of bimodal suspensions of 500 nm and 200 nm $BaTiO_3$ powders as a function of the ratio of coarse:fine particles (40 vol. % total solids at pH 9.3) and (b) the viscosity (at 299.3 s^{-1}) of bimodal suspensions of 500 nm and 100 nm $BaTiO_3$ powders as a function of the ratio of coarse:fine particles at pH 9.3.

Slip Casting Samples

The use of bimodal or multimodal powder mixtures offers the potential for enhanced particle packing during green state consolidation. This particle packing advantage is notable during the slip casting of concentrated suspensions. As shown in Figure 6(a), the addition of 100 nm powder to 500 nm powder increases the slip cast green density by up to 10% of the theoretical density relative to the coarse powder alone. Interestingly, the mixture yielding the optimum green density consists of 70% coarse powder, while one with 85% coarse was shown to result in suspensions of the lowest viscosity. This disparity can possibly be ascribed to the multimodal nature of both powders, thus leading to a complicated packing situation. Upon sintering, the superior density offered by powder mixing remains, although the magnitude of the benefit is somewhat diminished. From Figure 6(b), it is apparent that slip cast bodies using powder mixtures exhibit a sintered density that is roughly 3 % greater than that of the purely coarse powder. As with the green density, for sintering response a mixture with 70 vol. % coarse is shown to be the best for particle packing and densification behavior.

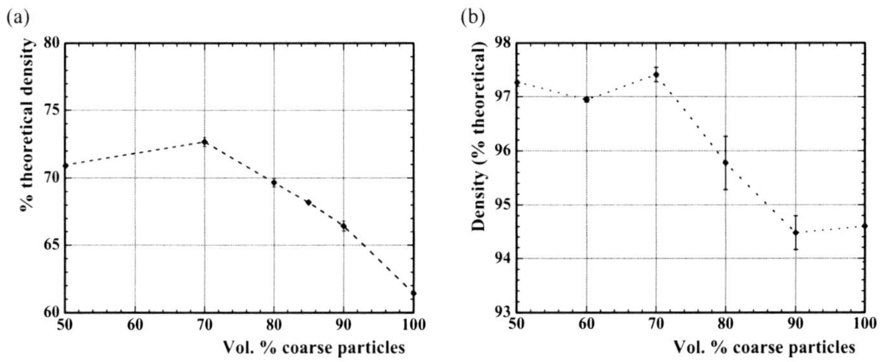

Figure 6. (a) The effect of coarse:fine particle ratio on the green density of slip cast preforms. (b) The density of slip cast bimodal suspensions (50 vol. % total solids), sintered at 1300°C for 2 hours.

Combined Agar/Suspension Rheological Testing

While the previous sections have outlined colloidal characterization and slip casting of $BaTiO_3$ suspensions, it is often favourable to be able to cast components into complex moulds or machine them in the green state, and consequently a suitable strengthening aid must be incorporated. In the present work this has been achieved through the addition of agar, which is a food grade, polysaccharide biopolymer. A more comprehensive discussion of agar rheology and gelling behavior, both alone and with $BaTiO_3$ powder, is described in a recent publication.[15] The evolution of viscoelastic properties during gelation of combined agar/ceramic suspensions was studied via oscillatory rheometry, and the results of these tests are represented by the curves of Figure 7. For the range of temperatures studied, the storage modulus, G', is greater than the loss modulus, G", indicating that combined suspensions behave more solid like than liquid like (Figure 7(a)). The rapid rise in both moduli signifies a gelation temperature below 30°C. This transition temperature is confirmed by examining the normal force response, which highlights a contraction of the gelling body perpendicular to the rheometer plates (Figure 7(b).

From individual plots of storage moduli similar to the representative curves of Figure 7, G' was taken as an indication of "gel strength". The effect of agar precursor concentration on these gel strengths is presented in Figure 8(a) for two different temperatures. These strengths were noted *immediately* upon cooling to the given temperatures at 10°C/min, although storage moduli were generally seen to increase by an average of 25 % in the 10 minutes following the initial oscillatory tests. The use of higher concentration agar precursor solutions, that incorporate less additional water, leads to stronger gelled structures when the overall agar content is constant with respect to solids loading. While it may seem advantageous to use as concentrated an agar precursor as possible, the inclusion of less water in the combined suspension increases apparent viscosity and thus may negatively affect the ability of the suspension to fill the mould, especially when complex features are present. The loss modulus, G", such as that plotted in the representative curve of Figure 7(a), gives an indication of the apparent viscosity of the combined suspension. Gel casting needs to take place at a temperature above the gelation point, and so the magnitude of the loss modulus at 55°C has been plotted against agar precursor concentration in Figure 8(b). As shown, increasing the precursor concentration from 2 wt. % to 8 wt. % leads to an increase in loss modulus of more than two orders of magnitude. Indeed, suspensions prepared from 8 wt. % precursors were essentially impossible to pour while those utilizing 2 wt. % precursors were able to flow readily when poured.

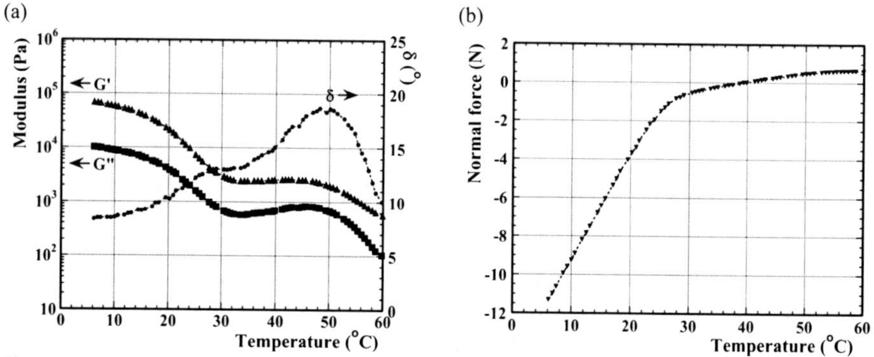

Figure 7. (a) Evolution of the storage modulus (G'), loss modulus (G") and phase angle (δ) upon cooling a gel cast suspension (55 vol. % bimodal $BaTiO_3$ with 4 wt. % agar solution) at 10°C/min. (b) Evolution of the normal force response upon cooling the gel cast suspension shown in (a).

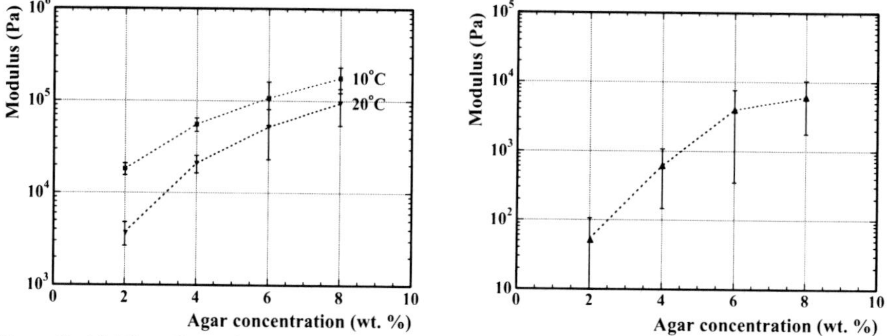

Figure 8. (a) The effect of agar precursor concentration on the storage moduli (G') of gel casting suspensions (prepared with 55 vol. % $BaTiO_3$ and a final agar content of 0.5 wt. % as a function of the total solids content). (b) The effect of agar precursor concentration on the loss moduli (G") of gel cast suspensions at 55°C (prepared with 55 vol. % $BaTiO_3$ and a final agar content of 0.5 wt. % as a function of the total solids content).

Gel Casting Behavior

Figure 9(a) compares the green densities of dried gel cast discs with an overall agar content of 0.5 wt. % of solids to the concentration of the agar precursor used (prepared from a bimodal mixture of 85 vol. % 500 nm and 15 vol. % 100 nm powder). The achievable densities increased slightly with agar precursor concentration, but were ultimately much lower than the ~68 % of theoretical density that was achieved via slip casting of the same suspensions (shown in Figure 6(a)). More notably, gel casting using higher agar precursor solutions limited drying shrinkage, as shown in Figure 9(b). The inclusion of less water in the gel casting suspension leaves less to be removed during consolidation, and thus inhibits shrinkage and enhances density. Linear shrinkages below 5% were consistently possible for castings utilizing an 8 wt. % precursor. It is notable that the sintered densities for the gel cast compositions were essentially the same as for the slip cast samples prepared with an 85:15 vol. % ratio of coarse to fine particles (Figure 9(a)), demonstrating the potential for utilising gel casting without significantly affecting sintering behavior; it is clear this could be further improved through the use of a 70:30 vol. % blend, based on the data from Figure 6(b).

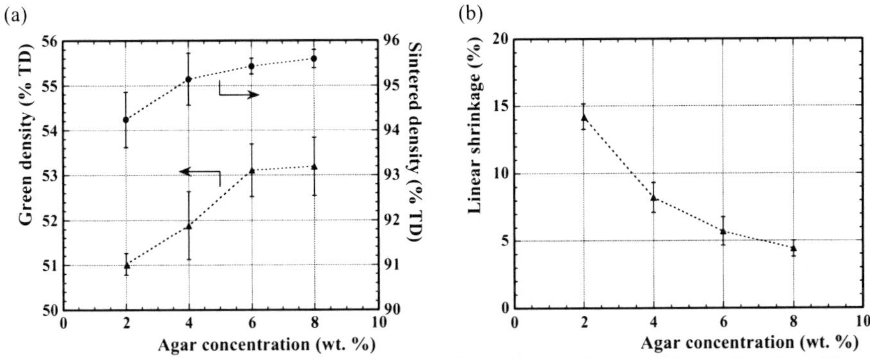

Figure 9. (a) The measured green and sintered densities of dried, gel cast solids as a function of initial agar concentration in suspension. (b) The drying shrinkage of gel cast solids as a function of initial agar concentration in suspension. All samples were prepared with 55 vol. % $BaTiO_3$ and a final agar content of 0.5 wt. % as a function of the total solids content.

Casting into complex moulds proved successful in creating difficult to form ceramic pieces. A moderately thin walled tube (4 mm wall thickness) with 0.5 wt. % agar was found to be strong enough both for demoulding and to maintain its shape during drying. Low drying shrinkage led to no noticeable cracking or warping (Figure 10(a)). It is clear that during casting, the suspension flowed well and evenly, filling the heated Teflon mould entirely.

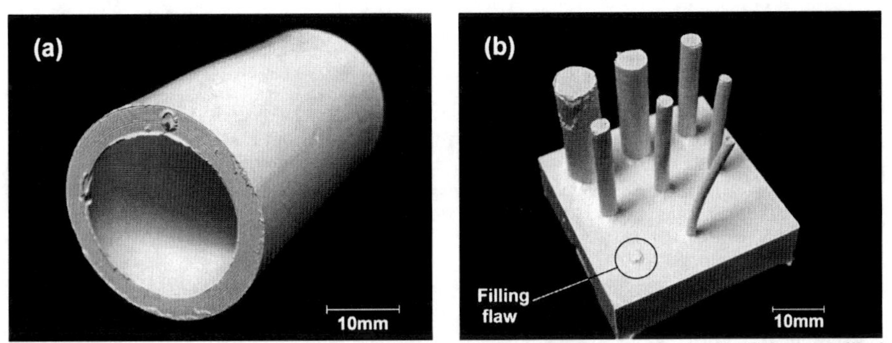

Figure 10. (a) Gel cast and dried tube (0.5wt% agar from 6wt% precursor). (b) Gel cast 'pillar' structure (1wt% agar from 6wt% precursor solution) immediately after demoulding.

The ability of the gel casting process to form pillars of material, such as is used in 1 3 piezoelectric composites for ultrasound applications,[16] was also studied. Casting of this structure required a higher agar concentration to impart the necessary strength for demoulding, namely 1 wt. % of total solids. Such a casting is shown immediately after demoulding in Figure 10(b). The successfully cast pillars exhibited a maximum achievable aspect ratio of ~13.5:1. As shown, not all pillars were strong enough to support their own weight at the given agar concentration, and so the component needed to be inverted during drying.

Green Machining

Dried, gel cast solids proved strong enough to undergo CNC machining, allowing the creation of intricate geometries. Cutting speeds up to 12 mm/s were possible without noting any chipping. Internal features, in the form of a hexagonal array of through holes (1/16th inch), were machined in a 2 mm thick disc containing agar in a concentration of 0.5 wt. % of total solids, added as a 4 wt. % precursor. This disc, after sintering, is pictured in Figure 11(a). External features, i.e. a series of 10 thin fan blades (tip aspect ratio of ~15:1), required twice the agar content (1 wt. % of total solids) as the machining of internal features. These sintered fan blades are shown in Figure 11(b). All of the blades were fully formed with the exception of those containing air bubbles and those damaged due to a communication error between computer and milling machine.

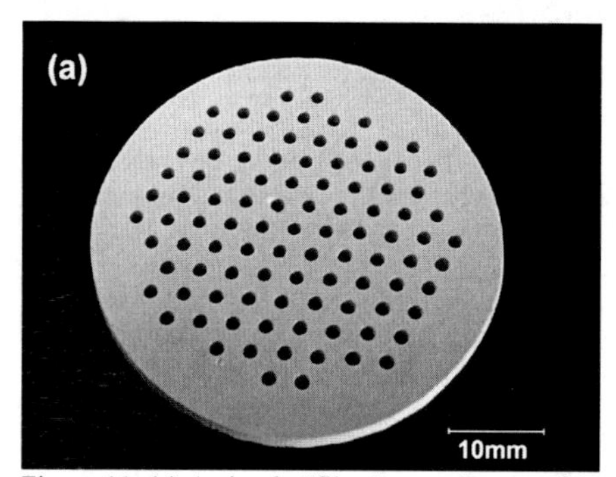

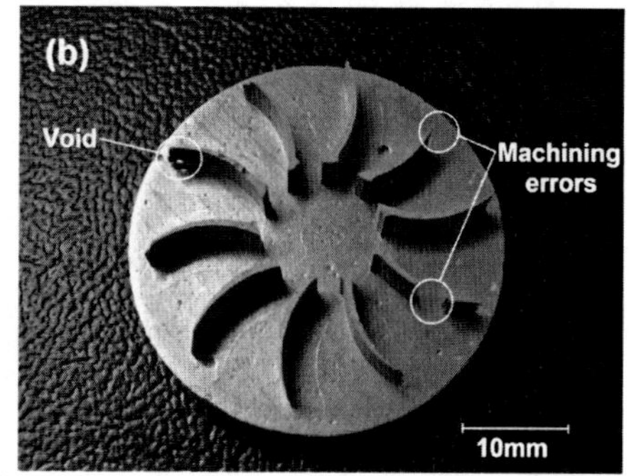

Figure 11. (a) A simple "filter" type disc, machined from a gel cast billet (0.5 wt. % final agar content from an initial 4 wt. % precursor solution), sintered at 1300°C for 2h. (b) Fan blades successfully machined from gel cast puck (1 wt. % final agar content from an initial 8 wt. % precursor solution), subsequently sintered at 1300°C for 2h.

Qualitatively, it is clear that more intricate structures, such as the fan blades, require higher agar content (1 wt. %) to impart the necessary strength for machining. It is possible, however, to cast bodies strong enough for machining less complicated internal structures using low agar content (0.5 wt %). It is therefore recommended that the gel casting formulation be tailored to the situation, in order to keep drying shrinkage to a minimum, and to maximize green density.

CONCLUSIONS

$BaTiO_3$ powders can be stabilized in an aqueous environment using low concentrations of a polyelectrolyte surfactant, PMA NH_4. Stabilization is best achieved under basic conditions, and the stabilization mechanism in this situation is electrosteric. Particles of a relatively spherical morphology lead to rheological properties amenable to colloidal consolidation, namely largely shear thinning behavior with shear thickening limited to very high shear rates. Suspensions comprised of two different sized particles present processing benefits such as reduced viscosity and enhanced particle packing, however only when a significant difference in particle size exists between the two fractions. These advantages allow the creation of higher solids content slips and yield consolidated bodies of enhanced green density. One consolidation technique in which it is favourable to maximize solids loading, and thus limits the liquid volume that needs to be removed later, is gel casting using an agent such as agar. Using this technique it is possible to cast structures that would be otherwise difficult to form, and to impart sufficient strength to the green body to permit machining without chipping or cracking.

ACKNOWLEDGEMENTS

The authors would like to thank the Natural Sciences and Engineering Research Council of Canada for the provision of funding through the Discovery Grants program. We are also grateful for the support of the Canada Foundation for Innovation, the Atlantic Innovation Fund, and other partners who helped fund the Facilities for Materials Characterization, managed by the Dalhousie University Institute for Materials Research, who provided access to the FE SEM. Dr. Gianfranco Mazzanti (Department of Process Engineering and Applied Science, Dalhousie University) is also gratefully acknowledged for his valuable comments and suggestions regarding rheological testing. We would also like to thank Dr. Mark Filiaggi (Department of Applied Oral Sciences) for access to the surface area analysis facility.

FOOTNOTES

* Address any correspondence to this author at: kevin.plucknett@dal.ca

REFERENCES

[1]G.H. Haertling, Ferroelectric Ceramics: History and Technology, *J. Am. Ceram. Soc.*, **882** [4] 797 818 (1999).
[2]J.A. Lewis, Colloidal Processing of Ceramics, *J. Am. Ceram. Soc.*, **83** [10] 2341 59 (2000).
[3]Z.C. Chen, T.A. Ring and J. Lemaitre, Stabilization and Processing of Aqueous $BaTiO_3$ Suspension with Polyacrylic Acid, *J. Am. Ceram. Soc.*, **75** [12] 3201 08 (1992).
[4]D.H. Yoon and B.I. Lee, Processing of Barium Titanate Tapes with Different Binders for MLCC Applications Part II: Comparison of the Properties, *J. Eur. Ceram. Soc.*, **24** [5] 753 61 (2004).
[5]Y. Hu, D. Zhou, D. Zhang and W. Lu, PTCR Characteristic of Gelcast $BaTiO_3$ Ceramic Thermistor, *Sensor. Actuat. A Phys.*, **88** [1] 67 70 (2001).
[6]H. Yaseen, S. Baltianski and Y. Tsur, Cathodic Electrophoretic Deposition of Barium Titanate Films from Aqueous Solution, *J. Mater. Sci.*, **42** [23] 9679 83 (2007).
[7]B. Velamakanni and F.F. Lange, Effect of Interparticle Potentials and Sedimentation on Particle Packing Density of Bimodal Particle Distributions During Pressure Filtration, *J. Am. Ceram. Soc.*, **74** [1] 166 72 (1991).
[8]J. L. Shi and J.D. Zhang, Filtration, Compaction and Sintering Behavior of Bimodal Alumina Powder Suspensions by Pressure, *J. Am. Ceram. Soc.*, **83** [4] 737 42 (2000).
[9]C.D. Munro and K.P. Plucknett, Aqueous Colloidal Characterization and Forming of Multimodal Barium Titanate Powders, submitted to *J. Am. Ceram. Soc.*.
[10]M.C. Blanco Lopez, B. Rand and F.L. Riley, The Isoelectric Point of $BaTiO_3$, *J. Eur. Ceram. Soc.*, **20** [2] 107 18 (2000).
[11]M.C. Blanco Lopez, G. Fourlaris and F.L. Riley, Interaction of Barium Titanate Powders with an Aqueous Suspending Medium, *J. Eur. Ceram. Soc.*, **18** [14] 2183 92 (1998).
[12]D.A. Anderson, J.H. Adair, D. Miller, J.V. Biggers and T.R. Shrout, Surface Chemistry Effects on Ceramic Processing of $BaTiO_3$ Powders, pp. 485 92 in *Ceramic Powder Science II, A: Ceramic*

Transactions, Vol. 1 (Eds. G.L. Messing, E.R. Fuller Jr. and H. Hausner), The American Ceramic Society, Westerville, OH (1988).
[13]J. H. Jean and H. R. Wang, Dispersion of Aqueous Barium Titanate Suspensions with ammonium Salt of Poly(methacrylic acid, *J. Am. Ceram. Soc.*, **81** [6] 1589 99 (1998).
[14]H.A. Barnes, Shear Thickening (Dilatancy) in Suspensions of Non aggregating Solid Particles Dispersed in Newtonian Liquids, *J. Rheology*, **33** [2] 329 36 (1989).
[15]C.D. Munro and K.P Plucknett, submitted to *Int. J. Appl. Ceram. Tech.*.
[16]V.F. Janas, T.F. McNulty, F.R Walker, R.P. Schaeffer and A. Safari, Processing of 1 3 Piezoelectric Ceramic/Polymer Composites, *J. Am. Ceram. Soc.*, **78** [9] 2425 30 (1995).

DISPERSION AND FLUIDITY OF AQUEOUS ALUMINIUM TITANATE SLURRY BY ADDITION OF TITANATE AQUEOUS SOLUTION

Seizo Obata, Yoshiyuki Iwata and Hisanori Yokoyama
Gifu Prefectural Ceramics Research Institute
3 11 Hoshigadai
Tajimi 507 0811, Japan

Osamu Sakurada and Minoru Hashiba
Faculty of Engineering, Gifu University
1 1 Yanagido
Gifu 501 1193, Japan

Yasutaka Takahashi
Daiken Chemical Co., Ltd.
2 7 19 Nishi hanaten Joto ku
Osaka 536 0011, Japan

ABSTRACT

A water soluble titanate compound (TNB Lac) prepared by direct reaction of titanium tetra *n* butoxide (TNB) with lactic acid (Lac) in water showed characteristics of a polyanion and acted as a dispersant for preparing aqueous aluminum titanate (Al_2TiO_5) slurries. The isoelectric point of Al_2TiO_5 particles shifted to lower pH with increasing concentration of TNB Lac. The rheological behavior of aqueous Al_2TiO_5 slurries with TNB Lac was examined and evaluated. The dispersion and fluidity of aqueous Al_2TiO_5 slurry were enhanced by the addition of TNB Lac at pH 10, and the thickening limit of the well dispersed and fluidized slurry with 0.075 mmol g^{-1} TNB Lac was found at a solid content of 79 wt%. A high green density of 63% for the sintered powder was obtained through the slip casting process using slurries with 0.075 mmol g^{-1} TNB Lac. The maximum relative density value of the sintered bodies was 96.5% when the green body was fired at 1500°C. The sintered body had a homogeneous grain growth and dense microstructure without pores or abnormal grain growth.

INTRODUCTION

Aluminum titanate (Al_2TiO_5) is an excellent refractory and thermal shock resistant material due to its relatively low thermal expansion coefficient ($\sim 1.0\times 10^{-6}$ °C^{-1}) and high melting point (1860°C)[1,2]. This material is mainly used in the fabrication of the components of industrial kilns, for example, as crucibles, ladles, nozzles, and sleeves[3]. For fabrication of these Al_2TiO_5 components with high uniformity, high performance, and high reliability, it is important to minimize defects, such as hard agglomerates and pores. The forming process for these materials and the preceding Al_2TiO_5 powder preparation process are very important in the fabrication of ceramic parts, as defects introduced during these processes will normally remain in the product even after successful sintering. Colloidal processing is a promising method to produce complex shaped parts with a reduction in the number and size of defects and increased reliability[4,5]. This colloidal process is suitable for fabricating near net shape components for which little machining is required. In this method, enhancement of the dispersion and fluidity of the Al_2TiO_5 slurry is a key technique for fabricating structural Al_2TiO_5 parts.

A good dispersant is required to enhance the dispersion and fluidity of a slurry. Polyelectrolyte dispersants, such as polyacrylic ($[CH_2CH(CO_2H)]_n$) and polymethacrylic acids ($[CH_2C(CH_3)(CO_2H)]_n$)[6,7] as polyanions and polyethylamine ($[CH_2CH_2NH]_n$, PEI)[8] as polycations have dispersing effects along with fluidizing effects, and are widely used. On the other hand, the effects of polymers that do not adsorb onto the particle surface have attracted a great deal of attention. Poly L lysine ($[NHCH[(CH_2)_4NH_2]CO]_n$) as a cationic polyelectrolyte[9] and nanoparticles of zirconium oxide[10] act as depletants. We have examined the effective use of some zirconium(IV) oxy salts (*e.g.*, $ZrOCl_2$) as fluidizing materials for ceramic slurries[11,12], and developed a new direct casting method based on aqueous acidic high solid loading of alumina slurries stabilized with zirconium(IV) acetate ($(CH_3CO_2)_4Zr$)[13].

Ohya *et al.* reported that very stable aqueous titanate solution could be prepared by direct reaction of titanium(IV) tetraisopropoxide ($Ti[OCH(CH_3)_2]_4$, TIP) with α hydroxycarboxylic acids ($HOCR_1R_2\ CO_2H$), such as lactic acid ($CH_3CH(OH)CO_2H$, Lac) in water, and that (004) oriented anatase film with a high refractive index of 2.54 was obtained from this stable titanate solution by the spin coating method[14]. Furthermore, they reported that lactic acid acted as a very effective stabilizer for TIP to yield a stable aqueous solution of titanates even with a molar ratio of Lac/TIP = 1. These solutions are very valuable as starting materials for other titanium compounds, including titanium oxide, because they have higher titanium content and lower carbon dioxide emission on firing to titanium oxide. In addition, the titanate solutions are free from hazardous halogen, nitrogen, sulfur, and metal ions other than titanium. Titanium and zirconium both belong to Group 4 in the periodic table because they have a similar valence shell that holds four electrons. Previously, we reported that water soluble titanate compound (TIP Lac) prepared by direct reaction of TIP with Lac in water showed characteristics of a polyanion and acted an effective dispersant for preparing alumina and titania slurries[15, 16]. On the other hand, we also reported that titanium aqueous solution could be prepared by direct reaction of titanium(IV) tetra *n* butoxide ($Ti[O^nC_4H_9]_4$, TNB) with Lac, and a well dispersed and fluidized barium titanate slurry could be obtained by addition of the aqueous solution (TNB Lac)[17].

In this study, the water soluble titanium compound (TNB Lac) was selected as a dispersant for Al_2TiO_5 slurries. In addition, the present study was performed to elucidate the optimum preparing conditions to obtain a dispersed, fluidized, and thickened aqueous Al_2TiO_5 slurry with TNB Lac. We evaluated the following processing factors: (i) the dispersibility of Al_2TiO_5 particles based on the zeta potential; (ii) the fluidizing and thickening of the slurry with TNB Lac; and (iii) the densities of sintered bodies.

EXPERIMENTAL

Materials

Aluminum titanate powder was synthesized from commercial α alumina (A 160SG3; Showa Denko K. K. Tokyo, Japan) with an average particle size of 0.6 μm and rutile (TM 1; Fuji Titanium Industry Co., Ltd., Osaka, Japan) with an average particle size of 0.66 μm. An equimolar mixture of both powders was ground in a ball milled in water using alumina pot and ball for 24 h. The dried and crushed powder was fired at 1550°C for 4 h in air. On X ray diffraction, compounds other than Al_2TiO_5 were not measurable in the synthesized powder. This synthesized Al_2TiO_5 powder was finely ground in water by ball milling for 168 h. The average particle size and specific surface area of the Al_2TiO_5 powder were 1.3 μm and 17 $m^2\ g^{-1}$, respectively. Water soluble titanate compound (TNB Lac) as a dispersant was prepared by directly mixing titanium tetra *n* butoxide (TNB) with lactic acid (Lac) and water in air at room temperature[17]. An exothermic reaction occurred instantly to yield a solid white mass. The solid product may be considered titanium hydroxide produced by the hydrolysis reaction. The solid product dissolved gradually with stirring and yielded a clear solution within 24 h, followed by addition of a suitable amount of water to adjust the titanium concentration to 1.0 *M*, after ridding the prepared solution of *n* butanol using a separating funnel. The molar ratio of Lac to TNB was 1. The titanium concentration of synthesized TNB Lac ($[Ti]_{TNB\ Lac}$) was determined gravimetrically by drying in air at room temperature and ignition of the dried TNB Lac to titanium oxide. The pH of the slurries was adjusted with tetramethyl ammonium hydroxide (TMAOH, $(CH_3)_4N(OH)$) and HNO_3.

Shaping processing and sintering

Slurries containing water and TNB Lac and Al_2TiO_5, which were added to the slurries in fixed amounts on a dry weight basis (dwb) for the Al_2TiO_5 powders, were prepared by ball milling using ZrO_2 balls for 24 h at room temperature. The water used to prepare all of the slurries was distilled and purified using a Milli Q system (Milli Q Plus; Millipore Co., Bedford, MA).

After ball milling, the slurries were degassed under vacuum and cast into gypsum molds. The solidified bodies were removed from the mold and dried at room temperature. Dried green bodies (80 × 10 × 5 mm) were subjected to sintering in air at 1400, 1450, 1500, and 1550°C for 2 h.

Measurements

The rheological behaviors of the slurries were measured at 25°C using a controlled stress rheometer (MARS II; Haake, Karlsruhe, Germany) with a parallel plate sensor 20 or 35 mm in diameter. The zeta potentials were measured by acoustophoretic spectrometry (DT 1200; Dispersion Technologies, Inc., Bedford Hills, NY) to determine the particle surface charge. Measurement of the zeta potential was carried out with 2 vol% suspensions at 25°C. Adsorption isotherms were determined from the amount of non adsorbed titanium in the supernatant of the slurry using an ICP AES (ICPS 7500; Shimadzu Co., Kyoto, Japan). The supernatant solutions were withdrawn after the 1 vol% slurries had been centrifuged at 10,000 × g (10,000 rpm) for 1 h. The densities of the green compacts were calculated from their size and weight. The sintered densities were measured by the Archimedes method. The sintered body was ground with an auto mortar for 30 min. The relative density of the sintered body was calculated based on the density of the powder, which was 3.64 g cm^{3} as determined with a helium pycnometer (Micromeritics Gas Pycnometer, Accupy 1330; Shimadzu Co., Kyoto, Japan) after drying in an oven at 150°C for 24 h. The microstructure of sintered bodies was observed by scanning electron microscopy (SEM, JSM 7001GC; JEOL Ltd., Tokyo, Japan).

RESULTS AND DISCUSSION

Electrokinetic properties of Al_2TiO_5 particles

Figure 1 shows the electrokinetic behavior of aqueous Al_2TiO_5 slurries with various amounts of TNB Lac as a function of pH. The zeta potential is strongly dependent on the pH of the slurry. The Al_2TiO_5 particles have an isoelectric point (IEP) at around pH 6.4 without TNB Lac. At pH > 6.4, the particles exhibit a negative charge. As a result, the mobility values decrease as the pH values increase. As the pH value increased from 6.4 to 11.2, the absolute value of the zeta potential increased. This was attributed to the adsorption of the OH ions onto the particle surface, which enhanced the electrostatic repulsive force. In the acidic range, a similar phenomenon was observed but with the adsorption of positive H^+ ions in the slurry. As the pH value decreased from 6.4 to 3.2, the value of the zeta potential increased. However, with the addition of more H^+ ions to the slurries, *i.e.*, when the pH value decreased from 3.2 to 2.2, the large number of positive ions resulted in a reduction of the double layer thickness, and hence a reduction of the repulsive force between the particles. At acidic pH, the measured zeta potential appeared to have a much lower absolute magnitude than that at basic pH. On the other hand, examination of the zeta potential of Al_2TiO_5 slurries with TNB Lac showed that the IEP of Al_2TiO_5 shifted to lower pH with addition of increasing amounts of TNB Lac. These phenomena could be attributed to the adsorption of the negatively charged TNB Lac as mentioned above on the surface of Al_2TiO_5 particles.

Fluidity and thickening of the Al_2TiO_5 slurry with TNB Lac

Figure 2 shows the effects of TNB Lac concentration on the apparent viscosity of 75 wt% Al_2TiO_5 slurries at pH 10. The apparent viscosity value of the Al_2TiO_5 slurries with TNB Lac

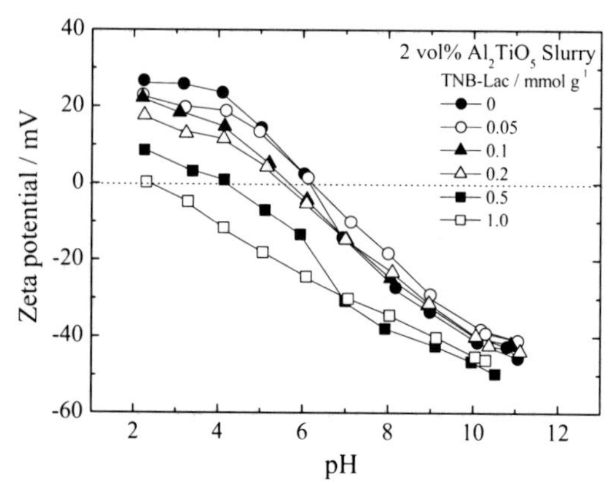

Fig. 1 Zeta potentials of Al_2TiO_5 particles with various amounts of TNB Lac in the slurries as a function of pH.

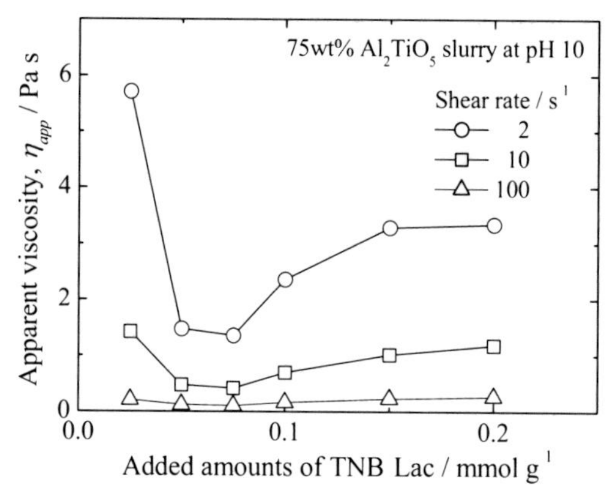

Fig. 2 Apparent viscosities of 75 wt% Al_2TiO_5 slurries with various amounts of TNB Lac.

decreases abruptly up to 0.05 mmol g ^{1}, and has a minimum value at the addition of 0.05 0.075 mmol g ^{1} TNB Lac, above which it gradually increases with increasing amount of TNB Lac.

The optimum amount of TNB Lac, which was required for the enhancement of fluidity of the Al_2TiO_5 slurries, was around 0.075 mmol g ^{1} (dwb) for the Al_2TiO_5 powders.

We reported previously that the optimum concentration of TIP Lac for preparing well dispersed and fluidized alumina slurries in the alkaline range is 0.5 1×10^{2} M[15]. This TIP Lac concentration was equivalent to 0.005 0.01 mmol g ^{1} calculated as added TIP Lac amount dwb for the alumina powder. In addition, the fluidity of titania slurry containing 0.01 mmol g ^{1} TIP Lac dwb for the titania powders was enhanced [10]. However, in the case of Al_2TiO_5 slurries, the optimum amount of Al_2TiO_5 slurry with TNB Lac was much greater than those of alumina or titania slurries. The apparent viscosity of the 70 wt% Al_2TiO_5 slurry with 0.05 mmol g^{1} TIP Lac was 2.5 Pa s at a shear rate of 10 s^{1}. The apparent viscosity of the 70 wt% Al_2TiO_5 slurry with 0.05 mmol g^{1} TNB Lac was 0.057 Pa s at a shear rate of 10 s^{1}. Comparison of the Al_2TiO_5 slurry including TIP Lac with that including TNB Lac, to which was added the same amount of titanium as Al_2TiO_5 powder, the apparent viscosity values of the slurries with TNB Lac were much lower than those of the slurries with TIP Lac. It should be noted that the difference in fluidity of the slurries with TIP Lac or TNB Lac was based on alcohol concentration in aqueous titanium solution as a dispersant. This flocculation was probably due to the isopropanol contained in the TIP Lac, thus reducing the dielectric constant in the media.

A thickened slurry with good fluidity was preferable for colloidal processing. The fluidity of electrostatically stabilized Al_2TiO_5 slurry was dependent on solid content and powder characteristics, such as particle size, size distribution, and aspect ratio. Aqueous Al_2TiO_5 slurry with good fluidity could be prepared by addition of 0.075 mmol g^{1} TNB Lac and TMAOH at pH 10.

Figure 3 shows the apparent viscosities of Al_2TiO_5 slurries with 0.075 mmol g^{1} TNB Lac as a function of solid loading at pH 10. TNB Lac was added to 0.075 mmol g^{1} dwb for Al_2TiO_5. The apparent viscosity at each share rate increased gradually up to 79 wt% Al_2TiO_5 followed by an abrupt increase above this value. We found that the thickening limit of the slurry with good fluidity, which was preferable for slip casting, was at a solid content of 79 wt%.

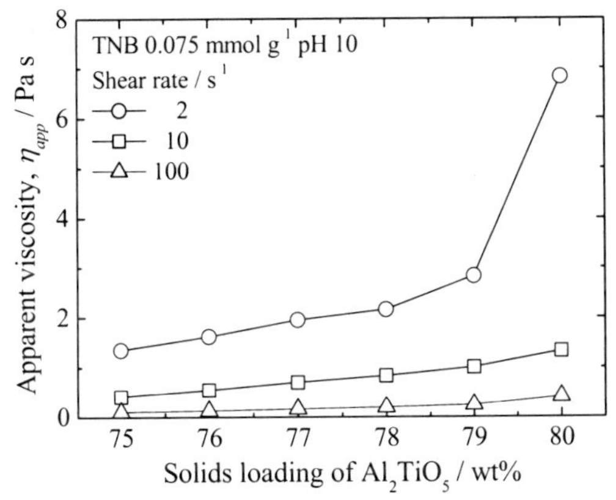

Fig. 3 Apparent viscosities of Al_2TiO_5 slurries with 0.075 mmol g^{1} TNB Lac as a function of solid content at pH 10.

Adsorption isotherm of TNB Lac on Al_2TiO_5

Figure 4 shows the adsorption isotherm of TNB Lac on Al_2TiO_5 surface as a function of the initial TNB Lac concentrations at pH 10. For comparison, the dotted line represents complete adsorption (100% adsorption) of added TNB Lac on the Al_2TiO_5 surface.

The adsorption was almost 100% when the initial concentration was very low. The amount of TNB Lac adsorbed increased with increasing TNB Lac concentration until it reached a plateau (saturated

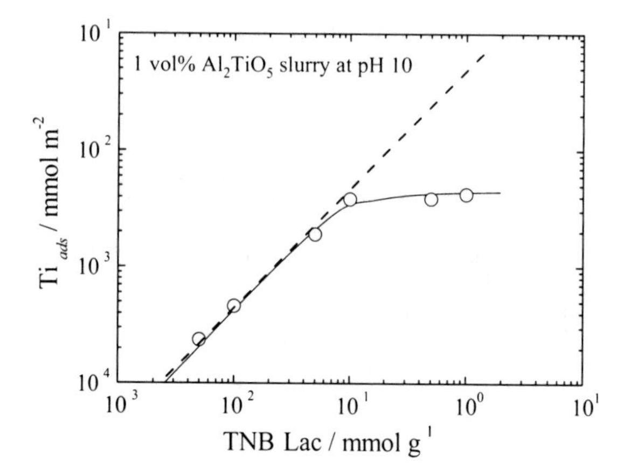

Fig. 4 Adsorption isotherm of TNB Lac on Al_2TiO_5 surfaces as a function of the initial TNB Lac concentration.

adsorption).. The isotherm adsorption was 4×10^{3} mmol m^{2} when more than 0.1 mmol g^{1} was used. This observation indicated the adsorption of anionic TNB Lac on the Al_2TiO_5 surface.

Evaluation of green and sintered bodies

The green bodies were obtained by the slip casting process using high solid loaded Al_2TiO_5

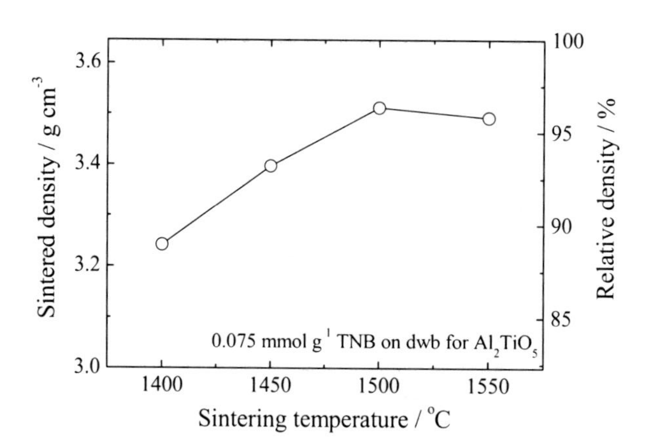

Fig. 5 Effects of sintering temperature on sintered density for Al_2TiO_5 bodies

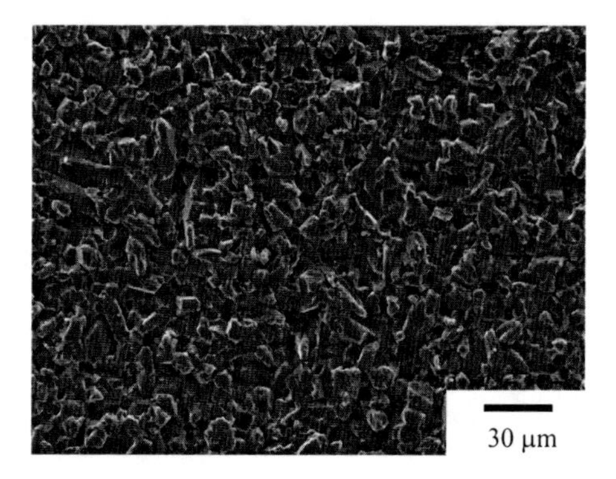

Fig. 6 Microstructure of a sintered body fired at 1500°C.

slurry with good fluidity by addition of 0.075 mmol g^{-1} dwb for Al_2TiO_5. Figure 5 shows the densities of the sintered bodies as a function of sintered temperatures. The relative density values of the green bodies were around 63% of the density of Al_2TiO_5 sintered powder (3.64 g cm^{-3}) due to the high degree of packing in the green body. The relative density values of the sintered bodies increased gradually with increasing sintering temperature, and the maximum value was 96.5% when the green body was fired at 1500°C. This value indicated that a highly dense sintered body was obtained through slip casting using Al_2TiO_5 slurries with TNB Lac, compared with the density of commercial Al_2TiO_5 materials (~3.4 g cm^{-3})[18].

Figure 6 shows the microstructure of the sintered body fired at 1500°C. The sample showed a dense microstructure without pores. In addition, the sintered body could be sintered homogeneously without abnormal grain growth.

CONCLUSIONS

The water soluble titanate compound TNB Lac prepared by direct reaction of titanium tetra *n* butoxide (TNB) with lactic acid (Lac) in water showed characteristics of a polyanion and acts as a dispersant for preparing aqueous aluminum titanate (Al_2TiO_5) slurries. The isoelectric point of Al_2TiO_5 particles shifted to lower pH with increasing concentration of TNB Lac. The dispersion and fluidity of aqueous Al_2TiO_5 slurry was enhanced by the addition of TNB Lac at pH 10, and the thickening limit of the well dispersed and fluidized slurry with 0.075 mmol g^{-1} TNB Lac was found at a solid content of 79 wt%. A high dense green body was obtained through the slip casting process using 79 wt% slurries with TNB Lac. The relative density values of the sintered bodies increased gradually with increasing sintering temperature, and the maximum value was 96.5% when the green body was fired at 1500°C. The sintered body had a homogeneous grain growth and dense microstructure without pores and abnormal grain growth.

REFERENCES

[1] R. G. Duan, G. D. Zhan, J. D. Kuntz, B. H. Kear, and A. K. Mukherjee, "Processing and Microstructure of High Pressure Consolidated Ceramic Nanocomposites," *Script. Mater.*, **51.** 1135 9 (2004).

[2] L. Stanciu, J. R. Groza, L. Stoica, and C. Plapcianu, "Influence of Powder Precursors on Reaction Sintering of Al_2TiO_5," *Script. Mater.*, **50.** 1259 62 (2004).
[3] H. Okuda, T. Hirai and O. Kamigaito, "Fine Ceramics Technology Series Vol. 6. Engineering Ceramics", Ohmsya Ltd., Tokyo, Japan (1987) pp. 194 7.
[4] F. F. Lange, "Powder Processing Science and Technology for Increased Reliability," *J. Am. Ceram. Soc.*, **72.** 3 15 (1989).
[5] J. A. Lewis, "Colloidal Processing of Ceramics," *J. Am. Ceram. Soc.*, **83.** 2341 59 (2000).
[6] J. Cesarano III and I. A. Aksay, "Processing of Highly Concentrated Aqueous α Alumina Suspensions Stabilized with Polyelectrolytes," *J. Am. Ceram. Soc.*, **71.** 1062 7 (1988).
[7] M. Itoh, O. Sakurada, M. Hashiba, K. Hiramatsu and Y. Nurishi, "Extensions of polyacrylic acid ammonium salts in the adsorption layer to fluidize alumina slurries," *J. Mater. Sci.*, **31.** 3321 4 (1996).
[8] F. Tang, T. Uchikoshi, K. Ozawa and Y. Sakka, "Dispersion of SiC Suspensions with Cationic Dispersant of Polyethylenimine," *J. Ceram. Soc. Japan*, **113.** 584 7 (2005).
[9] K. Furusawa, M. Ueda and T. Nashima, "Bridging and depletion flocculation of synthetic latices induced by polyelectrolytes," *Colloid Surf. A*, **153.** 575 81 (1999).
[10] V. Tohver, A. Chan, O. Sakurada and J. A. Lewis, "Nanoparticle Engineering of Complex Fluid Behavior," *Langmuir*, **17.** 8414 21 (2001).
[11] O. Sakurada, Y. Nakanishi and M. Hashiba, "Effect of zirconium acetate on the fluidity of acidic aqueous alumina slurries with high solids loading," *J. Mater. Sci. Lett.*, **20**. 929 31 (2001).
[12] O. Sakurada and M. Hashiba, "Depletion stabilization of ceramic suspensions with high solids loading in the presence of zirconium oxy salts," *Stud. Surf. Sci. Catal.*, **132**. 375 8 (2001).
[13] N. Adachi, O. Sakurada and M. Hashiba, "Direct Coagulation Casting of Alumina Slurries Stabilized through Zirconium Acetate Using an Enzyme Catalyzed Reaction," *Trans. Mater. Res. Soc. Jpn.*, **29**. 2037 40 (2004).
[14] T. Ohya, M. Ito, K. Yamada, T. Ban, Y. Ohya and Y. Takahashi, "Aqueous Titanate Sols from Ti Alkoxide .ALPHA. Hydroxycarboxylic Acid System and Preparation of Titania Films from the Sols," *J. Sol Gel Sci. Tech.*, **30**. 71 81 (2004).
[15] O. Sakurada, M. Saito, T. Ohya, M. Hashiba and Y. Takahashi, "Dispersion and Fluidity of Aqueous Aluminum Oxide Suspension with Titanate Aqueous Solution," *J. Ceram. Soc. Japan*, **115**. 846 9 (2007).
[16] O. Sakurada, M. Komaba, S. Obata, M. Hashiba and Y. Takahashi, "Electrophoretic Deposition on Anodes from Aqueous Titania Suspensions with Titanate Solution," *Key Engineering Materials* **412.** 313 6 (2009).
[17] O. Sakurada, M. Hashiba, N. Adachi, S. Obata, T. Ohya, Y. Takahashi and A. Harada, "Manufacture of titanium containing composite oxides, dielectric materials," *Jpn. Kokai Tokkyo Koho* (2007), JP 2007161502.
[18] The Ceramic Society of Japan, "Handbook of Ceramics, 2nd ed.," Gihodo Shuppan Co., Ltd., Tokyo, Japan (2002) pp. 1256 7.

Author Index